Heiau, ʻĀina, Lani

Heiau, ʻĀina, Lani

The Hawaiian Temple System in Ancient Kahikinui and Kaupō, Maui

Patrick Vinton Kirch and Clive Ruggles

With the collaboration of Andrew Smith

University of Hawaiʻi Press | Honolulu

© 2019 University of Hawai'i Press
All rights reserved
Printed in the United States of America

24 23 22 21 20 19   6 5 4 3 2 1

Library of Congress Cataloging-in-Publication Data

Names: Kirch, Patrick Vinton, author. | Ruggles, C. L. N. (Clive L. N.), author. | Smith, Andrew B. (Andrew Brown), contributor.

Title: Heiau, aina, lani : the Hawaiian temple system in ancient Kahikinui and Kaupo, Maui / Patrick Vinton Kirch and Clive Ruggles ; with the collaboration of Andrew Smith.

Description: Honolulu : University of Hawaii Press, [2019] | Includes bibliographical references and index.

Identifiers: LCCN 2019001987 | ISBN 9780824878276 (cloth : alk. paper)

Subjects: LCSH: Maui (Hawaii) —Antiquities. | Excavations (Archaeology)—Hawaii—Maui. | Temples—Hawaii—Maui. | Archeoastronomy—Hawaii—Maui.

Classification: LCC DU628.M3 K564 2019 | DDC 996.9/2102—dc23

LC record available at https://lccn.loc.gov/2019001987

Cover art: Aerial view of Loʻaloʻa Heiau in Kaupō, Maui. Photo by David Waynar.

University of Hawai'i Press books are printed on acid-free paper and meet the guidelines for permanence and durability of the Council on Library Resources.

Dedicated to the memory of
Charles Pili Keau,
respected *kupuna* and scholar of Hawaiian antiquity

CONTENTS

List of Illustrations ix
List of Tables xv
Preface xvii
Acknowledgments xix

Part I. The *Heiau* of Kahikinui and Kaupō: Analysis and Synthesis 1

1 Introduction 3

2 *Heiau*: Ethnohistorical and Archaeological Perspectives 13

3 The *Heiau* of Kahikinui and Kaupō: Architecture, Typology, Distribution, Chronology, and Function 40

4 *Heiau*, Landscape, and Sky: Viewsheds and Orientations 75

5 Summary and Conclusions 129

Part II. Catalog of *Heiau* Sites of Kahikinui and Kaupō 139

6 *Heiau* of Auwahi to Alena, Kahikinui *Moku* 143

7 *Heiau* of Kīpapa to Manawainui, Kahikinui *Moku* 199

8 *Heiau* of Kaupō *Moku* 285

Epilogue 335

Glossary of Hawaiian Words 337
References 343
Index 351

ILLUSTRATIONS

Figures

1.1. View of the Kahikinui landscape 5
1.2. Map of Kahikinui and Kaupō Districts 6
1.3. View of the Kaupō landscape 8
2.1. Ahuʻena Heiau in Kona, Hawaiʻi Island 22
2.2. Mean monthly rainfall at Auwahi, Kahikinui, and at Keʻeke, Kaupō 29
3.1. Typical *heiau* wall construction using stacked ʻaʻā cobbles 41
3.2. A large slab of conglomerate beach rock set as an upright at site LUA-36 43
3.3. Rank-order plot of Kahikinui and Kaupō *heiau* in ascending order of size 48
3.4. Interior of site AUW-11, a *koʻa* with an upright Kūʻula stone 50
3.5. A branch coral head in the wall of site ALE-140 51
3.6. Aerial view of site KIP-77, a notched *heiau* 53
3.7. Aerial view of site KIP-1156, a square enclosure with internal dividing walls 55
3.8. Aerial view of site KIP-1, an elongated double-court *heiau* 56
3.9. Aerial view of site KIP-1010, with smaller and larger twin notched enclosures 59
3.10. Map of Kahikinui and Kaupō, showing the distribution of *heiau* sites in this study 60
3.11. Frequency distribution of *heiau* by elevation above sea level 61
3.12. OxCal plot of radiocarbon dates obtained by M. Kolb from *heiau* excavations in Kahikinui 64
3.13. OxCal plot of radiocarbon dates obtained by P. V. Kirch and A. Baer from *heiau* sites in Kahikinui and Kaupō 65
3.14. OxCal plot of radiocarbon dates obtained by M. Kolb from excavations in site KAU-994 67
3.15. OxCal plot of radiocarbon dates obtained by M. Kolb from excavations in site KAU-324 67
3.16. Frequency histogram of the oldest ^{230}Th coral dates from temple sites 69
3.17. Frequency distribution of ^{230}Th coral dates derived from surface and architecturally integral contexts 70

3.18. Fishhooks and worked bone from Kahikinui and Kaupō *heiau* sites 73

4.1. Part of a DTM-generated horizon profile from site LUA-4 89

4.2. Radar plot showing the mean azimuths of Kahikinui and Kaupō temples in all four directions 101

4.3. Radar plot showing the mean azimuths of Kahikinui and Kaupō temples in the main direction of orientation 101

4.4. Radar plot showing the mean azimuths of upland Kahikinui and Kaupō temples in the main direction of orientation 102

4.5. Radar plot showing the orientations of "canoe-prow corners" 102

4.6. Cumulative probability histogram of the azimuths of Kahikinui and Kaupō temples in the direction of orientation and in all four directions 104

4.7. Cumulative probability histogram of the declinations of Kahikinui and Kaupō temples in the main direction of orientation 107

4.8. Cumulative probability histogram of the declinations of Kahikinui and Kaupō temples in all four directions 108

4.9. The east-northeastern horizon at *heiau* ALE-140 117

4.10. View from the platform of site NAK-30 119

4.11. DTM-generated visualizations of the northeastern and east-northeastern horizons at upland Auwahi *heiau* 121

4.12. DTM-generated profile of the eastern horizon at *heiau* WF-AUW-403 122

4.13. DTM-generated profile of the eastern horizon at *heiau* facing striking topographic features 123

4.14. Upright stone at KIP-1146, with Ka Lae o Ka 'Ilio directly behind it 124

5.1. Hierarchy of temples 133

6.1. Plan of *heiau* AUW-9 145

6.2. Plan of *heiau* WF-AUW-100 151

6.3. Plan of *heiau* WF-AUW-176A 153

6.4. DTM-generated profile of the western and northern horizon at *heiau* WF-AUW-176A 155

6.5. Plan of *heiau* WF-AUW-338 156

6.6. Plan of *heiau* WF-AUW-391 161

6.7. View of *heiau* WF-AUW-391 162

6.8. Plan of *heiau* WF-AUW-403 163

6.9. Eastern horizon at site WF-AUW-403 165

6.10. Plan of *heiau* LUA-1 169

6.11. View eastward along the N wall of LUA-1 172

6.12. Plan of *heiau* LUA-3 174

6.13. View of *heiau* LUA-3 175

6.14. Plan of *heiau* LUA-4 177

6.15. Plan of *heiau* LUA-29 179

6.16. Aerial view of *heiau* LUA-29 and the notched *pānānā* wall 181

6.17. Horizon to the WNW at *heiau* LUA-29 181

6.18. DTM-generated profile horizon to the W and WNW at *heiau* LUA-29 182

6.19. Plan of *koʻa* LUA-39 185

6.20. View of *koʻa* LUA-39 186

6.21. Plan of *heiau* ALE-1 187

6.22. View of *heiau* ALE-1 188

6.23. Plan of Wailapa Heiau (ALE-4) 191

6.24. View of Wailapa Heiau (ALE-4) 192

6.25. Plan of *koʻa* ALE-121 194

6.26. View of *heiau* ALE-140 195

6.27. Plan of *heiau* ALE-140 196

7.1. Aerial photo of *heiau* KIP-1 200

7.2. Plan of *heiau* KIP-1 201

7.3. View from an enclosure of *heiau* KIP-1 203

7.4. Map of the Heiau Ridge complex in Nakaohu, upland Kahikinui 204

7.5. Aerial photo of a portion of Heiau Ridge 205

7.6. Plan of *heiau* KIP-405 208

7.7. Branch coral offerings on the altar at *heiau* KIP-405 209

7.8. DTM-generated profile of the eastern horizon at *heiau* KIP-405 210

7.9. Plan of *heiau* KIP-75 211

7.10. Plan of *heiau* KIP-115 214

7.11. Plan of *heiau* KIP-77 216

7.12. Aerial view of *heiau* KIP-77 after excavation 216

7.13. Plan of site KIP-80 220

7.14. View of the main terrace facade at KIP-80, with phallic stone and *kī* plant 220

7.15. Plan of *heiau* KIP-273A and KIP-273B 224

7.16. Aerial photo of *heiau* KIP-273A 225

7.17. Plan of *heiau* KIP-275 227

7.18. Horizon to the ENE at *heiau* KIP-275 229

7.19. Plan of *heiau* KIP-307 231

7.20. View from *heiau* KIP-307, showing its alignment upon Ka Lae o Ka ʻIlio   233

7.21. Plan of *koʻa* KIP-330   234

7.22. View of the main part of *koʻa* KIP-330   235

7.23. Plan of *heiau* KIP-414   239

7.24. Plan of *heiau* KIP-567   242

7.25. Plan of *heiau* KIP-728, modified as a postcontact house site   244

7.26. View of the postcontact house enclosure within *heiau* KIP-728   245

7.27. Plan of *heiau* KIP-1010   247

7.28. View from an enclosure at *heiau* KIP-1010   251

7.29. DTM-generated profile to the E at *heiau* KIP-1010   251

7.30. Plan of *heiau* KIP-1025   252

7.31. Plan of site KIP-1146   255

7.32. Upright stone at KIP-1146, with Ka Lae o Ka ʻIlio behind it   255

7.33. DTM-generated profile of the eastern horizon at KIP-1146   256

7.34. Plan of *heiau* KIP-1151   258

7.35. Aerial photo of *heiau* KIP-1151   258

7.36. View from the upper leveled space at site KIP-1151, overlaid on a horizon profile   259

7.37. Plan of *heiau* KIP-1156   261

7.38. Plan of *heiau* KIP-1307   265

7.39. View of shrine KIP-1317   267

7.40. Plan of agricultural shrine KIP-1398   269

7.41. View from enclosure NAK-27, overlaid on a horizon profile   271

7.42. Plan of *heiau* NAK-29   273

7.43. View of *heiau* NAK-29, showing its landscape setting   274

7.44. Plan of *heiau* NAK-30   276

7.45. Plan of *heiau* MAH-231   279

7.46. DTM-generated profile of the northern horizon at *heiau* MAH-231   281

7.47. Plan of Kamoamoa Heiau (MAH-363)   282

8.1. Plan of site NUU-78   287

8.2. Plan of *heiau* NUU-79   288

8.3. View of an exterior wall face of NUU-79   289

8.4. Plan of *heiau* NUU-81   292

8.5. View of interior wall face and test excavation at site NUU-81   293

8.6. Plan of Halekou Heiau (NUU-100)   295

8.7. Basalt core tool, possibly an awl or graver, from Halekou Heiau (NUU-100) 295

8.8. Digitally generated profile of the western horizon at Halekou Heiau (NUU-100) 297

8.9. Plan of *heiau* NUU-101 299

8.10. The eastern horizon at *heiau* NUU-101 300

8.11. Plan of site NUU-153, a small agricultural *heiau* or men's house 302

8.12. Basalt adze preforms and a finished adze from site NUU-153 303

8.13. Plan of *heiau* NUU-175 305

8.14. View of the interior wall faces of site NUU-175 305

8.15. Plan of *heiau* NUU-188 307

8.16. Plan of Pōpōiwi Heiau (KAU-324) 310

8.17. DTM-generated profile of the northern horizon at Pōpōiwi Heiau (KAU-324) 314

8.18. Plan of Opihi Heiau (KAU-333) 316

8.19. Plan of *heiau* KAU-411 318

8.20. Aerial view of Loʻaloʻa Heiau (KAU-994) in its landscape setting 320

8.21. Plan of Loʻaloʻa Heiau (KAU-994) 322

8.22. View of the stepped terraces at Loʻaloʻa Heiau (KAU-994) 323

8.23. DTM-generated profile of the horizon to the ENE and E at Loʻaloʻa Heiau (KAU-994) 326

8.24. Plan of Kou Heiau (KAU-995) 327

8.25. View of a wall exterior at Kou Heiau (KAU-995) 328

8.26. Altar and high wall of the main enclosure at Kou Heiau (KAU-995) 330

8.27. Plan of Keanawai Heiau (KAU-999) 332

TABLES

2.1. Some attributes of the major Hawaiian deities 18

2.2. The Hawaiian calendar 27

3.1. Summary data on Kahikinui and Kaupō *heiau* sites 44

3.2. Artifacts recovered from *heiau* excavations 72

3.3. Faunal material from *heiau* excavations 74

4.1. Prominent cinder cones within the Kahikinui–Kaupō landscape 77

4.2. Known Hawaiian asterisms 79

4.3. Mean declination of the center of the sun at approximately 23-day (1/16-year) intervals through the year in AD 1600 83

4.4. Declination of the center of the moon at the "lunar standstill limits" around AD 1600 84

4.5. Primary and secondary indicators of the *kapu* direction of Kahikinui and Kaupō *heiau* 92

4.6. Directionality of notched corners and canoe-prow corners in relation to the identified *kapu* direction 96

4.7. Azimuths of Kahikinui and Kaupō temples in the direction of orientation 98

4.8. Orientation of canoe-prow corners at Kahikinui and Kaupō temples 102

4.9. Declinations of Kahikinui and Kaupō temples in the direction of orientation 105

4.10. Azimuths of prominent cinder cones, other peaks, and coastal promontories 111

4.11. Summary of topographic and astronomical influences on Kahikinui and Kaupō temple orientation 125

5.1. ^{230}Th-dated temple sites with attributes of morphology, orientation, and location 131

5.2. *Heiau* types and orientations 134

6.1. ^{230}Th dates on coral specimens from site AUW-9 147

7.1. Cultural content of KIP-75 excavations 212

7.2. Cultural content of KIP-77 excavations 217

7.3. Dates of sunrise in directions of possible significance at site KIP-1146 256

7.4. Cultural content of KIP-1156 excavations 262

7.5. Cultural content of KIP-1307 excavations 265

PREFACE

As is often the case in science, we did not begin with a neatly thought-out research design to investigate the role of *heiau* in Hawaiian astronomy and the Hawaiian calendar. Indeed, we did not even start out working together. Our joining forces in this endeavor came about almost by accident, evolving gradually into a collaboration that has now spanned fifteen years.

An archaeologist of Hawai'i and the Pacific islands, Kirch launched a study of ancient Hawaiian settlements in the Kahikinui District of Maui Island in 1995 (Kirch 2014). His research design included an intensive settlement pattern survey of large sections of this ancient *moku*, or political unit, as well as systematic excavations in residential and agricultural sites. As the project progressed, Kirch and his students mapped and recorded *heiau* they encountered, but this was not a primary focus in the early stage of his work. However, as accurate survey data on dozens of *heiau* in Kahikinui began to accumulate, Kirch became intrigued by what appeared to be a systematic pattern of temple orientations. Drawing upon data from a sample of twenty-three *heiau*, Kirch (2004b) outlined the case for three major groups of *heiau* orientations, each associated with one of the principal Hawaiian gods. Meanwhile, Kirch began to collaborate with geochronologist Warren Sharp, applying high-precision ^{230}Th dating to branch coral offerings found on *heiau*, in order to establish a chronology for the Kahikinui temples (Kirch and Sharp 2005).

An archaeoastronomer initially trained in both astrophysics and archaeology, Ruggles had studied the relationship between ancient and indigenous cultures and the heavens in several parts of the world, including prehistoric Britain and Ireland, continental Europe, sub-Saharan Africa, and Peru. His initial foray into the Hawaiian field came about as a result of an encounter, at a conference in Bulgaria in 1993, with retired architect Francis X. Warther, a Kaua'i resident with an intense interest in and respect for the Hawaiians' knowledge of the skies. Ruggles commenced his own systematic studies of *heiau* locations and orientations on Kaua'i and O'ahu in 1999, first visiting Hawai'i Island in 2000 and Maui in April 2001 (Ruggles 1999b, 2001). In the main, the results were disappointing because of the scattered and disparate nature of the extant *heiau* sites, the poor state of preservation of many of them, and the dearth of information about their archaeological and landscape context. Then, in January 2002, Ruggles reached Kahikinui and immediately recognized its importance as an area where many *heiau* existed within an exceptionally unspoiled landscape, and he discovered that the area was also the subject of an ongoing intensive archaeological study by Kirch and his team. The potential of this area for a systematic study of temple orientations was immediately evident. Ruggles visited Berkeley in March 2002 to familiarize himself with the archaeological survey work undertaken by Kirch, and embarked on his own study in April of that year (Ruggles 2007).

Having first met in Berkeley in March 2002 and deciding that a collaborative project would allow us to leverage our respective expertise, we agreed to spend an initial week in the field together in March 2003. We visited sites in the Auwahi, Alena, and Kīpapa areas of Kahikinui; many of these *heiau* had been previously mapped by Kirch. We made total station surveys of several sites, obtaining high-precision data on wall orientations, and also made extensive observations on landscape viewsheds from each site. In March 2008 a second joint trip extended the recording of sites to the Kīpapa and Mahamenui areas of Kahikinui and to Nuʻu in Kaupō District. In addition to working on sites at Nuʻu, we investigated the large *heiau* of Kou and Loʻaloʻa in Kaupō. In November 2011 we went to the field for a third time to look at *heiau* in the Auwahi, Lualaʻilua, and Alena areas of Kahikinui, an additional site in Nuʻu, and the large temple of Pōpōiwi in Kaupō. Following these collaborative field trips, Kirch has continued to conduct fieldwork intermittently in Kahikinui and particularly in Kaupō, where Alex Baer carried out a large-scale survey project for his Berkeley doctoral thesis (Baer 2015, 2016), under Kirch's direction. This latest fieldwork has allowed us to incorporate information on several additional *heiau* in Kaupō District.

ACKNOWLEDGMENTS

Initial archaeological fieldwork in Kahikinui was supported by Grants SBR-9600693 and SBR-9805754 from the National Science Foundation Archaeology Program, with Kirch as principal investigator. Later work was funded through Grant BCS-0119819 from NSF's Biocomplexity in the Environment Program, as part of the Hawai'i Biocomplexity Project. Additional funding was provided by the National Geographic Society's Committee for Research and Exploration (Grant 5692-96) and by the Stahl Fund of the Archaeological Research Facility, University of California, Berkeley. The Class of 1954 Professorship endowment at Berkeley also regularly provided financial support for our fieldwork.

Permission to carry out fieldwork on the lands of Kahikinui was provided by the Department of Hawaiian Home Lands, and for the Auwahi area, by 'Ulupalakua Ranch owners Pardee Erdman and Sumner Erdman. Access to sites in Kaupō was granted by Nu'u Mauka Ranch owners Bernie and Andy Graham and by Kaupō Ranch co-owner Jimmy Haynes.

We are especially grateful to the members of Ka 'Ohana o Kahikinui who over the years have shared their *aloha* and assisted us in many ways. Most of all we thank Mo Moler, whose hospitality and endless *kōkua* will never be forgotten.

Ruggles thanks the British Academy, the Royal Society, and the Royal Astronomical Society for grants that supported the fieldwork in 2002, 2003, and 2008; the School of Archaeology and Ancient History, University of Leicester, for the loan of the Leica TCR-705 Total Station used in the surveys; Lauren Lippiello for her assistance during survey work in 2002; and Laurel "Seeti" Douglass for support and encouragement during initial reconnaissance work in that year.

We also thank Masako Ikeda at the University of Hawai'i Press for her enthusiasm for this project and her careful editorial stewardship.

We are especially grateful to Jimmy Haynes for providing a generous grant-in-aid to the University of Hawai'i Press, making publication of this volume possible.

Patrick Vinton Kirch
Clive Ruggles

I

The *Heiau* of Kahikinui and Kaupō

Analysis and Synthesis

1

Introduction

No one knows the precise origins of the word *heiau*; its etymology is obscure. Elsewhere in Polynesia, temples and ritual spaces are generally called *marae*. At some point in their history, however, the Hawaiians began to call their places of prayer and ritual performance *heiau*. The great scholar of Hawaiian religion, Valerio Valeri, thought that *heiau* originated from *hai*, meaning "to sacrifice" (1985, 173). Indeed, *haiau* is a variant spelling of *heiau* (Pukui and Elbert 1986, 47). The offering of sacrifices, to gods great and small as well as to the collective ancestors (*'aumākua*), was a central function of *heiau*, whose types were as varied as the many deities themselves. These sacrifices ranged from the morsels of taro and *'awa* root offered in simple household shrines within the *mua*, or men's eating houses (one of the simplest forms of *heiau*), to the pigs baked in the earth ovens of agricultural temples to ensure the fertility of the land, to the ultimate of all sacrifices—a human body—presented exclusively in the great state temples of Kū, god of war. Through such sacrifices, presented in the context of stylized and strict ritual protocols, the Hawaiian priests entreated the gods and ancestors to look favorably upon the chiefs and the people, to bring rain and flowing waters upon which the life of the land depended, and to defend the kingdom against its enemies, within and without.

The first and foremost function of *heiau* was to serve as sacred places of sacrifice and prayer; this is well documented in the classic ethnohistorical texts (e.g., 'Ī'ī 1959; Kamakau 1964, 1976; Malo 1951). But it will be our contention—and a central theme of this book—that *heiau* had additional functions, ones that have been largely overlooked in much previous archaeological and ethnohistorical research. We present a body of empirical evidence to support the hypothesis that *heiau* were also places for the systematic observation of the heavens. We will also argue that *heiau* were places where men regularly gathered at ritually specified times—not only to worship and offer sacrifices but also to carry out a variety of material tasks, some of which may have required ritual (*kapu*) protocols.

Astronomical knowledge was highly developed among and essential to the Polynesian cultures. The very discovery of isolated Hawai'i around AD 1000 by voyaging Polynesians reflects their mastery of the art of wayfinding using the stars, wind patterns, and other noninstrumental aids, as did the later two-way voyages between Hawai'i and Tahiti by the famed Mo'ikeha, Kila, La'amaikahiki, Kaha'i, Pā'ao, and others, as encoded in the *mo'olelo*, or oral histories (Fornander [1878–1885] 1996). But long after the inter-archipelago voyages had ceased, after Hawai'i had become isolated in the ocean vastness of the North Pacific, astronomical knowledge continued to be essential to the populous and highly stratified society that arose in this fertile archipelago. Among the Hawaiian priestly orders were *kāhuna* known as *kilo hōkū*, "those who studied the stars," or as *papa kilokilo lani*, "those who could read the signs, or omens, in the sky" (Kamakau 1964, 8). These experts memorized the names of scores, and perhaps hundreds, of *hōkū*—stars, collections of stars (constellations, clusters), and other heavenly bodies—knew their movements, and could predict the timing of their coming and going across the heavens. The catalog of Hawaiian star names compiled by Johnson, Mahelona, and Ruggles (2015) lists over 450 distinct recorded names of stars and asterisms, plus variants, together with others for pathways across and regions of the sky.

Above all else, the priests relied on their astronomical knowledge to maintain the Hawaiian calendar, on which the orderly functioning of society depended. In chapter 2 we describe this calendar in greater detail, including the division of the year into two periods (one sacred to Lono and the other to Kū), the lunar calendar and the "nights-of-the-moon" into which each lunar month was segmented, and the intercalation and correlation of lunar and solar phenomena. The key activities of Hawaiian society, secular and sacred, agricultural and martial, commoner and chiefly, were regulated by this calendar, year after year. It was, we contend, from their vantage points in the many *heiau* carefully positioned across the landscape that the priests kept watch over the heavens, ensuring that the economic, social, and ritual life of the kingdom unfolded in a predictable and proper progression.

When archaeological research in Hawai'i commenced in the first few decades of the twentieth century, *heiau* were the center of attention. The often imposing and monumental stone foundations of *heiau* were thought to hold a key to the Hawaiian past. Pioneering Bishop Museum archaeologist John F. G. Stokes surveyed *heiau* sites on Hawai'i and Moloka'i Islands in 1906 and 1909, respectively (Flexner and Kirch 2016; Flexner et al. 2017; Stokes [1909], 1991). This was followed, in the 1920s and 1930s, by surface surveys of *heiau* on Maui, Moloka'i, Kaho'olawe, O'ahu, and Kaua'i Islands (Walker [1930]; Phelps [1937]; McAllister 1933a, 1933b; Bennett 1931). By the 1950s, however, when Hawaiian archaeology was rejuvenated with the application of stratigraphic excavations by Kenneth P. Emory and his colleagues (Emory, Bonk, and Sinoto 1959), the study of *heiau* had fallen out of favor. Indeed, with a few exceptions such as Ed Ladd's excavation and restoration of several *heiau* on Hawai'i and O'ahu (Ladd 1969, 1973, 1985) or the work of Michael Kolb on certain Maui *heiau* (Kolb 1991, 1992, 1994, 1999, 2006), the study of Hawaiian temple foundations remained a neglected aspect of research in Hawaiian archaeology. Several recent studies, however, suggest renewed interest in the archaeological study of temple sites (e.g., Baer 2015, 2016; Kikiloi 2012; M. D. McCoy 2008, 2014; M. D. McCoy et al. 2011; Mulrooney and Ladefoged 2005; Thurman 2015).

This book presents the results of the first intensive survey and detailed analysis of a large number of *heiau* distributed across a broad region of southeastern Maui, encompassing the ancient districts of Kahikinui and Kaupō. Throughout this region, we have studied 78 *heiau* structures. Some of these had been previously documented—although often poorly or incompletely—but 47 are reported here for the first time. We have brought to bear several methods of detailed mapping (plane table with telescopic alidade, theodolite, GPS) with selected excavation, combined with both radiocarbon and ^{230}Th dating to achieve a chronological framework. Our results have convinced us that renewed study of *heiau* holds great promise for advancing our knowledge and understanding of the Hawaiian past.

Kahikinui and Kaupō: Lands of La'amaikahiki and Kekaulike

Second largest of the main Hawaiian Islands, Maui was formed by the coalescence of two massive shield volcanoes. The eastern and higher of these mountains, Haleakalā (House of the Sun), soars to 3,055 m (10,023 feet) above sea level. First emerging above the ocean about one million years ago, Haleakalā remains dormant, having last erupted in the seventeenth century (Sherrod et al. 2007). The southeastern flank of Haleakalā, lying in the rain shadow created by the huge mountain mass, is an arid landscape characterized by relatively young, unweathered lava slopes, interrupted here and there by steep-sided gulches incised by intermittent, torrential rains that fall seasonally during the winter months (the *kona* rains). To the Hawaiians, this was an *'āina malo'o*, a dry land, where water was scarce and much hard work was required to wrest a living through the cultivation of sweet potatoes (*'uala*) and some dryland taro, augmented by fishing and shellfish gathering along the rocky, cliff-bound and wave-thrashed coast. This was a *kua'āina* (literally, back of the land) and its inhabitants were likewise known as *kua'āina* folk, country people, sunburned and hardened from their rugged existence. Even today, this part of Maui is known to locals as Backside, a thinly populated land of fenced-off ranches traversed only here and there by eroded four-wheel jeep tracks, used mostly by itinerant fishermen and hunters (fig. 1.1).

FIGURE 1.1. View of the Kahikinui landscape, with lava flow slopes sweeping up to the ridgeline of Haleakalā Volcano, in the Kīpapa-Nakaohu area. The enclosure walls of *heiau* site KIP-1010 are visible in the middle distance. (Photo by P. V. Kirch)

In ancient times Maui was divided into 12 districts (*moku*), each under the control of a district chief (*ali'i 'ai moku*) and—since roughly the end of the sixteenth century—all integrated into a single kingdom. Two of these *moku*, Kahikinui and Kaupō, make up the greater part of southeastern Maui, along with Kīpahulu District to the east (fig. 1.2). Kahikinui is the more westerly of the two, extending some 15 km from its western border with Honua'ula District (at Kanaio) to Wai'ōpai, where it shares its eastern boundary with Kaupō. Kaupō then runs another 10 km to the east, where it meets its eastern border with Kīpahulu, at Kālepa Gulch.

The name "Kahikinui" resonates with a deep Polynesian past. Literally translating as "Great Tahiti," the name was probably bestowed by early voyagers who recognized the close resemblance between Maui and the twin-volcanic island of Tahiti, where the southern part of the larger volcano is called Tahiti Nui. "Kaupō" is a toponym likewise found elsewhere in Eastern Polynesia, as with the renowned volcanic lake and geothermal springs at Taupō in New Zealand. Hawaiian oral traditions (*mo'olelo*) link Kahikinui and Kaupō to early voyagers from central Eastern Polynesia. The nineteenth-century Hawaiian scholar Samuel Kamakau (1991, 112) related the following tradition:

> According to the *mo'olelo* of Kāne and Kanaloa, they were perhaps the first who kept gods ('*o laua paha nā kahu akua mua*) to come to *Hawai'i nei*, and because of their *mana*

FIGURE 1.2. Map of Kahikinui and Kaupō Districts, showing the location of major areas mentioned in the text.

they were called gods. Kahoʻolawe was first named Kanaloa for his having come there by way of Ke-ala-i-kahiki ["The road to Tahiti"]. From Kahoʻolawe the two went to Kahikinui, Maui, where they opened up the fishpond of Kanaloa at Lua-laʻi-lua, and from them came the water of Kou at Kaupō.

Kahikinui is also associated in Hawaiian traditions with the voyaging chief Laʻamaikahiki, descendant of the famed Māweke, who ruled Oʻahu in the early thirteenth century, and son of Moʻikeha, who had sailed from Hawaiʻi to Tahiti, where he had an affair with a local chiefess named Kapo. Years later, Moʻikeha sent another son, Kila, to Tahiti to fetch Laʻamaikahiki, who then spent some time in Kauaʻi with his father. Laʻamaikahiki "moved over to Kahikinui in Maui" eventually, however: "This place was named in honor of Laamaikahiki. As the place was too windy, Laamaikahiki left it and sailed for the west coast of the island of Kahoolawe, where he lived until he finally left for Tahiti" (Fornander 1916–1920, 4:128). A unique notched stone wall near the southernmost point of Kahikinui, the *pānānā* (sighting wall) of Hanamauloa, may be a memorial to Laʻamaikahiki (Kirch 2014, 88–97; Kirch, Ruggles, and Sharp 2013).

Kaupō, for its part, is closely linked with Kekaulike, who ruled the island between roughly AD 1710 and 1730. A descendant of the Piʻilani dynasty of rulers, who had made their seat in the *moku* of Hāna, Kekaulike moved the royal center to Kaupō and rededicated the massive *luakini* war temples of Loʻaloʻa and Pōpōiwi (Kamakau 1961, 66). From these temple platforms overlooking the bay at Mokulau, where his war canoes were launched, Kekaulike gazed across the ʻAlenuihāhā Channel to Hawaiʻi Island, which he unsuccessfully attempted to conquer.

Although they are contiguous regions along the leeward slopes of Haleakalā, Kahikinui and Kaupō differ in important respects. Kahikinui's lava slopes range from about 225,000 years old (in the eastern part of the district) to a mere 10,000 years in the Alena area. Only a few narrow gulches dissect this terrain of undulating lava flows. In contrast, Kaupō District straddles a massive depositional fan of lava and mud flows emanating from late-stage Hāna volcanic series eruptions within Haleakalā Crater, which poured out through an erosional valley known as Kaupō Gap (fig. 1.3). Some of these flows are 50,000–140,000 years old, but many are as young as 750–1,500 years (e.g., the Puʻu Nole basanite). The youthfulness of this landscape and the high nutrient content of its soils, combined with an optimal range of annual rainfall for sweet potato cultivation (from about 750 to 1500 mm west to east across the Kaupō fan), made Kaupō ideal terrain for the intensive cultivation of sweet potatoes (Maunupau 1998; Handy 1940). Not surprisingly, the population of Kaupō at the time of the 1832–1833 missionary census, some 3,220 persons, was considerably higher than that of Kahikinui, with only 517 persons (Schmitt 1973, 18). The greater agricultural productivity of Kaupō is doubtless one reason why Kekaulike chose the *moku* as his royal seat. The archaeological legacy of this agricultural productivity is an intensive dryland field system, a grid of stone walls and earth-and-stone embankments that indelibly inscribes the landscape (Kirch, Holson, and Baer 2009).

Archaeological Research in Kahikinui and Kaupō

Until recently, Kahikinui and Kaupō remained relatively unstudied and unknown from an archaeological perspective. Winslow Walker conducted a fleeting survey through the region in 1929 as part of a Bishop Museum–Yale University research fellowship, making brief descriptions

FIGURE 1.3. View of the Kaupō landscape, taken from the Pauku-area lowlands looking up into Kaupō Gap. (Photo by P. V. Kirch)

and in some cases rough maps of about 20 *heiau* sites shown to him by Native Hawaiian informants from Kaupō. Walker's manuscript ([1930]) remains unpublished but, along with his field notes, is an important source of information on temple sites in southeastern Maui. In 1966, Peter Chapman directed a Bishop Museum field team in a settlement-pattern survey of parts of Kahikinui, recording about 500 stone structures. Kirch, then a Punahou School student, was a member of the 1966 field team (Kirch 2014, 18–26). No report of the 1966 survey was completed, although a map of the sites was compiled.

Archaeological research in Kahikinui recommenced in the mid-1990s, when Kirch launched the University of California, Berkeley, project, which concentrated primarily on the Kīpapa-Nakaohu region, building on the initial results of the 1966 Chapman survey. At the same time, Dixon et al. (2000) conducted a cultural resources management survey of the higher-elevation areas of Kīpapa and Nakaohu, which were being divided into lots for lease to Native Hawaiian occupants by the Hawaiian Homes Commission. And Michael Kolb spent two field seasons investigating several *heiau* sites in Kahikinui, mostly sites that had been identified by Walker in 1929. Initial results of all three projects were summarized in a volume edited by Kirch (1997b). While the Dixon and Kolb projects ended by 1999, Kirch and his students continued their fieldwork in Kahikinui for another decade (Coil 2004; Holm 2006; Kirch 2004b, 2007, 2011a; Kirch and O'Day 2002; Kirch et al. 2004, 2010; Kirch and Sharp 2005; Kirch, Mertz-Kraus, and Sharp 2015; Van Gilder 2005). Most recently, large parts of the *ahupua'a* (territory under control of a chief) of Auwahi, in the westernmost part of Kahikinui, were archaeologically investigated in conjunction with the development of a wind energy project (Shapiro et al. 2011). Information on all of the sites recorded during these surveys was compiled in a Geographic Information System (GIS) database at the Oceanic Archaeology Laboratory, Berkeley.

Kirch (2014) synthesized the results of all this Kahikinui archaeology, as well as the postcontact history of the *moku*.

In spite of these substantial field projects, Kahikinui *moku* has by no means been completely archaeologically surveyed. Kirch (2014, table 1) estimates that 25.2 km^2 has been intensively surveyed, with 3,058 sites recorded. This is about 20 percent of the total land area of Kahikinui, although it is closer to 40 percent of the area below about 900 m elevation, where most sites are concentrated. Nonetheless, large sections of Kahikinui remain archaeologically terra incognita.

Kaupō has received even less archaeological attention than Kahikinui. In 2003, at the invitation of the late Charles P. Keau, Kirch began a long-term archaeological study in the *ahupua'a* of Nu'u, a large land area on the western side of Kaupō. Kirch and his students have continued to work at Nu'u, where approximately 450 sites have now been recorded and several test-excavated. Berkeley doctoral student Alex Baer carried out a large-scale survey across much of the core of Kaupō, using a combination of aerial photography and ground survey integrated in a GIS database and recorded about 500 sites, along with a series of test excavations and radiocarbon dates (Baer 2015). As in Kahikinui, however, many parts of the district remain to be surveyed.

Theoretical Approach: Settlement Patterns, Landscapes, and Skyscapes

Cultural landscape studies within Kahikinui and Kaupō, first instigated in 1966 by Peter Chapman and pursued more intensively since the 1990s by Kirch and colleagues, have followed an integrated "settlement pattern" approach (Kirch 1997a, 10) focusing not only on recognizable "sites" but on the overall distribution of a range of human remains and environmental indicators studied in their geographic and archaeological context. This progress beyond the exclusively site-based inventories such as those compiled by Walker, Stokes, and Emory earlier in the twentieth century reflects a broader trend within archaeology that has seen regional analyses focusing exclusively on "sites" give way to broader landscape studies and to cultural interpretations increasingly informed by anthropological perspectives on people's interactions with and cognition of the landscape (e.g., Ashmore and Knapp 1999; Ingold 2000).

Yet sites functioned within the lived-in landscape; *heiau*, in particular, were the focus of ritual activities within the cultural landscape in the Hawaiian Islands, as were temple sites elsewhere in Polynesia. As has been increasingly recognized in interpretive approaches since the 1990s, this implies that we should be open-minded regarding a range of possible symbolic relationships that conveyed meaning to those dwelling within and moving through the landscape, including the location and design of monuments in relation to visible topographic features (e.g., Tilley 1994; Scarre 2005).

The very terms "landscape" and "landscape archaeology" mask the fact that the perceived environment includes not only the terrestrial realm but also the sky and imposes a Western dichotomy (deriving from Linnaean taxonomy) that implicitly excludes the latter (Ruggles and Saunders 1993). Tangible links between human constructions and celestial objects could also have been of cultural significance—for example, in relation to calendar, navigation, and cosmology. For Hawai'i, the importance of the calendar—and of seasonally constrained rites, including processions through the landscape such as the famous circuits of the "long god" (*akua loa*) and the "short god" (*akua poko*) (see chapter 2)—suggests an obvious need for calendrical timing devices using the sun or stars. Archaeoastronomy provides a set of tools and practices available to archaeologists wishing to investigate tangible links to objects and events in the sky (Ruggles 2014a,c,e,f; Prendergast 2014). The term "skyscape" is gaining popularity as a means of ensuring that such enquiries are well integrated within broader archaeological investigations rather than being pursued as a separate "interdiscipline" (e.g., Silva and Campion 2015).

As Ruggles proposed in his first paper on Hawaiian temples in 1999: "The main agenda for the future…must surely be to develop archaeoastronomical investigations as part of coordinated programs addressing questions of landscape use and cognition" (Ruggles 1999b, 79). The collaborative project that has resulted in this book represents just such a sustained effort in pursuit of this goal, offering a case study that we hope illustrates, in broad terms, how this could be achieved in other contexts.

Research Methods

Eighteen of the *heiau* sites discussed in this book were originally recorded by Walker in 1929 (Walker [1930]); a few others had been located during the 1966 Bishop Museum survey in the Kīpapa area of Kahikinui. All of the other sites have been found through intensive field survey, primarily carried out by Kirch and his Berkeley students in the Kīpapa, Nakaohu, Mahamenui, Manawainui, and Nuʻu areas since 1995 and by the Pacific Legacy field team in Auwahi (Shapiro et al. 2011). In these surveys, field teams walked in systematic transects across the landscape, usually 5 to 10 m apart, recording all stone structures visible. While very small or low structures were undoubtedly missed in some areas where the vegetation was dense (see Holm and Kirch 2007), it is unlikely that any significant *heiau* were overlooked in the areas that we systematically covered. As noted above, some 25.2 km^2 of Kahikinui *moku* have been surveyed in this manner.

Several different site-numbering systems have been applied in Hawaiian archaeology over the decades. Walker ([1930]) simply gave sequential numbers beginning with 1 to all of his sites. In the 1960s, the Bishop Museum devised a system in which the island, district, and *ahupuaʻa* were designated by codes, followed by sequential site numbers within each *ahupuaʻa*. Chapman used this system during the 1966 Bishop Museum survey in Kīpapa, each site number being prefaced with "MA-A35-" (MA, Maui; A, Hāna District; 35, Kīpapa). After the State of Hawaiʻi Historic Preservation Division was established, a new system indexing sites to the US Geological Survey 7.5-minute topographic maps was adopted; Kolb and Radewagen (1997) and Dixon et al. (2000) used this state system to number sites in Kahikinui. Owing to recurrent problems with maintenance of the state's site inventory (the Historic Preservation Division's GIS system was unavailable for a number of years), Kirch began to use his own site numbering system for the Kahikinui and Kaupō surveys. This system follows the widespread practice in Polynesian archaeology of designating sites within geographic sections of islands. The following three-letter abbreviations are used in this book: AUW, Auwahi; LUA, Lualaʻilua; ALE, Alena; KIP, Kīpapa (including Nakaohu); NAK, Nakaaha; MAH, Mahamenui; MAW, Manawainui; NUU, Nuʻu, Kaupō; KAU, Kaupō other than Nuʻu. Site numbers are assigned sequentially within each area. For Auwahi, a separate prefix, "WF-," is used to designate those sites in Auwahi recorded during the Pacific Legacy wind energy survey project (Shapiro et al. 2011).

Several techniques were used to map the *heiau* sites of Kahikinui and Kaupō. A few smaller sites were mapped with compass and tape, but most structures were mapped by Kirch using a

W. & L. E. Gurley optical alidade and plane table, usually at a scale of 1:100 or, in the case of some larger structures, at 1:200. The great advantage of plane-table mapping for Hawaiian stone structures is that one is able to artistically render architectural details directly onto the map in the field, something that cannot be done with a digital total station (Flexner and Kirch 2016). Plane-table maps were later scanned and digital plans prepared using Adobe Illustrator software. Sites were also recorded using GPS, for most sites using a Trimble GeoXT instrument or, more recently, a Trimble Juno 3B. The datum used for GPS mapping was WGS84. Geographic coordinates, where specified, are in the Universal Transverse Mercator (UTM) system, zone 4. Exterior and interior perimeters of all main walls, terraces, and platforms were traced using polylines, while individual features were recorded as points. The GPS shapefiles have been integrated into the Kahikinui-Kaupō GIS maintained at the Oceanic Archaeology Laboratory, Berkeley.

For accurately determining structural orientations, viewsheds, and the astronomical potential of *heiau* sites, surveys were undertaken wherever possible using a Leica TCR-705 Total Station (theodolite with electronic distance measurement [EDM] device) loaned from the University of Leicester (UK). Where this was impracticable, less accurate surveys were undertaken using handheld prismatic compasses and compass-clinometers. Details of our methods for determining the precise orientation of *heiau* structures are provided in chapter 4.

The analysis of the spatial orientations of *heiau*, including their relationship to the surrounding landscape and to viewsheds, depends on determining what we call the "axis of orientation" of a structure. Polynesian ritual spaces, including Hawaiian *heiau*, were regarded by their users as having an axis progressing from less to increasingly sacred; this is evident from ethnohistorical descriptions of *heiau* and of the rituals conducted within them (see chapter 2). The most sacred part of a *heiau* was the sector where wooden images of the gods were arrayed, sometimes with an *'anu'u* tower of wood and barkcloth immediately behind or next to such images, and with an offering platform (*lele*) in front of the images. From the perspective of an observer standing within the temple precincts, the axis or orientation would therefore be looking toward the images and/or the tower. (This is analogous to someone entering a Christian church and looking toward the altar.)

Given that the images, wooden towers, and other superstructures of *heiau* have long since vanished, how does the archaeologist infer the axis of orientation? The evidence is encoded in the architectural foundations themselves. *Heiau* are rarely perfectly symmetrical in plan; in most of the sites we have studied, one end is more elaborated than its opposite. Typically such elaboration involves higher walls, often buttressed with external terraces on the exterior. Elevation differences between internal courts or divisions also serve to indicate orientation, with the higher or elevated spaces indicating the axis of orientation—this reflecting the widespread Polynesian spatial association of height with rank and sanctity (Oliveira 2014). Additional evidence lies in internal features such as platforms, which were probably the bases for *'anu'u* towers, benches, or altars (some of which have branch coral offerings on them).

An example of how all of these features can be combined to infer the axis of orientation is provided by site KIP-1, a medium-sized temple in the Kīpapa uplands of Kahikinui (see part II for plan map and details of KIP-1). The eastern end of this site exhibits thick, high walls that are externally buttressed, while the western end has very low walls that almost grade into the surrounding terrain. Consistent with this, the eastern court is elevated above the level of the western court. Furthermore, near the eastern end of the higher court we find a low stone platform that was most likely the foundation for the ʻanuʻu tower. All three lines of evidence converge to indicate that the sacred end of this temple was on the east and that its axis of orientation was to the east.

In our *heiau* descriptions in part II of this book, we indicate for each site the probable axis of orientation, citing the relevant architectural evidence. To be sure, in some cases the evidence is ambiguous, so that two possible axes of orientation can be inferred. We have pointed out these cases wherever they occur.

Précis of This Volume

The long-term value of a study such as ours, in which now-vanished cultural practices are decoded from the patterns remaining in lichen-encrusted stone foundations, ultimately rests on the careful and systematic presentation and analysis of the archaeological details. These we present as fully as possible in part II of this book, where each of the 78 *heiau* that we have mapped and scrutinized are described and illustrated with maps and photographs. Additional information too extensive or too mundane to include in a printed volume may be found in a repository of supplementary online materials available at www.cultural-astronomy.com. The catalog of *heiau* presented in part II thus stands as an empirical record of these sites, independent of our interpretations.

Part I, consisting of chapters 2 through 5, draws upon the systematic observations presented in the site catalog (part II) to advance a series of arguments and interpretations regarding the place of *heiau* in the precontact society of Maui. We begin, in chapter 2, with a review of the ethnohistorical literature regarding *heiau* and *heiau* rituals in Hawaiian culture, followed by an overview of previous archaeological studies of *heiau* sites. In chapter 3 we turn to an analysis of the archaeological remains of *heiau* of Kahikinui and Kaupō, with a consideration of architectural patterns, types of *heiau*, and their geographic distribution over the landscape. We also review the evidence from radiocarbon and ^{230}Th dating for the chronological development of *heiau* in the region, as well as evidence for *heiau* function derived from test excavations in a number of sites. Chapter 4 is devoted to the place of the *heiau* in relation to the landscape and the sky: their orientation, their viewsheds, and the inferences that may be drawn regarding *heiau* as observation places for astronomical phenomena such as the rising of the Pleiades and the progression of the sun from solstice to solstice. Finally, chapter 5 summarizes our main findings and conclusions.

2

Heiau

Ethnohistorical and Archaeological Perspectives

With the abolition by Liholiho (Kamehameha II) of the *'ai kapu* system of gendered eating prohibition in November of 1819, and the subsequent order that the temples and images be destroyed (Kuykendall 1938, 68), *heiau* passed from being integral, functioning components of Hawaiian society into history. As the new Christian religion—introduced by the Congregationalist missionaries in 1820—took hold and began to replace the old ritual system as a new form of state religion (Sahlins 1992), knowledge concerning *heiau* and the rituals performed at these places was suppressed and gradually lost. By the time that the first archaeologists started to record and study *heiau* sites in the early twentieth century, five or six generations had passed; although some informants still knew the names of certain *heiau* and even some details of their function or history, no one living had directly participated in the pre-Christian *heiau* system. Consequently, information concerning *heiau* and their place in traditional Hawaiian society must be gleaned from the finite set of ethnohistorical sources. Valeri (1995, xvii–xxviii) succinctly summarized the main sources, which include the accounts of early voyagers and missionaries as well as the writings of several Native Hawaiians; a few remarks of our own are, nonetheless, in order.

A number of early European voyagers, beginning with Captain James Cook's third expedition in 1778–1779 (Beaglehole 1967) and continuing up to the time of Kamehameha I's death in 1819, contain valuable firsthand accounts of temples and temple rituals witnessed by these visitors, as well as priceless sketches and lithographs depicting *heiau* with their perishable superstructures and accoutrements (e.g., Joppien and Smith 1988). These sources, however, have two main shortcomings: they tend to focus on major royal temples (*luakini*) and the rites associated with the highest-ranked chiefs, neglecting lesser temples and commoner modes of worship; and they were for the most part constrained by the voyagers' lack of fluency in the Hawaiian language and limited understanding of Hawaiian culture. As Valeri remarks (1995, xxiii), the later missionary accounts are for the most part of little value, with the exception of that by Rev. William Ellis of the London Missionary Society, who toured Hawai'i Island for two months in 1823–1824 (Ellis [1827] 1963). Although the *kapu* system had been abolished by this time, Ellis (who spoke Tahitian and quickly gained fluency in Hawaiian) acquired valuable information, in particular from Hewahewa, the last *kahuna nui* (high priest) of Kamehameha I, and from Kēlou Kamakau, a Kona chief.

Without doubt, however, the most important ethnohistorical sources are those of several Native Hawaiian writers of the mid-nineteenth century, most of whom attended the missionary seminary at Lahainaluna, Maui, where they were encouraged by William Richards and Sheldon Dibble to collect and transcribe accounts of traditional Hawaiian culture and history, or *moʻolelo*. Foremost among these Hawaiian scholars is Davida Malo, who was born to a chiefly family in Kona, Hawaiʻi, in 1795 and became a pupil of ʻAuwae, "Kamehameha's bard, genealogist, and ritual expert" (Valeri 1995, xxiv). Malo's *Moolelo Hawaii*, translated by missionary son Nathaniel Emerson and published by the Bishop Museum (Malo 1951), is a critical source on Hawaiian religion and ritual practices by someone born and raised prior to the advent of Christianity. A second member of the Lahainaluna seminary group and a prolific writer on Hawaiian traditions was Samuel Manaiākalani Kamakau. Born in 1815, Kamakau was not an eyewitness to temple ritual but had access to many individuals who had been (two of his grandparents had been *kāhuna* [Chun 1993, 17]). Kamakau recounted what he had learned in a long series of articles in *Ka Nupepa Kuokoa* and several other Hawaiian-language newspapers; these were later translated by Mary Kawena Pukui and published by the Bishop Museum (Kamakau 1961, 1964, 1976, 1991). The third major source is John Papa ʻĪʻī, born in 1800 and raised in the royal court as an attendant to Liholiho (later Kamehameha II). ʻĪʻī (1959) provides a firsthand account of events in the royal household, including certain temple rituals. Other Native Hawaiian sources include Kepelino (who lived from approximately 1830 to 1878), whose *moʻolelo* was edited by Beckwith (1932) and contains valuable information on *heiau* as well as the Hawaiian calendar, and Kēlou Kamakau (born about 1773), a chief of Kona and an informant to William Ellis. Kēlou Kamakau's manuscript, probably written around the same time as Malo's (Valeri 1995, xxvi), was eventually translated and published as a part of the *Fornander Collection of Hawaiian Antiquities and Folk-Lore* (Fornander 1916–1920, 6:2–45).

We turn now to an overview of Hawaiian ritual places and practices with the objective of providing sufficient cultural context within which to situate our own investigation of *heiau* sites in southeastern Maui. This will lead us not only to a review of indigenous Hawaiian concepts regarding temples (i.e., their "emic" classification) but also to a consideration of the ritual cycle and its relation to the Hawaiian calendar. Finally, we conclude the chapter with a review of prior archaeological research on Hawaiian *heiau*, including their typology and chronology.

Defining *Heiau*

Archaeologists—and indeed many people in contemporary Hawaiʻi—tend to think of *heiau* simply as stone structures, often of monumental proportions, that were places for pre-Christian religious ceremonies. This view is only partly correct. While the majority of *heiau* sites do incorporate stone structures, architecture alone does not define a *heiau*. Rather, the word *heiau* denotes any place where worship was performed and offerings were made to one or more gods (*akua*) or ancestral spirits (*ʻaumakua*). The first Hawaiian dictionary, authored by Lorrin Andrews, offers as definitions of *heiau* both "a large temple of idolatry" and "one of the six houses of every man's regular establishment—the house for the god" (1865, 146). The latter is explained by Malo as a place within the *mua*, or men's eating house: "The sanctuary where they worshiped was called *heiau*, and it was a very tabu place" (1951, 29). *Heiau* can thus also be natural stones or uprights (such as the *pōhaku o Kāne*) where prayers are recited and offerings made. Pukui and

Elbert (1986, 64) offer the broadest definition of *heiau* as simply a "pre-Christian place of worship, [a] shrine."

The word *heiau* is a unique Hawaiian lexical innovation. Virtually everywhere else in Polynesia the word for a temple or place of religious ritual is *malae* or *marae*, a word that can be confidently reconstructed to the Proto-Polynesian language (PPN **malaqe*; see Kirch and Green 2001, table 9.2). That the original Hawaiian word for a place of worship was *malae* is indicated by its retention in at least four instances as a place-name or as the proper name of a *heiau* (e.g., Malae, a large *heiau* near the Wailua River on Kaua'i Island; Pukui, Elbert, and Mookini 1974, 143). At some point in Hawaiian history—for reasons that will probably never be known—the word *malae* was replaced by the new term *heiau*. This lexical innovation may have been related to major changes in the structure of Hawaiian society, land tenure, and politics that occurred around the end of the sixteenth century and beginning of the seventeenth, with the rise of archaic states on Hawai'i and Maui Islands (Hommon 2013; Kirch 2010, 2012). Valeri (1995, 173) avers that the etymology of *heiau* lies in the root *hai*, "to sacrifice" (Pukui and Elbert 1986, 47). Thus, in its essence, *heiau* or *haiau* designates any place where sacrifices are offered.

Heiau in Comparative Polynesian Context

The original Proto-Polynesian word for ritual place was **malaqe*, a term that can be confidently reconstructed from its 25 cognate terms in as many different Polynesian languages and dialects (Kirch and Green 2001, table 9.2). Throughout both Western and Eastern Polynesia, these variants of the ancient word **malaqe* all refer to various kinds of ritual or ceremonial spaces, even though such spaces vary greatly in architectural form and in function.

Drawing upon multiple lines of evidence from historical linguistics, comparative ethnography, and archaeology, Kirch and Green (2001, 254–256) reconstructed the ancient form of the **malaqe* complex or ritual space as it would have been manifest in Ancestral Polynesian culture during the mid-first millennium BC, describing it in this way (255):

> We infer these to have been architecturally simple affairs, consisting of an open, cleared space (**malaqe*) lying seaward of a sacred house (**fale-{qatua}*), the latter constructed upon a base foundation (**qafu*). The sacred house may sometimes have been the actual dwelling of the priest-chief (**qariki*), and may at times have contained the burials of ancestors (**tupunga* or **tupuna*). But we are confident that one or more posts (**pou*) within the sacred house were ritually significant.

From this simple ritual space centered around the dwelling of the priest-chief with an open ceremonial space on the seaward side, different forms of Western Polynesian *malae* and Eastern Polynesian *marae* gradually developed over time.

In Western Polynesia, the *malae* typically consisted of an extensive ceremonial plaza—where kava ceremonies were performed—fronting either a mound, a god house, or a row of upright stones (Emory 1943, 13–15; Kirch 1990). At the important ceremonial center of Mu'a in Tongatapu, for instance, the broad *malae* was flanked by burial mounds (*langi*) containing the bones of deceased ancestors of the sacred Tu'i Tonga line. Each mound was topped with a small, thatched god house (Kirch 1984, 227–230; McKern 1929). In Futuna and Alofi Islands, ritual plazas (*malae*) were flanked by rows of upright stone or limestone slabs; behind and above the slabs sat the house of the priest-chief (Kirch 1994). A simple form of *marae* is exemplified in the Polynesian Outlier of Tikopia, consisting of an open space on the seaward side of the chief's house (under the floor of which chiefly ancestors were buried), bounded around three sides of its perimeter with upright stones marking the "seating places" (*noforanga*) of ancestral deities (Firth 1967). These Western Polynesian and Outlier examples demonstrate the essential combination of elements making up a *malae:* court, uprights, and sacred house.

When Polynesian voyagers expanded eastward out of the Western Polynesian homeland in the late first millennium AD, they carried some version of the early *malae* complex with them. However, in the newly discovered archipelagoes of central Eastern Polynesia (Society Islands, Cook Islands, Marquesas, Tuamotus, Australs, and Mangareva), older Ancestral Polynesian concepts of ritual space were re-created with innovations particular to this Eastern Polynesian region. The first innovation was the formal delineation of a rectangular court, usually by enclosing the space with a stone wall or curbing. The term *marae* now applied to this architecturally delineated space. Second, the sacred house that had originally stood on the landward side of the ritual space in ancient Polynesia was supplanted by a platform or altar erected at one end of the *marae* enclosure. The Eastern Polynesian word for this altar, *ahu*, derives from the Proto-Polynesian term **qafu*, which originally signified the foundation of the ancient sacred house that fronted the ceremonial plaza. The sacred house itself was often now miniaturized into a small portable container of wood. In the Society Islands, for example, miniature *fare atua* (literally, god houses) held the sennit images of gods (Oliver 1974, 96, fig. 3-13). Third, the sacred posts (**pou*) of the ancient god house—which represented the ancestral deities whose bodies were frequently interred within the **qafu* foundation—were in the new Eastern Polynesian *marae* replaced by stone uprights set either along the *ahu* platform or in front of it. A further innovation was to replace some or all of these posts with anthropomorphic images of deities or deceased ancestors, carved in stone or wood; such images were termed *tiki, ti'i*, or in the Hawaiian case *ki'i*. The greatest elaboration of anthropomorphic images, upon such *ahu* platforms, arose in the case of the *moai* statues of Easter Island (Kirch 2000, 272–273).

These innovations in Eastern Polynesian ritual architecture evidently developed within the first few centuries following the discovery and settlement of the central Eastern Polynesian archipelagoes, because they were subsequently carried by voyagers who extended the Polynesian diaspora as far as Easter Island, New Zealand, and Hawai'i. We can infer that the early Eastern Polynesian *marae*—introduced to Hawai'i around AD 1000—were fairly simple and small affairs. The relatively simple *marae* of the Tuamotu Islands, described thoroughly by Emory (1934, 1947), in his view retained many of the basic characteristics of these earliest Eastern Polynesian *marae*. Likewise, the simple platform temples with uprights and anthropomorphic *tiki* of Nihoa and Mokumanamana Islands were thought to be representative of this earlier phase in Hawaiian temple architecture (Emory 1928). However, as with all aspects of Polynesian culture and society, *marae* were not static. In each island group, *marae* builders developed distinctive innovations and architectural forms, in part responding to local changes in religious belief and ritual practice. Just as *marae* architecture evolved in the Society Islands (Kahn and Kirch 2014) or on Easter Island, so temple architecture changed over time in the Hawaiian Islands, along with the change in the name for a sacrificial place, from *marae* to *heiau*.

Hawaiian oral traditions frequently mention specific *heiau* in relation to the deeds of prominent *aliʻi* (members of the chiefly class) and contain hints as to the development of *heiau* over time. Samuel Kamakau (1976, 135) wrote that the first *heiau* were constructed at Waolani, in Nuʻuanu, Oʻahu: "That was where Wakea *ma* lived. At that place was the first heiau built in this archipelago, and there at Waolani were named the sacred and the consecrated places within the heiau." Kamakau recites a chant that incorporates the names of the principal components of a *heiau*, such as the pavement (*ʻiliʻili*), *ʻanuʻu* tower, *mana* house, drum house, offering stand, and principal image (*ka moʻi*). According to Kamakau, this "ancient style" of *heiau* was later modified by combining its features "with the styles of [the *kahuna*] Paʻao and [the prophets] Makuakaumana, Luhaukapawa, Luahoʻomoe, and Maihea *ma*" (135). In his account of the early Oʻahu chief Māʻilikūkahi, Kamakau (1991, 56) alludes to an earlier period of *heiau*, when human sacrifice was not practiced.

The traditions attribute a major change in Hawaiian religion to the arrival of Pāʻao, a priest who came from "Kahiki," probably the Society Islands, possibly toward the end of the fourteenth century (Kirch 2012, 204). According to Fornander ([1878–1885] 1996, 35–37), Pāʻao built Wahaʻula Heiau in Puna District and Moʻokini Heiau in Kohala District on the island of Hawaiʻi. He is also credited with introducing the worship of Kū and the practice of human sacrifice (Beckwith 1932, 58). Based on these traditions, Fornander was convinced that a major change had occurred in Hawaiian religious architecture around the time of Pāʻao's arrival (and cessation of the so-called voyaging period). Fornander argued that an earlier, "open truncated pyramidal structure" lacking enclosing walls was replaced by "the four-walled, more or less oblong, style of Heiau." In his view, this architectural change signaled a major shift in the nature of Hawaiian ritual practice:

> Under the old, the previous régime, the Heiau of the truncated pyramid form, with it presiding chief, officiating priests, and prepared sacrifice, were in plain open view of the assembled congregation, who could hear the prayers and see the sacrifice, and respond intelligently to the invocations of the priest. Under the innovations of this [later] period, the presiding chief, those whom he chose to admit, and the officiating priests, were the only ones who entered the walled enclosure where the high-places for the gods and the altars for the sacrifices were recited, the congregation of the people remaining seated on the ground outside the walls, mute, motionless, ignorant of what was passing within the Heiau until informed by the officiating priest or prompted to the responses by his acolytes. (Fornander [1878–1885] 1996, 59)

Fornander's theory prompted Bishop Museum director William T. Brigham to instruct pioneering archaeologist John F. G. Stokes to test the theory out by mapping the foundations of *heiau* throughout Hawaiʻi Island (Dye 1989); Stokes' data, however, did not support Fornander's theory.

Hawaiian Gods and Their Priests

A brief overview of the principal Hawaiian gods (*akua*) and of the priestly class (*kāhuna*) is essential before we turn to the problem of *heiau* classification and function.[1] The Hawaiians, like other Eastern Polynesians, were polytheistic, with four major deities dominating: Kū, Lono,

---

1. This section has been adapted from Kirch (2010, 53–58).

Kāne, and Kanaloa (Malo 1951, 81; Valeri 1985, table 1). Kū, god of war, was perforce the principal god of the king; his anthropomorphic images, covered in brilliant red feathers (Chun 2014), were carried into war as well as displayed at tribute collections of the king (*aliʻi nui*). But Kū had other functions beyond warfare, including fishing, canoe building, and sorcery; it was to him alone that human sacrifices were offered.

Lono, who divided the annual ritual calendar with Kū, was the god of dryland agriculture, fertility, and birth; his signs were clouds, thunder, and the rain essential for sweet potato growth. Lono's ritual season commenced once the Pleiades star cluster ("Makaliʻi" in Hawaiian) became visible immediately after sunset, which happened in late November (Valeri 1985, 199) and continued for the next four lunar months. This was the Makahiki (whose significance is discussed in the section "Traditional Hawaiian Calendar"). During the Makahiki, the cult of Kū was in abeyance and warfare was forbidden.

Kāne embodied the male power of procreation but was also the god of flowing waters and as such was the patron deity of irrigation and fishponds; his special crop was *kalo* (taro, *Colocasia esculenta*). Finally, Kanaloa was a deity of death, of the subterranean world, and of the sea (Valeri 1985, 15).

Each of these principal gods was associated with particular directions, colors, plants, animals, and astronomical phenomena, some of which are summarized in table 2.1. Kū was associated with the direction north, high mountains, old-growth forest, and trees such as *koa* and *ʻōhiʻa*. Lono had directional attributes of high and leeward (*kona*, from which direction the winter rains come), but most important was his link to the Pleiades (Makaliʻi), which rises in the east-

TABLE 2.1. Some attributes of the major Hawaiian deities.

	DEITY			
ATTRIBUTE	KŪ	KĀNE	LONO	KANALOA
Directions	North, high, right	East, right	High, leeward	West, south, left
Landscape and natural phenomena	High mountains, old-growth forest, *ʻōhiʻa* and *koa* trees, sky	Light, sun and the sunrise, fresh waters and streams	Thunder, rain, *kukui* tree, black pig	Seawater, ocean, setting sun
Time of day	—	Dawn	—	Sunset
Time of month	First 3 days of lunar month	27th–28th days of lunar month	—	23rd–24th days of lunar month
Season	Season of *kapu pule* (temple ritual)	Summer, sun's most northerly path	Visibility of Pleiades immediately after sunset, Makahiki season	Winter, sun's most southerly path
Functions	War, canoe building, farming, fishing (Kūʻula), sorcery	Male power of procreation, irrigated agriculture	Dryland agriculture, medicine	Fishing, navigation, death, healing

Note: Data are abstracted in part from Valeri (1985, table 1).

northeast and, from most viewpoints in Kahikinui, would rise over the long, broad slope of Haleakalā. Kāne was strongly associated with the east and the rising sun, or, as Kepelino put it, "the great sun of Kane" ("*ka la nui o Kane*") (Beckwith 1932, 14–16, 80). Kanaloa, god of the sea and the underworld, had directional attributes of west, south, and left; he was associated with the sunset and death.

Kāne and Kanaloa are often paired in the Hawaiian *moʻolelo*, traveling the land together and opening springs for water to mix with their *ʻawa* (kava). Indeed, Samuel Kamakau (1991, 112) relates how Kāne and Kanaloa first "opened the waters" at Lualaʻilua in Kahikinui and at Kou in Kaupō. The two deities are also linked in the daily and seasonal movements of the sun. The dawn, the rising sun, and the eastern sky belong to Kāne, while the afternoon, the setting sun, and the western sky are associated with Kanaloa. The northernmost path of the sun through the sky, at the summer solstice, is called "the black shining road of Kāne" (*ke alanui polohiwa a Kāne*) while its southernmost path, at the winter solstice, is called "the black shining road of Kanaloa" (*ke alanui polohiwa a Kanaloa*) (Johnson, Mahelona, and Ruggles 2015, 30, 179–180).

Each of these major gods had its own cults, its special priests, and its particular temples and rituals. The king concerned himself primarily with the annual ritual cycle alternating between Kū and Lono. While the commoners, the *makaʻāinana*, also participated in the state rituals and were required to furnish substantial quantities of sacrificial offerings (pigs, dogs, coconuts, and other foodstuffs) necessary for major temple rites, these commoners practiced—on a daily basis—an entirely separate, more private religious cult. Their practice focused on the collective body of ancestors, the *ʻaumākua*. The word *ʻaumākua* derives from the prefix *ʻau-*, designating a group, and *mākua*, the word for "parents" (Pukui and Elbert 1986, 32). *ʻAumākua* were literally then the collective body of parents, grandparents, and ancestors of particular family groups of *makaʻāinana* who—having no proper names because deep genealogies were not kept by the common people—took the form of various natural creatures, such as sharks, owls, octopuses, eels, shellfish, or plants. Prayers were offered to the *ʻaumākua* in the *mua*, the men's eating house, along with offerings of kava root and taro or other foods placed at a simple shrine (the *heiau* within the *mua* [Malo 1951, 29]). This distinction between the daily worship of the collective, nameless *ʻaumākua* in the men's houses of the commoners and the ostentatious public ceremonies carried out with great formality on the temple platforms dedicated to the state gods attests to the strong class distinctions that permeated late precontact Hawaiian society (Hommon 2013; Kirch 2010, 2012).

The Hawaiian priesthood was elaborated well beyond anything known elsewhere in Polynesia (Handy 1927). The general term for priest is *kahuna* (plural *kāhuna*), derived from the Proto-Polynesian term **tufunga*, or expert (Kirch and Green 2001, tables 8.7 and 9.2). In Hawaiʻi, *kāhuna* were full-time specialists. There were several major classes of *kāhuna*, the most important being the *kahuna pule*, the *kahuna lapaʻau*, and the *kahuna ʻanāʻanā*. The priests who officiated at temples controlled by the king and major chiefs were the *kāhuna pule* (*pule* means "prayer"). These were subdivided into a number of specific orders or cults, especially those pertaining to Kū and Lono (*moʻo Kū* and *moʻo Lono*). Priests were drawn from high-ranking elite families, typically of *papa* rank (in which the person's mother comes from one of the three highest ranks). The most important priest of the order of Kū was the *kahuna nui*, or high priest (Valeri [1985] calls him the "royal chaplain"), who was responsible for the king's religious duties and looked after his temples and main gods.

Kāhuna lapaʻau were medicinal priests or curing experts. Kamakau (1964, 98) lists eight specific varieties, including those who specialized in diagnosis, others who used sorcery or magic in treatment, and priests who were particularly knowledgeable in the application of herbal medicines. Greatly feared were the *kāhuna ʻanāʻanā*, or sorcerers, who tended to perform their work outside of, and indeed in opposition to, the main state cults (Valeri 1985, 138). *Kāhuna ʻanāʻanā* were thus somewhat marginal figures, although they could wield substantial power and influence. Another marginal category is that of the *kāula*, or prophets, a category that can also be traced back to Ancestral Polynesia (PPN *taaula*; Kirch and Green 2001, table 9.2). These individuals were thought to have direct relationships with particular gods and would enter into trance states in which they spoke the god's oracle. Although *kāula* did not participate in the state cults, they were frequently sought out by kings for their advice. A famous example from the immediate postcontact era is that of Kapoukahi, a prophet of Kauaʻi who was consulted by the chiefess Haʻaloʻu, on behalf of Kamehameha, to ascertain what acts would assure the latter's ascendancy over all of Hawaiʻi (Kamakau 1961, 149–150). Kapoukahi directed Kamehameha to build the great *luakini heiau* of Puʻukoholā. Another famous *kāula* was Lanikāula, who dwelt in a remote *kukui* (candlenut) grove on eastern Molokaʻi.

Types of *Heiau* and Their Hierarchy

The ethnohistorical sources regarding *heiau* offer a variety of names for different types or categories of *heiau*; some terms appear to be synonyms for the same functional type of temple (e.g., *luakini* and *poʻokanaka*). For the most part, these Hawaiian categories signify *functional* types rather than architectural types; indeed, it is not certain that there were even names for different architectural forms such as terraced platforms or walled enclosures. Kepelino, however, does remark that *heiau* dedicated to Kāne were square (Beckwith 1932, 58). The following is an overview of the main categories, beginning with the largest royal temples reserved for the king and proceeding down the temple hierarchy to the simplest forms of ritual space.

War Temples (*Luakini* or *Poʻokanaka*)

The primary royal temples reserved exclusively for the king (*aliʻi nui*) and dedicated to the war god Kū were the *luakini*, the only kind of temple in which human sacrifices were offered. These were also known as *poʻokanaka* (literally, human head), possibly because of "the custom of placing the skulls of human victims on top of poles erected around the temple" (Valeri 1985, 181). Malo (1951, 159–176) provides a lengthy description of the ceremonies involved in constructing a *luakini*. He writes: "The luakini was a war temple, *heiau-wai-kaua*, which the king, in his capacity as ruler over all, built when he was about to make war upon another independent monarch or when he heard that some other king was about to make war against him; also when he wished to make the crops flourish he might build a *luakini*" (161). This latter remark suggests that although Kū was the principal deity worshipped at *luakini*, the agricultural deities Lono and Kāne may also have at times been supplicated at these temples. Kamakau concurs that *luakini* temples "were only for the paramount chief, the *aliʻi nui*, of an island or district (*moku*)," adding, "Other chiefs and *makaʻainana* could not build them; if they did, they were rebels" (1976, 129).

Malo mostly concerns himself with the rites performed at *luakini* but does offer some brief comments on the physical layout of these war temples, noting that older temples were at times renovated or refurbished while at other times an entirely new temple would be constructed. The temple architect was known as the *kahuna kuhi-kuhi-puʻu-one*, a class of priests "thoroughly educated in what concerned a *heiau*" and "acquainted with the *heiau* which had been built from the most ancient times" (Malo 1951, 161). Malo says that these priest-architects had studied the old temples "on the ground" and "had seen their sites and knew the plans of them all." In proposing a new temple, the *kahuna kuhi-kuhi puʻu-one* built a sort of model of the structure on the ground, showing it to the king and obtaining his approval prior to construction.

According to Malo's description, a central feature of the *luakini* was a three-tiered structure of *ʻōhiʻa* wood, known as the *lana-nuʻu-mamao*, which the high priest and king climbed up on during certain temple rites: "In front of the *lana-nuu-mamao* stood the idols, and in their front, a pavement (*kipapa*) and the *lele* on which the offerings were laid" (1951, 162). In front of the *lele*, or offering platform, was another pebble pavement where offerings were laid prior to being offered up (*hai*) onto the *lele*. Continuing down the temple court farther from the *lele* were several thatched houses:

> In front of the *lele* was a house called *hale-pahu*, with its door facing the *lele*, in which the drum was beaten. At the back of the *hale-pahu* stood a larger and longer house, called *mana*, its door also opening toward the *lele*. To the rear again of the *hale-pahu* was another house which stood at the entrance of the *heiau*. In the narrow passage back of the drum house (*hale pahu*) and at the end (*kala*) of the house called *mana* was a small house called *waiea*, where the *aha* cord was stretched.
>
> At the other end (*kala*) of the *mana* house was a house called *hale-umu*, in which the fires for the *heiau* were made. The space within the *pa*, or enclosure, was the court, or *kahua*, of the *heiau*. Outside of the *pa*, to the north, was a level pavement (*papahola*), and to the south and outside of the *pa*, stood the house of Papa. At the outer borders of the *papahola*, crosses were set in the ground to mark the limits of the *heiau*. (Malo 1951, 162)

ʻĪʻī (1959, 35), in his account of *kapu loulu* rites held by Kamehameha I at Papaʻenaʻena Heiau on the slopes of Lēʻahi (Diamond Head) around 1804, offers a description of the layout of a *luakini* temple that largely corroborates Malo's account. Here ʻĪʻī uses the term *ʻanuʻu* for a tower that served as the priest's oracle place (presumably the same kind of structure called the *lana-nuʻu-mamao* by Malo), which was fronted by a "row of idols." Between the images and the drum house stood the *lele* altar, or offering platform. ʻĪʻī says that the *mana* house was long, "like a *halau*." There were as well the oven house (*hale umu*) and the *hale waiea*. Of the latter, ʻĪʻī writes: "Two images stood before it on either side of the opening, and the king and kahuna conducted their *ʻaha* services at the right side of the opening, in the dark of the night before the birds began to twitter" (ibid.).

Kamakau (1976, 130) adds a few other details regarding the layout of a *luakini*, noting that "first the foundation, the *kahua*, was nicely leveled." Kamakau refers to "seven raised pavements, *kipapa nuʻu*, and also seven *paehumu* fences," above (or beyond) which were evidently the *mana* house and then the *lele* offering stand.

Some of the principal features of a *luakini*—in this case the temple of Ahuʻena at Kailua, Kona, used by Kamehameha I and reconstructed based on early nineteenth-century drawings and descriptions—are visible in figure 2.1.

FIGURE 2.1. Ahuʻena Heiau in Kona, Hawaiʻi Island, with superstructures reconstructed based on ethnohistorical descriptions and sketches. Note the wooden fence, the offering platform, the *ʻanuʻu* tower wrapped in white barkcloth, and the wooden images. (Photo by Thérèse Babineau)

Fertility or Agricultural Temples, *Heiau Hoʻoulu ʻAi* (*Hale o Lono, Mapele, Unu*)

Certainly the largest class of *heiau*, and that for which there is the most confusion regarding names and possible subtypes, is the fertility temples, or *heiau hoʻoulu ʻai*. The word *hoʻoulu*—from the root *ulu*, meaning "to grow, increase, spread; growth"—signifies "to grow, sprout, propagate, to cause to increase" (Pukui and Elbert 1986, 369). Hence, these were temples for increasing food. This kind of temple is sometimes referred to as a *hale o Lono* (Valeri 1985, 177–179). (There were also temples called *heiau hoʻoulu ʻia* and *heiau hoʻoulu ua* for increasing fish and for inducing rain, respectively.)

Malo discusses fertility temples only briefly, as an aside in his long treatment of *luakini* (1951, 160), using the terms *mapele* and *unu o Lono* for such temples:

> The *mapele* was a thatched *heiau* in which to ask the god's blessing on the crops. Human sacrifices were not made at this *heiau*; pigs only were used as offerings. Any chief below the king in rank was at liberty to construct a *mapele heiau*, an *unu o Lono*, or an *aka*, but not a *luakini*.... The *mapele*, however, was the kind of *heiau* in which the chiefs and the king himself prayed most frequently.

Malo notes that "if the king worshipped after the rite of Lono," he would erect either a *mapele* or an *unu o Lono*. Such temples were also outfitted with a *lana-nuʻu-mamao* tower, but in this case it would be constructed of *lama* (*Diospyros sandwicensis*) wood, and the temple house was thatched with *kī* leaves (*Cordyline fruticosa*). "There were also idols," Malo adds.

Kamakau (1976, 129–131) uses slightly different terms for these agricultural temples, calling them *waihau ipu-o-Lono* (or simply *waihau*) or *heiau ipu-o-Lono*, although he does mention the term *unu* in passing as well. He calls these "comfortable heiaus" in that they did not require human sacrifices: "The chief would build one first; afterward the people, the *makaʻainana* and the *kanaka*, of the land would build them so that the land might 'live' (*no ke ola o ka ʻaina*). The offerings were easy ones" (129). Further on, Kamakau adds: "The ruler, *moʻi*, also had *waihau* and *unu* heiaus, but chiefs could build these heiaus too. Pigs, bananas, and coconuts were the sacrifices at these heiaus" (132). Kamakau refers to such agricultural or fertility temples in the following passage (132):

> Some of their heiaus were large, and some were small. Some were surrounded by wooden tabu enclosures, *paehumu laʻau*, or sometimes stone walls, with a single house within. Outside of the gable end of the house was an *ʻanuʻu* tower, and a *kuapala* offering stand and *lele* altar where bananas were placed. The dedication of these heiaus belonging to the people called for pig eating, and when it was heard that some prominent person had a heiau to dedicate, *hoʻokapu*, and that many pigs had been baked, the *poʻe paʻa mua akua* came from all about to the pig-eating.

Kepelino is spare with details of *heiau*, but he does comment, "The heiau of commoners were small with just a circular fence and an elevation where the god was placed. A little lower was a place for the person who offered the sacrifice" (Beckwith 1932, 58). This seemingly describes a lower court where a supplicant was seated and an elevated terrace or court where the image or other representation of the deity resided. As will be seen in chapter 3, such vertical divisions within *heiau* are typical of temples throughout Kahikinui and Kaupō. Kepelino specifies that coconuts, red fish, and bananas were characteristic offerings at such commoner *heiau* (Beckwith 1932, 60).

ʻĪʻī (1959, 56) described a *hale o Lono* at Honolulu in the early years after Kamehameha's conquest of Oʻahu:

> Wooden female images stood outside of [the] enclosure, with *iholena* and *popoʻulu* bananas in front of them. There were *maoli* bananas before the male images at the *lele* altar inside of the enclosure of *lama* wood. Back of the male images of wood was an *ʻanuʻu* tower, about 8 yards (*iwilei*) high and 6 yards wide. It stood on the right side of the house, and was covered with strips of white *ʻoloa* tapa attached to the sticks resembling thatching sticks. The *opu* tower was just as tall and broad as the *ʻanuʻu*, and was wrapped in an *ʻaeokahaloa* tapa that resembled a *moelola* tapa. The small *lama* branches at its top were like unruly hair, going every which way. The *opu* stood on the left side of the house, facing the images and the *ʻanuʻu*. Between the two towers and extending from one to the other was a fine pavement of stones. In line with the middle of the pavement were the gate and the house which was called the Hale o Lono, where Liholiho was staying.

In his discussion of the hierarchy of Hawaiian temples, Valeri (1985, 175–177) devotes several paragraphs to the *unu* category of *heiau* and to how it overlaps or differs from the other categories of *māpele* and *waihau*. Valeri concedes that there is considerable confusion in the literature but thinks that *unu* referred to an architectural form, in the shape of an oval or a horseshoe; his conclusion in this regard is not convincing to us. It is evident, however, that the term *unu* was used with respect to both small shrines (such as fishing or agricultural shrines) and larger *luakini* temples and might refer to an altar on either of those kinds of *heiau*, rather than to the *heiau* itself. Emerson (1915, 202), for example, refers to *unu kupukupu* as an altar upon which fishermen laid their offerings and before which they uttered their prayers. It is interesting that the term *unu* also occurs in both the Society Islands and the Tuamotu Islands, where it refers to carved boards set upright along the back of the altar (*ahu*) and also on the court (Emory 1933, 15–16; 1947, 17). This suggests that the word *unu* may have derived from a Proto-Eastern Polynesian word indexing an altar.

Fishing Shrines, Koʻa

Koʻa, or fishing shrines, were a special category of *heiau* where fishermen made offering to Kūʻula, their patron deity. Kamakau writes: "*Heiau koʻa* were close to the beach or in seacoast caves, on lands with cliffs. The purpose of the *heiau koʻa* was important. The *koʻa* brought life to the land through an abundance of fish; there was no other purpose for the *koʻa* but this" (1976, 133). Kamakau adds that most *koʻa* were small and simply "rounded heaps of stones with a *kuahu* altar," although "some consisted of a house enclosed by a wooden fence."

Hale Mua

As noted earlier, the most fundamental kind of *heiau* was that which was a part of every household, or *kauhale* (literally, group of houses): the *hale mua* or simply *mua*, the men's eating house. Malo (1951, 29) noted that the *mua* was a kind of *heiau* and a *kapu* place. Likewise, Kamakau (1976, 133) wrote that "the *heiau ipu-o-Lono* constantly maintained by the populace was the *hale mua*, the men's eating house, which every household had." Valeri (1985, 173–174) discusses *mua*, noting that the *mua* of the king and other prominent chiefs were prominent structures with altars and images of gods.

In this book we will not concern ourselves with archaeological structures that functioned as *mua*, for this is better treated as an aspect of household archaeology. A number of sites with what we interpret as *mua* were studied and excavated by Kirch in Kahikinui; these will be described in a later book on household excavations.

Pōhaku o Kāne

A final kind of sacred place at which offerings and prayers were made is the *pōhaku o Kāne*, an upright stone (either entirely natural or a stone set in place) whose phallic shape represents the male procreator deity. Kamakau describes these as "single stones set up in commemoration of Kane," noting that "families would set up these stones at their residences, and to them they would go and make offerings and sacrifices" (1976, 130).

Heiau Architectural Components

While the archaeological study of *heiau* is largely confined to their stone foundations, as the descriptions above make clear, *heiau* when in use included several different kinds of houses, structures, and images, all made of perishable materials. Here we briefly summarize the main kinds of perishable superstructures (see also Valeri 1985, 237–253; Buck 1957, 519–527).

Fences (*paehumu*). Wooden fences often enclosed *heiau*; sometimes the wooden posts were themselves carved with anthropomorphic images of deities. Early European visitors described human skulls (of sacrificial victims) as being fixed atop poles that were part of *luakini heiau* fences. There were apparently also internal fences within some *heiau*, marking spatial divisions.

Towers (*'anu'u, lana-nu'u-mamao*). A tower constructed of wooden poles with three stages or platforms seems to have been an essential component of *luakini heiau* and probably of many (if not all) *heiau ho'oulu 'ai* as well. The kind of wood used varied between *luakini* and *heiau ho'oulu 'ai*. Towers were sometimes covered with white *'oloa* barkcloth. Such towers seem to have been located at the most sacred end of the *heiau* court (what Valeri calls the "sanctum sanctorum"), fronted by the image or images of the gods. The most commonly used term for the tower is *'anu'u*, although Malo (1951, 162) calls them *lana-nu'u-mamao*. 'Ī'ī (1959) mentions a second tower, called *ōpū*, at the *hale o Lono* in Honolulu. During temple ceremonies the priest would climb up on the *'anu'u* tower, where it appears he would receive the oracle of the deity being addressed. Valeri, however, opines that the towers might also have been used as "an observatory for noting the positions of the stars and other heavenly signs of divine will" (1985, 238).

Houses (*hale mana, hale pahu, hale umu, hale waiea*). As described earlier, four functionally distinct kinds of thatched structures were contained within the enclosure of a war temple (*luakini heiau*). The largest house, the *hale mana*, was used to keep cult paraphernalia and images, including the feathered war gods (Valeri 1985, 239). The functions of the drum house (*hale pahu*) and of the oven house (*hale umu*) are made clear by their names. (The more common variant of the Hawaiian word for an earth oven is *imu*, but Kamakau [1976, 138] used the alternate, *umu*, which is also the more widespread Polynesian term.) The *hale waiea* (literally, water of life) was quite small but considered very important as the locus for the *'aha* rite.

Altars and offering stands (*kīpapa, kuahu, lele*). Sacrificial offerings were placed on three different kinds of structures within *heiau*. The first were simple pavements (*kīpapa*), often of waterworn pebbles or gravel (*'ili'ili*), in front of images or offering stands. Wooden stands called *lele*, consisting of four sticks or posts supporting a wooden platform, typically were situated in front of temple images to receive offerings such as pigs, bananas, coconuts, and so forth. A third kind of altar is the *kuahu*, which seems to indicate a stone platform or an elevated terrace. As will be seen in chapter 3 and in the various *heiau* descriptions in part II, such stone terraces or benches are quite common in southeastern Maui *heiau*.

Refuse pits (*lua, lua-kini*). Pits called *lua* or *lua-kini*, used for the disposal of sacrificial offerings, were a component of *heiau*. Malo (1951, 152) states that the pit in a *luakini heiau* was situated within (or beneath?) the oracle tower, but it is likely that pits were located in other parts of temples as well. Pits are present in a number of the temples described in part II of this book.

Images and carved slabs (*ki'i*). Images called *ki'i*, typically in wood but sometimes in stone or in wickerwork covered with feathers, were regarded as representations of the deities and as receptacles for the gods when these descended to receive their offerings during temple rites. Fixed images were permanently set up within temples, their bases set into image holes in the stone foundations of the temple court. Portable images and wicker, feather-covered images were also brought out for use during ceremonies (Chun 2014). Hawaiian images are extensively described and illustrated by Buck (1957, 467–502) and by Cox and Davenport (1974).

Heiau Orientations

Archaeologists, beginning with Stokes (1991, 35–36), have stated that *heiau* do not show preferential orientations. The comments of McAllister (1933b, 9) regarding the orientations of Oʻahu *heiau* are typical: "Orientation apparently depended only upon the slope of the land. The heiaus face in all directions of the compass. The only generalization that can be made is that most of them face the sea." In chapter 4, we will demonstrate this claim to be false in the case of the Kahikinui and Kaupō temples. The Native Hawaiian writers, however, provide some hints that *heiau* were laid out according to specific directions. Malo (1951, 162), for example, in writing about the layout of a *luakini* temple, states:

> The plan of the *luakini* was such that, if its front faced west or east, the *lana-nuu-mamao* would be located at the northern end. If the *heiau* faced north or south, the *lana-nuu-mamao* would be located at the eastern end; thus putting the audience either in the southern or western part of the *luakini*.

Valeri (1985, 253–256) reviewed in some detail the ethnohistorical evidence for cultural principles underlying the orientation of *heiau* and of component structures such as the ʻanuʻu tower within *heiau*. Referring to the case of Puʻukoholā temple at Kawaihae, Hawaiʻi, where the ʻanuʻu tower and the main altar are situated in the southern part of the temple, Valeri raises the following important question: "Can we suppose that in any *luakini* inaugurated during a time of war the tower and the altar for the human sacrifice are oriented in the enemy's direction?" (256). This is a question we will return to later in this book, when evidence is presented that a number of *heiau* in Kahikinui and Kaupō have features apparently oriented to enemy territory, especially to Hawaiʻi Island across the ʻAlenuihāhā Channel.

The Ritual Cycle

The rituals performed in *heiau* by the Hawaiian priests followed a prescribed annual cycle. This ritual cycle was divided into two main periods: the first—of four lunar months' duration—was called the Makahiki and dedicated to the god Lono; the second, lasting eight lunar months, was dedicated to the *luakini* temple rituals of the king, with Kū being the principal (but not exclusive) deity to whom sacrifices were offered. This latter period (which had no formal name) was also the time during which war could be promulgated.

Traditional Hawaiian Calendar

The traditional Hawaiian calendar in use at the time of European contact was based on a sequence of 12 lunar months (table 2.2). This protohistoric Hawaiian calendar had been inherited from the much older Proto-Polynesian lunar calendar of Ancestral Polynesia, which Kirch and Green (2001, 267–273) reconstructed in some detail. (Although most of the Hawaiian names for the specific lunations had changed over time, two names clearly retain their Proto-Polynesian form: Makaliʻi from PPN *Mataliki, and Hinaia-ʻeleʻele from PPN *Siringa-kelekele [ibid., table 9.7].) And as in Ancestral Polynesian times, determining the onset of the new year (and synchronizing the 354-day lunar cycle with the 365-day solar year) was based on careful observation of

TABLE 2.2. The Hawaiian calendar.

SEASON	LUNAR MONTH	RITUALS	COMMENT
Kau	ʻIkuwā	Makahiki rituals	Acronychal rising of Pleiades (Makaliʻi)
Hoʻoilo	Welehu	Makahiki rituals	
Hoʻoilo	Makaliʻi	Makahiki rituals	
Hoʻoilo	Kāʻelo	Luakini temple rituals	Aku fishing season
Hoʻoilo	Kaulua	Luakini temple rituals	Aku fishing season
Hoʻoilo	Nana	Luakini temple rituals	
Hoʻoilo	Welo	Luakini temple rituals	
Kau	Ikiiki	Luakini temple rituals	
Kau	Kaʻaōna	Luakini temple rituals	
Kau	Hinaia-ʻeleʻele	Luakini temple rituals	ʻŌpelu fishing season
Kau	Māhoe-mua	Luakini temple rituals	ʻŌpelu fishing season
Kau	Māhoe-hope	Luakini temple rituals	ʻŌpelu fishing season

Note: Modified after Valeri (1985, 199).

the rising of the Pleiades, called Makaliʻi in Hawaiian (a continuance of the old Proto-Polynesian name *Makaliki, or Little Eyes), as well as by other names such as Huhui or Huihui, meaning "cluster" (Johnson, Mahelona, and Ruggles 2015, 161, 211).

The Hawaiian calendar was described by all of the main nineteenth-century Native Hawaiian scholars, including Malo (1951, 30–33), Kamakau (1976, 13–19), and Kepelino (Beckwith 1932, 82–113). These sources agree in the main but have some differences in the naming, ordering, and timing within the seasonal year of the various months (see Johnson, Mahelona, and Ruggles 2015, 216 for a tabular summary), which reflect regional variations between the islands and the fact that by the time of their writing (especially for Kamakau and Kepelino) the traditional calendar had been supplanted by the missionary-introduced Gregorian calendar. Like Valeri, we take Malo's account—being the earliest—as the most accurate; the following synopsis is based on Valeri's careful consideration of all the sources (1985, 194–199).

While an annual calendar of 12 or 13 lunations is pervasive across all Polynesian cultures—and hence can be reconstructed back to Ancestral Polynesian times—a major innovation occurred in Eastern Polynesia with the development of a set of 30 "nights-of-the-moon" names (Kirch and Green 2001, 276). Thus in Hawaiʻi, each phase of the moon through its waxing and waning of a complete lunar cycle has a specific name, beginning with the night of Hilo and ending with the night of Muku (Malo 1951, 31–33). Unlike the naming and sequencing of the months themselves, the naming of the nights-of-the-moon is remarkably consistent among the Hawaiian sources (Johnson, Mahelona, and Ruggles 2015, 217). Malo, Kamakau, and Kepelino all describe in detail how each named phase was regarded as being propitious for certain kinds of activities, such as planting or fishing.

More important from our perspective of heiau rituals, however, there were four periods of two or three nights duration each, known as *kapu pule*, which were reserved for ritual activity and during which the men (including commoners) were required to spend their nights within the temple precincts. As Kamakau (1976, 18) states: "There were four tabu periods in the month that were made sacred and treated with great reverence." According to Malo (1951, 32–33), the four *kapu* periods were Kū (nights of Hilo, Koaka, and Kū-kāhi), Hua (nights of Mōhalu and Hua), Kaloa (nights of 'Ole-pau and Kāloa-ku-kāhi), and Kāne (nights of Kāne and Lono). Malo states that these *kapu* periods were kept during the eight months outside of the Makahiki period (33). That the men were required to spend their nights within the temple precincts is of considerable interest (see chapter 3). There is ample evidence that male activities such as adze production, wood carving, and the manufacture and repair of fishing gear took place within these temples.

In addition to the lunar months and the named nightly moon phases, the Hawaiian calendar was divided into two main ritual periods (see table 2.2). The four lunar months beginning with 'Ikuwā and ending with Kā'elo were known as the Makahiki, a period dedicated to the rituals of Lono. (Note that Kamakau [1976, 15] has the Makahiki period commencing with Welehu and ending with Kaulua.) Malo writes: "In the month of Ikuwa the signal was given for the observance of Makahiki, at which time the people rested from their prescribed prayers and ceremonies [i.e., the *kapu pule* days and nights] to resume them in the month of Kau-lua. Then the chiefs and some of the people took up again their prayers and incantations, and so it was during every period in the year" (1951, 33).

The "signal" referred to by Malo for the onset of the Makahiki was the acronychal rise of the Pleiades—that is, their first appearance immediately after sunset. This is the apparent rather than the true acronychal rise—the rising of the Pleiades (or any asterism) at sunset—which would not be visible to the naked eye, because of the bright sky (Ruggles 2014b, 464–465). In practice, as this season approached, the Pleiades would be seen to rise early in the night, progressively earlier each night. Eventually, the brightness of the sky following sunset would prevent the actual rise from being seen, and as the sky darkened, the Pleiades would appear already in the sky.

As with the nights-of-the-moon sequence, the Makahiki (or **mata-fiti* in Proto–Central Eastern Polynesian) was an innovation in Eastern Polynesia (Kirch and Green 2001, 276; see also Makemson 1941, 82–83). John Papa 'Ī'ī described this event as follows: "In the month of October, Ikuwa by Hawaiian count, the king declared a single kapu night, called Kuapola, in the *luakini*. The kapu extended from evening far into the night. They all gathered at the *kuahu* altar and waited for the appearance of the Huhui over the forest or mountain top" (1959, 72). The observation of the Pleiades so as to keep the lunar calendar in sync with the sidereal year is a pan-Polynesian phenomenon (see Kirch and Green 2001, table 9.4); it must be a very old practice, traceable back to Ancestral Polynesian times. As Valeri notes, however, there are hints that in Hawai'i (as in Mangareva; see Kirch 2004a) the changing positions of the sun were also observed and that "the solar year was recognized as well" (1985, 197). One outcome of our research is that many of the *heiau* of Kahikinui and Kaupō were sited and oriented for the purposes of solar observation (see chapter 4). This strongly reinforces the view that both sidereal and solar observations were used in the annual recalibration of the Hawaiian lunar calendar.

Finally, the Hawaiian year was also divided into two named seasons of six months each: Kau (from Ikiiki to 'Ikuwā) and Ho'oilo (from Welehu to Welo). Kau begins when the Pleiades set at dawn and roughly corresponds to the dry season, especially on the leeward sides of the large islands of Maui and Hawaiʻi. Hoʻoilo, which incorporates the Makahiki, is the wet season of *kona* rains in the leeward districts, and hence the main growing season for sweet potato.

The correlation between the strongly seasonal rainfall patterns of the leeward regions of the islands and the divisions of the traditional calendar bears further comment. Figure 2.2 displays the mean monthly rainfall recorded in historic times at Auwahi in Kahikinui and at Keʻeke in Kaupō. Although the Kaupō station has a slightly higher overall rainfall, both stations exhibit marked seasonality, with a wet period that begins around November and ends by April or May, followed by a distinct dry period. The five to six wet months, when southeastern Maui receives its *kona* (southerly) storms that bring the rain, is the period during which the sweet potato crop had to be grown. The onset of the Makahiki, marked by the acronychal rising of the Pleiades, thus occurred at approximately the same time that the first *kona* rains began to fall in Kahikinui

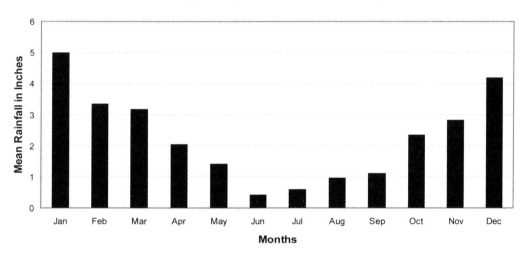

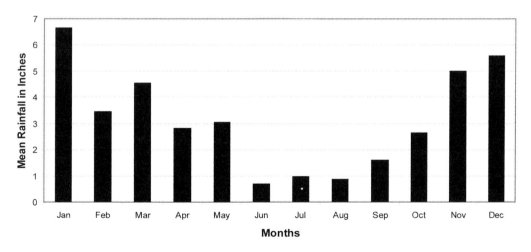

FIGURE 2.2. Mean monthly rainfall at Auwahi, Kahikinui, and at Keʻeke, Kaupō, in inches. (Data source: Giambelluca et al. 2013)

and Kaupō. Moreover, the four lunar months of the Makahiki season corresponded to the period of greatest rains and hence to the main growing season of the sweet potato crop. The earliest maturing varieties of sweet potato would have been ready for harvesting at the close of the Makahiki, when the *hoʻokupu*, or tribute, was collected from each of the *ahupuaʻa*. This timing may also be referenced in the Hawaiian proverb *Pua ka wiliwili nanahu ka manō* (When the *wiliwili* tree blooms, the sharks bite) (Pukui 1983, 295). The *wiliwili* (*Erythrina sandwicensis*), a dominant tree throughout Kahikinui and Kaupō, blooms around March to April, at the close of the Makahiki period. Hawaiian chiefs were metaphorically regarded as sharks, *manō*, who "traveled on the land" to collect their tribute (Kirch 2012; 2014, 45).

Makahiki Rituals

As noted earlier, the Hawaiian lunar calendar was divided ritually into two parts: a period of four lunar months collectively called the Makahiki, sacred to Lono, and a period of eight lunar months when the main temple rituals of Kū were practiced (Valeri 1985, 194–199).[2] The closing of one year and the signaling that a new one would soon commence began with a rite in which the king placed a rag before the main war temple of Kū (K. Kamakau in Fornander 1916–1920, 6:35); after this the king and the *kahuna nui* gathered at night in the temple to await the first visibility of the Pleiades in the darkening sky after sunset. When they became visible, the king performed the *kuapola* rite, in which he broke open green coconuts that signified life, purification, and renewal.

The ensuing Makahiki season has been called both a "New Year's Festival" (Valeri 1985) and a "harvest festival" (Handy and Handy 1972). The period was sacred to Lono, god of dryland agriculture, and during this time ritual activity centered on the *hale o Lono* temples, particularly the main *hale o Lono* of the king. Symbolically, Lono returned from Kahiki, the distantly remembered homeland beyond the horizon, to fertilize the land and in particular to bring the rains to the vast dryland field systems. It is thus no coincidence that the Makahiki corresponds to the *hoʻoilo*, the wet season of winter rains (brought by *kona* storms) that deposit most of the annual quotient of rainfall on the leeward sides of the islands, where the great sweet potato and dryland taro plantations were situated. The planting of sweet potatoes was presumably completed by the time that the Makahiki was declared; the Makahiki season corresponds closely to the main growing period for the crop. That war was forbidden during this interval, and indeed that the gardens themselves were declared to be *kapu* for a certain period, can be seen as a ritually inscribed means to assure that nothing adversely affected the new crop.

The Makahiki was not a uniquely Hawaiian concept but was the Hawaiian version of a more widespread Eastern Polynesian "first fruits" or "new year" celebration; the term can be reconstructed back to Proto–Central Eastern Polynesian *Mata-fiti (Kirch and Green 2001, 276, table 9.8). The earlier Proto–Central Eastern Polynesian *Mata-fiti was probably a simple first-fruits prestation to the gods and chiefs (257–260). By the time of contact with the West, however, the Makahiki had taken on a much more significant function with respect to the emergent Hawaiian state: it was the principal ritualized form of tribute exaction. Here we only briefly summarize the complex set of rites by which the king and his subsidiary *aliʻi* (chiefs) used the Makahiki to collect their respective *hoʻokupu* (tribute). Accounts are given by Malo (1951), ʻĪʻī (1959, 70–77), and Kamakau (1964, 19–21) and discussed in detail by Handy and Handy (1972, 327–371)

---

2. This section and the "*Luakini* Rituals" section have been adapted from Kirch (2010, 60–66).

and especially Valeri (1985, 200–233). Indeed, some aspects of the Makahiki rites were witnessed and described by Captain Cook's expedition at Kealakekua Bay in 1779 (Beaglehole 1967). Not the least of these was the ritual feeding of Cook himself, cum Lono, as described and analyzed by Sahlins (1985b) and famously illustrated by the expedition's artist, John Webber (Joppien and Smith 1988, 526).

Two distinct phases of tribute collection occurred during the course of the Makahiki. The first has been less remarked in the conventional anthropological summaries but is clearly described by Kēlou Kamakau (Fornander 1916–1920, 6:38–39):

> And when seven more nights had come to pass and on the day of *Laau-ku-lua*, the deities of all the lands were turned on this day. They were not to be stood up, as the annual restrictions prevailed, and the collectors of tributes [*na kanaka halihali waiwai*] from all over the land were near, and had brought a great collection of goods for the king's annuity, consisting of dogs, cloths, malos, fish and all other things and placed them before the king, all the districts paying tribute this day. And in the night of *Laau-pau* (the 20th) the collection was displayed and the king's feather deity and the lesser priests came to distribute the offerings this night. This was a very sacred night, no fires burning, and no noise to be heard. They offered prayers this night and then went to sleep.

This passage describes a kingdom-wide collection of prestige foods (dog in particular) and wealth items (especially barkcloth) brought specifically for the king and, indeed, placed in front of the feathered image of his personal god, Kūkā'ilimoku. As Valeri (1985, 204) points out, during the process of collection of the tribute (*waiwai*, or wealth), the "gods of each land (*akua 'āina*) cannot remain upright but are placed in a horizontal position, doubtless as a sign of homage to the king." The assembled wealth was then redistributed to the various priests, *ali'i*, warriors, and other favorites of the king; notably, nothing is given to the commoners who produced the goods.

The second phase of tribute collection under the aegis of Lono has been much more widely remarked and described, both by Native Hawaiian writers and by ethnographers. This is the famous circuit of the "long god" (*akua loa*) and shorter circuit of the "short god" (*akua poko*) through the territorial units (*ahupua'a*) of the kingdom. The *akua loa* consisted of a wooden shaft tipped with a carved human head, with a crossbeam attached from which were suspended streamers of white barkcloth and the white pelts of *ka'upu* birds. The image was brought before the king, "who cried for his love of the deity," placing a *niho palaoa* (whale tooth pendant) necklace on the image and feeding pork, taro, and coconut pudding (*kūlolo*) and kava to the man who carried the image (thus representing its mouth). Properly anointed and supplicated by the king, Lono was sent across the land to do his work: "After this the long god was carried forth on a circuit of the land. The different lands paid tribute to the deity in cloth, pigs, feathers, chickens and food" (K. Kamakau, in Fornander 1916–1920, 6:42). The long god took 23 days to complete his right-handed (clockwise) circuit, whereas the short god took only four days on a left-handed counterclockwise circuit, apparently restricting his visit to the particular lands (*'ili kūpono*) of the king. Kamakau (1964, 20–21) gives a vivid description of the work of the long god as it made its circuit throughout the land:

Much wealth was acquired by the god during this circuit of the island in the form of tribute (*hoʻokupu*) from the *mokuʻaina*, *kalana*, *ʻokana*, and *ahupuaʻa* land sections at certain places and at the boundaries of all the *ahupuaʻa*. There the wealth was presented—pigs, dogs, fowl, poi, tapa cloth, dress tapas (*ʻaʻahu*), *ʻoloa* tapa, *paʻu* (skirts), malos, shoulder capes (*ʻahu*), mats, *ninikea* tapa, *olona* fishnets, fishlines, feathers of the *mamo* and the *ʻoʻo* birds, finely designed mats (*ʻahu pawehe*), pearls, ivory, iron (*meki*), adzes, and whatever other property had been gathered by the *konohiki*, or land agent, of the *ahupuaʻa*. If the tribute presented by the *konohiki* to the god was too little, the attendant chiefs of the god (*poʻe kahu aliʻi akua*) would complain, and not furl up the god nor twist up the emblems and lay him down. The attendants kept the god upright and ordered the *ahupuaʻa* to be plundered. Only when the keepers were satisfied with the tribute given did they stop this plundering (*hoʻopunipuni*) [in the name] of the god.

Although perishable foodstuffs such as poi (pounded taro root) were evidently consumed by the crowd accompanying the Makahiki circuit, no doubt the other valuables listed by Kamakau were returned to the king's storehouses. While Hawaiʻi has generally been characterized as having a "staple finance" form of economy (Earle 1997), most of the goods assembled during the two Makahiki tribute collections consisted of various kinds of material manufactures, some of considerable prestige value (such as the fine barkcloth, mats, birds' feathers, and ivory mentioned by Kamakau). This indicates that the Hawaiian economy also incorporated a substantial element of "wealth finance" based on the control of circulation of these material items.

The Makahiki season concluded with a series of rites whereby the king wrested symbolic control of the polity from Lono (Valeri 1985, 210–213). Most dramatic of these was the *kāliʻi* rite, which took place after the Lono gods had returned to the principal *hale o Lono* temple. The king set forth in a canoe and landed in front of the temple. A sham battle ensued, with spears thrown at the king and/or his champion (an expert in spear dodging), who were obliged to parry them. Symbolically, the king thus reclaimed his polity. He then went to the *hale o Lono* to see the god images, sacrificing a pig in honor of Lono, who had done the hard work of visiting the land "that belongs to us both" (Valeri 1985, 212, translating from K. Kamakau). After this the Lono images were taken down, wrapped up, and stored in the *mana* house of the *luakini heiau* until the next year. In the concluding rite of the Makahiki a tribute canoe (*waʻa ʻauhau*), filled with foods of various kinds, was set to sea and allowed to drift off, carrying Lono back to Kahiki, from whence he would return again after eight lunar cycles.

Luakini Rituals

With the Makahiki season ended, the king—after various purification rites and the first open-sea trolling for *aku* (bonito) fish—prepared with his high priest to undertake a second, more complex set of rites dedicated to Kū. These rites were concerned with various stages of building and dedicating (or rededicating) a *luakini heiau*, or temple of human sacrifice, the exclusive prerogative of an *aliʻi nui*. In practice, new temples were not constructed every year but, rather, from time to time, especially in preparation for an intended war of territorial conquest (as with Kamehameha's construction of the great *luakini* temple of Puʻukoholā at Kawaihae in 1790). On an annual basis, however, the principal *luakini* of the king was refurbished, and some version of these rites performed. Malo (1951, 159–187), ʻĪʻī (1959, 33–45), and Kēlou Kamakau (Fornander 1916–1920, 6:2–45) all give indigenous accounts of the *luakini* temple rites; Valeri has synthesized and interpreted these and other sources extensively (1985, 234–339). In their full version, the rites began with the *haku ʻōhiʻa*, the selecting and carving of a special *ʻōhiʻa* log into an image of Kū. The *ʻōhiʻa* tree (*Metrosideros polymorpha*) was a manifestation of Kū, and the nectar of its flowers nourished the forest birds that provided the brilliant red and yellow feathers for the capes, cloaks, and helmets of the *aliʻi* and for the images of Kū. The *haku ʻōhiʻa* was the first occasion on which a human sacrifice would be offered, the body buried along with that of a sacrificial pig at the foot of a selected tree.

After the image had been carved and transported to the temple, various houses on the stone foundation had to be constructed, thatched, and consecrated (or refurbished and reconsecrated, as the case may be), coinciding with the *kauila nui* rite, a complex ritual sequence involving the thatching of the temple houses (Valeri 1985, 281–285). With these preliminaries accomplished, the new image of Kū was brought into the temple to be erected between the *ʻanuʻu* tower and the *lele* altar. Another human sacrifice was required for this *ʻaha hulahula* rite, during which the corpse was "thrust into the hole where the base of the god's image is to stand" (Valeri 1985, 289).

These rites were preliminaries to what Valeri calls the "final transformation of the god," which was marked by a "great sacrifice" (*ka haina nui*) requiring enormous quantities of pigs, bananas, coconuts, red fish, special barkcloth, and, once again, human victims. As Malo writes, these offerings were all in specific quantities of 400: "When the chiefs and the people had finished feasting on the pork, the king made an offering to his gods of 400 pigs, 400 bushels of bananas, 400 cocoanuts, 400 red fish, and 400 pieces of *ʻoloa* cloth; he also offered a sacrifice of human bodies on the *lele*" (Malo 1951, 174). In a peculiar rite the following day, the king's ritual double (Kahoʻāliʻi) consumed the eye of one of the human victims. These rites also involved the binding and wrapping of the images—for example, in *kapa* (barkcloth) loincloths. The concluding rites of the entire sequence took place, not at the *luakini* temple, but at the *hale o Papa* temple of the female *aliʻi*, officiated by the priest of Papa (a principal female deity); this involved the participation of the king's wife. Valeri (1985, 330) interprets these final rites as the "desacralization" of the male congregation, allowing them once again to enter into contact with women and the society at large.

The foregoing précis of the Hawaiian state cults gives some sense of how the Hawaiian rulers and their priests used the ritual cycle both to cement their power and to extract surplus from the commoner class. The Makahiki rituals were carried out in Lono's name by his priests, accompanied by various celebrations and games. The common people were in various ways made to feel that Lono was "their" god, who brought prosperity to the land. But there can be little doubt that it was the king and chiefs who benefited from the tribute collections that swept the lands for the collective *waiwai*, "wealth" that they could yield. The withholding of wealth was not advisable, as the threat of greater plundering was real. In contrast, the rituals of Kū were conducted by the king and his chief priest in order to establish the bond between the king and this most fearsome god and to reinforce the king's divinity. These rites also required large quantities of sacrificial items, such as hogs and other foodstuffs that had to be provided by the common people. Moreover, they required human sacrifice, either of those who had broken some *kapu* or, in times of war, the body of a vanquished chief or king. Witnessed by assembled males from special seating places outside the immediate temple walls, these terrifying rites reinforced the power of the king. Any false movement, cough, or noise by one of the participants or observers could bring instant death (ʻĪʻī 1959, 43–44). The threat of death—and real death (of the sacrificial victims)—permeated the *luakini* rites, underscoring the king's power over the people.

Commoner Rituals

In contrast with the extensive ethnographic accounts of both Makahiki and *luakini* rituals, relatively little information is available concerning the ritual practices of *makaʻāinana*. In an account of canoe making, however, Malo wrote that "the *kahuna* went at night to the *mua* [men's house] to sleep before his shrine, in order to receive a revelation from his deity" (1951, 126). Samuel Kamakau (1976, 133) provides a more detailed description of shrines within the *mua*:

> The *heiau ipu-o-Lono* constantly maintained by the populace was the *hale mua*, the men's eating house, which every household had. In it was a large gourd container, an *ipu hulilau*, with four pieces of cord for a handle. Inside the gourd there was "food" and "fish" (*ʻai* and *iʻa*) and outside, tied to the cord handle, was an *ʻawa* root. This gourd was called an *ipu kuaʻaha*, or an *ipu-o-Lono*, or an *ipu ʻaumakua*. Every morning and evening the householder prayed and offered food to the god; then he would take the ipu-o-Lono…and bring it to the center of the house, take hold of the *ʻawa* root on the handle, and pray to the god about troubles or blessings, and pray for peace to the kingdom, to the king, the chiefs, the people, his family, and to himself. When the praying was ended, he sucked on the *ʻawa* root, opened the gourd, and ate of the "food" and "fish" within.

The Archaeology of Hawaiian *Heiau*

History of *Heiau* Research

Interest in recording and collecting information about Hawaiian *heiau* began in the late nineteenth and early twentieth centuries. Fornander ([1878–1885] 1996) mentions various *heiau* in connection with the *moʻolelo*, or oral traditions of the chiefly lines, while Thrum (1906, 1909) in short articles published in the *Hawaiian Annual* compiled lists of *heiau* along with information obtained from Hawaiian informants. The first true archaeological survey of *heiau* structures, however, must be credited to John F. G. Stokes, who held the position of curator of Polynesian ethnology at the Bernice P. Bishop Museum. In 1906, William T. Brigham, the museum's first director, assigned Stokes the task of mapping the *heiau* ruins of Hawaiʻi Island; the objective was to test Fornander's assertion that there had been a temporal succession in *heiau* styles, with an original platform type of *heiau* supplanted with walled enclosures following the arrival of the legendary priest Pāʻao (Dye 1989). Stokes spent five months visiting more than 100 temple ruins on Hawaiʻi, preparing detailed plan maps of more than 40 structures (Stokes [1909]). This was followed in 1909 by a similar survey of temple foundations on Molokaʻi Island (Flexner et al. 2017). Stokes' reports on these surveys were not published in his lifetime, but those for the Hawaiʻi Island sites were eventually edited and published posthumously (Stokes 1991).

During the 1920s and 1930s, archaeological surveys of most of the main Hawaiian Islands were carried out under the auspices of the Bishop Museum, primarily by young researchers from various mainland universities (Hiroa 1945, 57). The first of these was a survey of structures (including some shrines) in Haleakalā Crater, Maui, by Kenneth P. Emory, followed soon after by Emory's survey of Lānaʻi Island and then of the remote islets of Nihoa and Necker, also known as

Mokumanamana (Emory 1921, 1924, 1928). Emory's surveys followed in the tradition of Stokes', with careful site mapping and description. A burst of activity took place between 1928 and 1931, with archaeological field surveys of Kaua'i by Wendell C. Bennett (1931), Maui by Winslow Walker ([1930]), O'ahu and Kaho'olawe by J. Gilbert McAllister (1933a, 1933b), and Hawai'i by Alfred E. Hudson ([1931]). The last such island-wide survey was of Moloka'i, conducted in 1937 by Southwick Phelps ([1937]). Although other kinds of sites were included, the main emphasis in all of these surveys was *heiau*, with the archaeologists following up on sites originally reported by Thrum or known to local informants. There was little or no effort at systematic survey or to locate smaller sites not already known to Native Hawaiian informants. More importantly, the work was limited to mapping and describing surface remains; no excavations were undertaken or evidently even contemplated. Nonetheless, this period in Hawaiian archaeology resulted in the compilation of information on approximately 816 temple sites (Valeri 1985, table 5), including valuable ethnographic details that might otherwise have been lost.

Beginning in the early 1950s with Emory's excavation at Kuli'ou'ou Rockshelter on O'ahu, a resurgence of archaeological research in Hawai'i was fueled by the recognition of stratification and by the direct dating of sites made possible by Willard Libby's invention of radiocarbon dating (Emory, Bonk, and Sinoto 1959). Numerous excavations were conducted throughout the main islands, primarily in stratified rockshelters and coastal dune sites, where artifacts such as adzes and fishhooks were plentiful. *Heiau* sites, however, were neglected in this resurgence of activity, largely because it was assumed that no chronological information or useful artifact assemblages would be obtained from them. The first attempt to excavate in *heiau* foundations was made by Ed Ladd at the 'Āle'ale'a and Hale o Keawe temple sites at Hōnaunau on Hawai'i Island (Ladd 1969, 1985), demonstrating that these structures were built in a succession of stages that could be architecturally and stratigraphically discriminated. Ladd's subsequent excavation of Kāne'ākī Heiau in Makaha Valley, O'ahu, similarly revealed six construction stages that began with a simple double terrace and ended with a large, complex *luakini* structure at the time of European contact (Ladd 1973).

Michael Kolb's investigation of 108 temple sites on Maui Island, with excavations at eight structures, conducted for his doctoral dissertation at the University of California, Los Angeles, was a significant contribution to the archaeology of *heiau* (Kolb 1991, 1992, 1994, 1999). Kolb's "processual" research paradigm sought to elucidate the power relations within chiefly societies, *heiau* being regarded as materializations of chiefly power. Kolb excavated at both smaller and very large *heiau* structures, including Lo'alo'a and Pōpōiwi Heiau in Kaupō District. Construction sequences for the excavated *heiau* were proposed using stratigraphic information combined with radiocarbon dates. Kolb continued his Maui *heiau* research in the mid-1990s with survey and excavation at a number of sites in Kahikinui District, supported by a grant from the National Science Foundation. To date, however, only a preliminary report (Kolb and Radewagen 1997) and a listing of radiocarbon dates (Kolb 2006) are available for this later project.

With Hawaiian archaeology turning in recent decades largely to development-driven "cultural resource management," there has been relatively little primary research on *heiau*. Important exceptions include Kikiloi's (2012) study of ritual structures on the small northwestern Hawaiian islands of Nihoa and Mokumanamana, Kirch's (2004b) and Ruggles' (2007) studies of Kahikinui *heiau* orientations, Kirch and Sharp's application of ^{230}Th dating of branch coral offerings from *heiau* structures (Kirch and Sharp 2005; Kirch, Mertz-Kraus, and Sharp 2015),

M. D. McCoy's work on ritual practice and religious authority (M. D. McCoy 2008, 2014; M. D. McCoy et al. 2011), P. C. McCoy's ^{230}Th dating of a high-altitude shrine on Mauna Kea (P. C. McCoy et al. 2009), Mulrooney and Ladefoged's (2005) study of temple distribution within the leeward field system of Kohala, Hawaiʻi, and Thurman's (2015) detailed mapping and test excavations at Maunawila Heiau on Oʻahu.

Heiau Typology

The island-wide surveys of *heiau* conducted in the early part of the twentieth century were largely descriptive and placed little emphasis on classifying *heiau* into types or groups, whether morphological or functional. Stokes (1991, 21) wrote of Hawaiʻi Island *heiau* that "based on the character of their foundations, the heiau would seem to fall into two classes—the platform and the walled enclosure," but he qualified that statement by adding that "there were many intermediate forms and combinations of the two." McAllister (1933b, 9) found that the variation in Oʻahu Island *heiau* defied his efforts at classification: "Classifying the heiaus remaining on Oahu into types is an arbitrary and unsatisfactory procedure. Not only are there too few of these structures, but no two heiaus, furthermore, are alike." Nonetheless, McAllister opined that the island's temples could be "classified generally" as walled, terraced, or terraced-and-walled structures.

Wendell Bennett's study of Kauaʻi Island *heiau* was the only effort at serious classification of *heiau* based on the morphology of the stone foundations (1931, 30–35). Based on the 122 *heiau* that he recorded, Bennett offered the following classificatory scheme:

I. Natural sites
II. Small heiau
   A. Open platforms
   B. Walled enclosures
   C. Terraced platforms
III. Large heiau
   A. Platforms
     1. On level ground or terraced against a slope
     2. Crowning the top of a hill or rise
     3. Two or more terraced divisions
   B. Walled enclosures
     1. Square and rectangular
     2. Divisioned enclosures
     3. Compound enclosures
   C. Terraced heiau
     1. Two terrace types
     2. Three terrace types
     3. Four terrace types
   D. Round heiau
   E. Unclassified heiau
     1. L-shaped
     2. Community houses
     3. Large structures

Bennett's classification would be considered a taxonomy (Dunnell 1971), even though it was evidently derived through a process of "grouping" of like structures rather than by systematic delineation of sets of ideational attributes. Bennett's first-order, rather arbitrary, distinction between "small" and "large" *heiau* was defined with a cutoff of "less than 50 feet on the longest side" for small *heiau* (Bennett 1931, 31).

The only modern, island-wide study of *heiau* is that of Kolb (1991, 1992, 1994) for Maui Island, based on data for 108 temple sites. Kolb built upon the basic distinctions in Bennett's scheme between enclosures, terraces, and platforms to derive a set of eight "Maui types" (1991, fig. 4.8). Kolb's classification of Maui *heiau* can be summarized as follows, with the percentage of each major category indicated:

Enclosure *heiau* (42%)
    Simple enclosure
    Notched enclosure
Terraced *heiau* (37%)
    Simple terrace
    Multiple terrace
    Notched terrace
    Walled terrace
Platform *heiau* (21%)
    Simple platform
    Notched platform

Within the Kolb's group of enclosure-type *heiau*, the notched enclosure predominates, with 39 examples (36 percent of the total).

Kolb examined the geographical distribution of these *heiau* types across four major "ecological zones" of Maui Island, using chi-square tests to evaluate statistically meaningful differences in the distribution patterns (1991, 118–123). This analysis demonstrated that whereas platform-type *heiau* are evenly distributed across the island, the terraced and enclosure forms are differentially concentrated. Terraced *heiau* are more common in the windward districts (including Hāna, Kaupō, and Koʻolau), whereas enclosures are more concentrated in the arid, leeward regions of Honuaʻula, Kahikinui, and Kula.

Heiau Chronology

Fornander's theory that an "open, pyramidal" or platform type of *heiau* was later superseded by *heiau* of a walled-enclosure type was the first formal statement of a putative chronological sequence for Hawaiian religious structures ([1878–1885] 1996). Fornander's theory, however, was not supported by Stokes' (1991) survey of Hawaiʻi Island *heiau*. Emory (1928, 1943) proposed that the earliest form of Hawaiian temple was that evidenced on remote Mokumanamana (Necker Island)—a simple, open platform with a low, raised altar at the rear upon which was erected a row of upright stone slabs. The Necker Island *"marae"* were similar to simple *marae* in both the Society Islands and the Tuamotu Archipelago, which Emory regarded as the immediate homeland of the first settlers of Hawaiʻi. Lending some support for this thesis is the presence of simple, *marae*-type shrines with rows of upright slabs—as, for example, on Mauna

Kea (Emory 1943, plate V, 1; see also P. C. McCoy et al. 2009) and elsewhere in the main Hawaiian Archipelago.

Dating of *heiau* construction sequences using radiocarbon or other methods remains relatively limited throughout the Hawaiian Islands, with the exception of Maui Island, due to the small number of *heiau* that have been stratigraphically excavated. Based on his excavations at seven *heiau* on Maui, Kolb made a number of chronological interpretations regarding Maui *heiau* (1991, 367–371; 1992; 1994). He noted that none of the sites he studied dated to earlier than AD 1235–1374; thus, the earliest forms of Maui temples, whatever these may have been, were not represented. The earlier phases at his excavated sites, however, circa AD 1300–1500, were primarily terraces (although there was one "stacked enclosure"). These early terrace phases were then stratigraphically buried under later rebuilding phases consisting of platforms in four instances and of core-filled enclosures in two instances (see summary diagram in Kolb 1992, fig. 8). Kolb concluded that "in general, terraced *heiau* appear to be chronologically older than enclosure *heiau*, and probably older than platform *heiau*" (1991, 367). Kolb also calculated the labor investments in *heiau* construction over time, arguing that the greatest period of *heiau* building (in terms of volume of stonework) occurred during what he termed the Consolidation Period, roughly coinciding with the fifteenth century AD (1994, fig. 6). As Kirch (2010, 233–234) pointed out, however, this interpretation depends almost exclusively on evidence from a small test pit at Piʻilanihale Heiau in Hāna District. Removing the bias created by that limited and questionable sample, the labor investment trend shows the greatest period of *heiau* building to have been in the Unification Period (ca. 1500–1650) and final Annexation Period (1650 to contact).

Kirch and Sharp (2005) applied the high-precision method of ^{230}Th dating of branch corals obtained from seven *heiau* in Kahikinui District, Maui. The branch corals, which were evidently collected live from the ocean, had been placed as offerings on the temple sites. The narrow range of dates obtained from the branch corals (AD 1580–1640) suggested that the temples had all been dedicated within a relatively short time span, a period coinciding with the consolidation of the Maui kingdom under high chief Piʻilani and his immediate successors, according to the oral traditions. These results prompted Kirch and Sharp to hypothesize that there had been a major phase of *heiau* construction corresponding with the imposition of a "ritual control hierarchy" by the newly established Maui archaic state (104).

In response to Kirch and Sharp's (2005) report on ^{230}Th dating of Kahikinui *heiau*, Kolb published a brief article reporting "a corpus of 73 new and 17 published [radiocarbon] dates collected from specific construction phases from 41 different temples in six different political districts" (2006, 657). In addition to the dates previously reported in his dissertation work (Kolb 1991), the new dates derived from Kolb's otherwise unpublished research on *heiau* in Kahikinui. No stratigraphic details other than the most basic contextual information (e.g., "firepit, pavement") were included for the new dates. More than two-thirds of the samples were run on unidentified wood charcoal (and no botanical identifications were presented for any of the dates), leaving unaddressed the thorny problem of "built-in age" for samples from older wood. Despite these shortcomings, this was at the time the largest corpus of radiocarbon dates from any sample of *heiau* in the Hawaiian Islands. Based on these radiocarbon dates, Kolb (2006, 663) concluded:

> The general trend of temple construction followed four phases between ~ AD 1200 and 1800, phases that correlate with some general sociopolitical trends distilled

from ethnohistory. These include (1) the formation of district-sized polities and the rise of chiefly prerogatives, (2) the expansion of the chiefly hierarchy and a bifurcation of the island into eastern and western kingdoms, (3) island unification and a shift in land tenure, and (4) interisland competition and eventual absorption into a larger incipient state. An important shift in temple construction and use coincided with island unification and a shift in land tenure and occurred ~ AD 1452–1625. Overall, the temple system followed a cycle of construction and use characteristic of incipient state development, coinciding with distinct periods of political tension when it was important to encourage and control social allegiances.

Most recently, Kirch, Mertz-Kraus, and Sharp (2015) expanded their sample of ^{230}Th high-precision dated corals from Kahikinui temples, to further test and refine the hypothesis of a phase of rapid *heiau* construction in the late sixteenth and early seventeenth centuries. An extensive sample of 46 ^{230}Th dates from 26 temple sites, including corals from both surface and "architecturally-integral" (e.g., interior wall) contexts, strongly supports the interpretation of a major phase of *heiau* construction around AD 1550–1700. This does not mean that no *heiau* were built prior to AD 1550, nor does it mean that *heiau* were no longer used after AD 1700. These dates, however, strongly support the hypothesis that a major phase of *heiau* construction was associated with the consolidation of the Maui archaic state under the rulers Pi'ilani and his immediate successors, Kiha-a-Pi'ilani and Kamalālāwalu (see Kirch 2010, 2012).

Elsewhere in the Hawaiian Islands, there have been a few recent efforts to date religious structures. For the small northwestern islets of Lehua, Nihoa, and Mokumananana (Necker Island), where Emory (1928) recorded small religious structures that he regarded as representing an early form of Eastern Polynesian ritual architecture, Kikiloi (2012) dated coral offerings with the ^{230}Th method. Kikiloi writes: "Two small shrines on Lehua had coral samples that were dated to A.D. 1470 ± 7 y and A.D. 1478 ± 6 y. Nihoa's chronology is based on thirty six ^{230}Th dates that span from a total of 110 years from A.D. 1496 ± 6 y to 1606 ± 7 y. These dates originate from both interior construction and surface contexts from 14 heiau temples and 2 shrines referencing building and use events" (247). For Mokumanamana, he notes, "two coral samples were located from a single ritual site which dated early to A.D. 677 ± 15 and A.D. 1420 ± 5" (250). The first coral date is older than the accepted date for the Polynesian settlement of the Hawaiian Islands; Kikiloi reasonably argues that this piece of coral may have been an "heirloom item" brought by voyagers from elsewhere in Polynesia. Kikiloi's dating of the Nihoa and Mokumanamana temples is important in establishing a time frame for these relatively simple ritual structures.

P. C. McCoy et al. (2009) used the ^{230}Th method to date a single piece of branch coral (presumed to be a dedicatory offering) on a simple shrine with uprights associated with an alpine-zone basalt adze quarry on Mauna Kea, Hawai'i. The coral returned a date of AD 1441 ± 3 (2009, table 1), within the same time frame as the simple ritual structures of the northwestern islands. Finally, Thurman (2015) obtained four radiocarbon dates from her test excavations at Maunawila Heiau on O'ahu Island, demonstrating that initial construction at this site began in the early sixteenth century, with later modifications in the mid- to late seventeenth century.

3

The *Heiau* of Kahikinui and Kaupō
Architecture, Typology, Distribution, Chronology, and Function

In chapter 2, we reviewed the Hawaiian ethnohistorical record regarding the varied kinds of *heiau*, their functions, and their role in the annual ritual cycle, as well as the contributions of previous archaeological studies of Hawaiian temple sites. We turn now to an analysis of the corpus of 78 *heiau* structures recorded by us in the *moku* of Kahikinui and Kaupō. Drawing upon the detailed descriptions of these sites provided in part II, we present here a synthesis of *heiau* architecture (including construction techniques, features, and ground plans), the main structural types of *heiau*, their distribution over the landscape, aspects of their function as revealed by excavations, and the chronology of their construction as indicated by radiocarbon and ^{230}Th dating. This is followed, in chapter 4, by an analysis of the orientations of the Kahikinui and Kaupō *heiau*, the viewsheds that they were positioned to overlook, and their likely relationships to astronomical phenomena. In both chapters 3 and 4, we make frequent reference to specific *heiau* sites; the reader will find the relevant details in the site descriptions provided in part II.

Heiau Architecture

It must be kept in mind that all that remains today of the ancient temples of Kahikinui and Kaupō are their stone foundations. As described in chapter 2, *heiau* when in use incorporated a variety of superstructures made of perishable wood, fiber, thatch, and barkcloth, including houses of varied sizes, offering stands, oracle towers, wooden fences, and carved images. None of these superstructural features has survived, although in some cases it is possible to infer their former presence through aspects of the stone architecture, such as *'ili'ili* house pavements, alignments of curbstones, or image holes.

Stonework: Building Materials and Techniques

The geologically youthful landscapes of Kahikinui and Kaupō strongly influenced the region's stonework. This landscape is dominated by massive *'a'ā* lava flows; consequently, most of the stone used in *heiau* construction consists of *'a'ā* cobbles, boulders, or clinkers. These rough and "gnarly" stones form on the surfaces and leading edges of slow-moving *'a'ā* lava flows. Geologists Gordon Macdonald and Agatin Abbott (1970, 26; figs. 18 and 19) offer the following concise description of ʻaʻā lava rocks:

> The fragments of clinker that cover aa flows are often fantastically jagged and spiny. The spines are so sharp that fragments sometimes cause painful cuts when handled, and the leather boots of a person crossing the flow are soon scarred with a multitude of deep gashes. In mapping the upper parts of Mauna Loa we found that boots would last only a week or two on this cruel terrain.

ʻAʻā rocks are also relatively porous and susceptible to breaking apart when smashed or dropped; hence, they could easily be broken up into smaller rubble that was used as wall fill and for paving the surfaces of terraces and platforms.

The majority of temple foundations in Kahikinui and Kaupō were constructed with ʻaʻā rocks. These stones offered one advantage in that their rough and spiny surfaces tightly mesh and interlock (fig. 3.1). We have more than once marveled at the ability of Hawaiian stone masons to handle and manipulate these rough stones—they surely endured countless gashes and wounds in the process of building the region's temple walls and platforms.

Less common but not entirely lacking in southeastern Maui are *pāhoehoe* lava flows, formed by fast-moving lava that quickly crusts over, leaving "smooth, billowy, or ropy surfaces" (Macdonald and Abbott 1970, 22). *Pāhoehoe* lavas fracture along cleavage lines parallel to the surface, resulting in angular blocks ranging from 10 to 30 cm in thickness that resemble bricks. These

FIGURE 3.1. Typical *heiau* wall construction using stacked ʻaʻā cobbles, in this case at site ALE-140 in Alena, Kahikinui. (Photo by P. V. Kirch)

pāhoehoe slabs can be stacked to form neat walls with nearly vertical sides (as at site LUA-3). Where available such *pāhoehoe* slabs were also sometimes set on edge to form the curbstones of house structures within *heiau* courts (as at site ALE-1) or to line small cists within *heiau* walls (as at WF-AUW-403).

In many parts of the Hawaiian Islands, temples were constructed using dense, waterworn boulders and cobbles, known in Hawaiian as *'alā* stones (Pukui and Elbert 1986, 16). Throughout Kahikinui, however, sources of such stones are restricted to a few small boulder beaches nestled at the bases of the steep coastal cliffs, difficult to access. In Kaupō, waterworn stones are more readily available, at the boulder or black sand beaches of Nu'u and Mokulau. The use of waterworn cobbles or boulders in the *heiau* of this region is therefore rare, although the remnant temple foundation of KAU-1001 at Mokulau, where there is an extensive boulder beach, makes use of such stones. However, individual *'alā* stones do occur on *heiau* sites, where they sometimes appear to have served as representations of deities or of *'aumākua* (as at sites AUW-11 and WF-AUW-574). More extensive use was made of waterworn gravel, or *'ili'ili* (Pukui and Elbert 1986, 98), also obtained from the coastal gravel beaches. Such *'ili'ili*, typically consisting of smooth, rounded or elliptical stones ranging from 1 to 10 cm in diameter, was laid down as paving for the floors of thatched house structures within *heiau* courts in several sites (e.g., ALE-1, ALE-140, KIP-1, KIP-273, KIP-1010, NUU-79, and KAU-995). Individual *'ili'ili* pebbles and small cobbles, ranging between about 3 and 10 cm in diameter, often displaying one or more scars where flakes were removed, also frequently occur on the region's *heiau*, often together with pieces of branch coral; these evidently were offerings.

The most extensive use of *'ili'ili* paving was at the massive *luakini* temple of Pōpōiwi in Kaupō (KAU-324), where the two main courts are covered in a thick deposit of waterworn gravel, hauled up to the site from the gravel beach at Mokulau. These paved courts cover an area of approximately 2,700 m^2; if originally paved with a layer of *'ili'ili* averaging 10 cm thick, roughly 270 m^3 of gravel would have been required. *'Ili'ili* is dense and heavy, with a cubic meter probably needing to be divided up into perhaps 20 basket loads. The work of hauling something like 5,000 or more baskets full of gravel up the steep slopes to Pōpōiwi speaks to the intensive labor involved in constructing the largest Hawaiian state temples.

Temple walls in Kahikinui and Kaupō are predominantly of the core-filled type, in which the parallel inner and outer faces of the wall were constructed of carefully stacked boulders and cobbles, with the interior space between these filled with much smaller rubble and clinker. The rough *'a'ā* stones used in most temples readily lent themselves to this core-filled technique. Boulders used in the base courses of such temple walls often exceed 1 m in length, while stones used in the higher courses typically range between 20 and 60 cm in diameter. The core-filled technique is not unique to temple structures; most precontact house walls also used this method. In the postcontact period, however, the core-filled technique was gradually abandoned and replaced by simple stacking, as seen both in residential sites and in boundary or cattle walls.

Core-filled temple enclosure walls are frequently buttressed on one or more sides by the addition of a narrow terrace (or, more rarely, two or even three terraces) abutting the exterior face of the wall. Such terraces add to the overall wall thickness, lending an impression of monumentality, especially when the *heiau* is viewed from a downslope perspective. A fine example of such exterior buttressing can be seen at KIP-1 in Kīpapa, where the north, east, and south walls of the eastern enclosure have a well-constructed terrace wrapping around them. Another example is at KIP-1010 where up to three exterior terraces help retain portions of the southern enclosure wall. Site NUU-101 has an impressive exterior terrace running the entire length of the eastern wall. By far the most elaborate example of exterior buttressing, however, is at Loʻaloʻa Heiau (KAU-994), where the eastern end of the massive platform is retained by a descending flight of three stepped terraces with a total height of nearly 7 m.

Exterior wall faces occasionally incorporate large, vertical slabs set upright, either singly or in rows. One such example is at LUA-36, which has an impressive slab of conglomerate beach rock set into its eastern facade (fig. 3.2). More visually impressive is LUA-4, at Kiakeana Point, where the southern and eastern exterior wall facades are made up of rows of adjacent upright slabs.

Stacked-stone walls, lacking a rubble core, are rare in Kahikinui and Kaupō *heiau*, although they do occur at a few sites (such as WF-AUW-574 or KIP-414). Stacked walls are generally associated with the small coastal fishing shrines known as *koʻa*.

For the most part, *heiau* enclosure walls are broadly orthogonal to each other, so that the junction of two walls forms (at least approximately) a right angle. At a number of sites, however, such wall junctions have been intentionally modified by elongating the walls and causing them to join at an acute angle, with the corner noticeably protruding. The corner formed by this kind of acute wall junction resembles the prow of a canoe, and we have therefore adopted the term "canoe prow" for these corners. (Whether the Hawaiian temple architects regarded these features in such a manner we do not know, although it is conceivable.) A particularly clear example of such canoe-prow corners is seen in the southeastern corners of both the smaller (western) and larger (eastern) enclosures at site KIP-1010. Such a canoe-prow shape also forms the southeast-

FIGURE 3.2. A large slab of conglomerate beach rock set as an upright in the exterior face of the eastern wall of site LUA-36, a *koʻa* (fishing shrine) at Wekea Point, Kahikinui. (Photo by P. V. Kirch)

ern corner of the massive enclosure wall of Pōpōiwi Heiau (KAU-324). This is likewise an aspect of the southeastern corners of three massive platform structures that we believe to have functioned as chiefly residences (AUW-20, KIP-394, and NUU-172).

Temples whose foundations consist of platforms (elevated above the ground on four sides) or terraces (elevated on three sides and flush with the slope on the uphill side) were constructed by carefully stacking several courses of retaining wall or facade (much like the outer face of a core-filled wall), beginning with massive base stones and gradually reducing the size of stones in the upper courses, then filling the space behind the retaining wall with rubble and clinker. The surfaces of such platforms and terraces were often carefully leveled and paved with fine, size-sorted ʻaʻā clinkers, pieces of ʻaʻā rubble that had been purposefully broken up for this purpose.

Size Variation in *Heiau*

The sizes of Kahikinui and Kaupō *heiau*—as measured in terms of their basal areas, or footprints—vary from a mere 4 m² in the case of one small shrine (KIP-1317) up to the massive Pōpōiwi temple (KAU-324), which covers approximately 7,500 m² (table 3.1). When *heiau* basal areas are plotted on a logarithmic scale, as in figure 3.3, a distinctive pattern emerges. At the low end of the scale are nine structures with distinctively smaller footprints. Almost all of these are agricultural or fishing shrines, the largest of which has an area of 58 m². There is then a nearly continuous size progression of *heiau* from 60 to 900 m² in basal area. Within this range are all the major types of *heiau* structures, including notched enclosures, square and U-shaped enclosures, platform and terraced forms, and elongated double-court structures (see the section "Types of *Heiau* in Kahikinui and Kaupō"). At the upper end of the distribution plot are five unusually large structures. In ascending order of size, these are (1) the double notched *heiau* KIP-1010 in the Kīpapa uplands of Kahikinui, at 1,385 m²; (2) the partially destroyed *heiau* at Nuʻu Landing (NUU-101), at 1,870 m²; (3) Kou Heiau (KAU-995), with an area of 2,500 m²; (4) Loʻaloʻa Heiau (KAU-994), with an area of 4,200 m²; and (5) Pōpōiwi Heiau (KAU-324), at 7,500 m². The last four of these are all located within the *moku* of Kaupō; both Loʻaloʻa and Pōpōiwi are known to have been state *luakini* temples associated with the late precontact Maui king Kekaulike.

TABLE 3.1. Summary data on Kahikinui and Kaupō *heiau* sites.

SITE NO.	TYPE	SIZE (M²)	ELEVATION (MASL)	ORIENTATION	DATE (AD)
AUW-6	Shrine	17	10	N	1545–1549
AUW-9	Notched; with later platform addition	225	10	N	1690–1698 1436–1529, 1544–1634
AUW-11	*Koʻa*	22	10	NNE	1771–1753, 1758–1752 1663–1710, 1717–1890
WF-AUW-100	Notched	144	190	NNE	—
WF-AUW-176A	Notched	60	290	N	—
WF-AUW-338	Notched	142	375	N	1331–1325

SITE NO.	TYPE	SIZE (M²)	ELEVATION (MASL)	ORIENTATION	DATE (AD)
WF-AUW-343	Notched	57	370	N or E	—
WF-AUW-359	Notched	160	410	E	1592–1584, 1709–1715
WF-AUW-391	Square enclosure	86	425	E	—
WF-AUW-403	Notched	75	405	E	—
WF-AUW-493	Square enclosure	100	490	E	—
WF-AUW-574	U-shaped enclosure	155	380	N	—
LUA-1	Elongated double-court	512	610	E	1436–1690, 1729–1810 1675–1778
LUA-3	Square enclosure	241	100	N	1588–1592
LUA-4	Notched	67	10	NNW	1694–1698
LUA-29	Notched; incorporates older *ko'a*	88	10	NNE	1612–1618 (*ko'a*) 1656–1662 (addition)
LUA-36	*Ko'a*	58	10	E	1653–1659
LUA-39	*Ko'a*	12	70	N	—
ALE-1	Elongated double-court	700	120	N	1490–1603, 1611–1706, 1720–1819
ALE-4	Square enclosure, modified to elongated double-court	455	20	NNW	1551–1559 1676–1777, 1799–1941
ALE-121	*Ko'a*	19	30	N	1697–1701
ALE-140	Notched	70	20	WSW	1589–1595
ALE-211	Elongated double-court	225	640	N	—
KIP-1	Elongated double-court	790	640	E	1322–1328 1297–1495, 1601–1615 1444–1649
KIP-75	Square enclosure	225	590	ENE	1667–1783, 1796–1891 1670–1780, 1798–1893
KIP-77	Notched	330	575	ENE	1521–1575, 1626–1684, 1736–1805 1524–1559, 1631–1684, 1736–1806
KIP-115	Irregular	220	580	ENE?	1482–1652 1442–1529, 1552–1634
KIP-188	Notched	290	620	E	1453–1645
KIP-273A	Notched	150	240	N	1666–1672
KIP-273B	U-shaped enclosure	70	240	ENE	1611–1625 1625–1633
KIP-275	Platform	80	20	ENE	1619–1631
KIP-306	*Ko'a*	14	5	N	1576–1582 1591–1599
KIP-307	Notched	230	10	NNW	1644–1650 1640–1646 1461–1691, 1729–1811

(*continued*)

TABLE 3.1. Summary data on Kahikinui and Kaupō *heiau* sites. *(continued)*

SITE NO.	TYPE	SIZE (M²)	ELEVATION (MASL)	ORIENTATION	DATE (AD)
KIP-330	*Koʻa*	25	20	ENE	1655–1661 1731–1735
KIP-366	Square enclosure	465	125	E	1642–1680, 1763–1802
KIP-405	Irregular	320	625	E	1594–1608 1438–1675, 1777–1800 1664–1708, 1718–1888
KIP-410	Notched	220	650	ENE	1297–1495, 1601–1615
KIP-414	Irregular	295	725	E	1571–1577
KIP-424	Square enclosure	160	440	N	—
KIP-567	Platform	125	425	ENE	1571–1577
KIP-728	Elongated double-court	620	540	N	1568–1576
KIP-1010	Notched	1,385	660	N	1570–1590 1445–1654 1432–1664 1522–1575, 1585–1590, 1625–1950
KIP-1025	Irregular (unfinished?)	520	90	E	—
KIP-1146	Irregular	100	290	SE	—
KIP-1151	Platform	190	310	ENE	—
KIP-1156	Square enclosure	203	350	ENE	1665–1894
KIP-1306	Shrine	16	475	E?	1667–1783, 1796–1891
KIP-1307	Irregular	90	475	NE?	1686–1731, 1808–1927 1665–1709, 1718–1786, 1793–1889
KIP-1317	Shrine (*pōhaku o Kāne*)	4	480	E?	996–1161
KIP-1398	Shrine	40	325	E	1642–1697, 1725–1815, 1835–1878
NAK-27	Notched	90	610	ENE?	—
NAK-29	Notched	200	590	ENE	1600–1606
NAK-30	Platform	95	580	ENE	—
NAK-34	Notched	102	650	N	1437–1666, 1785–1794
MAH-231	Platform	288	490	N	1445–1524, 1558–1632 1664–1707, 1719–1826
MAH-363	Platform	400	20	?	—
MAW-2	Notched	120	150	ENE or NNW	1691–1730, 1810–1924
NUU-78	Enclosure (square?)	400	20	?	—
NUU-79	Elongated double-court	627	20	E	1667–1709, 1717–1784, 1796–1890
NUU-81	Elongated double-court	700	27	WNW	1664–1707, 1719–1826, 1832–1884
NUU-100	Elongated double-court	850	200	W	1667–1709, 1717–1784, 1796–1890
NUU-101	Elongated double-court?	1,870	10	E or N	—

SITE NO.	TYPE	SIZE (M²)	ELEVATION (MASL)	ORIENTATION	DATE (AD)
NUU-151	Terraced platform	200	45	E or N	—
NUU-153	Terraced platform	140	90	N?	1485–1650
NUU-175	Notched	393	330	E	1489–1604, 1611–1654
NUU-188	Notched	300	250	N or E	—
NUU-424	Terraced platform	350	500	N	—
KAU-324	Irregular; rectangular double-court with additional features	7,500	30	NNW	1043–1104, 1118–1302, 1367–1382 1051–1082, 1129–1133, 1151–1417 1284–1438 1421–1646 1286–1492, 1602–1612
KAU-333A	Notched	230	160	N	1441–1530, 1531–1635
KAU-333B	Square enclosure	160	160	E?	1488–1604, 1610–1670, 1780–1799 1675–1778, 1799–1942 1682–1736, 1805–1935
KAU-411	U-shaped enclosure with attached notched structure	900	200	NNE	—
KAU-536	Notched	240	50	ENE?	1446–1524, 1559–1563, 1571–1631 1442–1522, 1575–1625 1456–1638
KAU-994	Terraced platform	4,200	100	ENE	1438–1675, 1777–1800 1442–1664 1443–1695, 1726–1814, 1838–1843 1674–1778, 1799–1942 1523–1572, 1630–1950 1669–1781, 1798–1945 1431–1683, 1736–1805 1669–1780, 1798–1945
KAU-995	U-shaped enclosure	2,500	10	ESE	1091–1107 1389–1403 1625–1643 1774–1784 1790–1798
KAU-996	Terraced platform	210	30	ENE	—
KAU-999	Irregular (elite residence?)	480	160	E?	1296–1477 1530–1538, 1634–1695, 1726–1814 1641–1697, 1725–1815, 1835–1878
KAU-1001	Terraced platform?	400	10	?	—

Note: ²³⁰Th date ranges are in italics; radiocarbon (¹⁴C) date ranges are in roman; and elevations are in meters above sea level (masl).

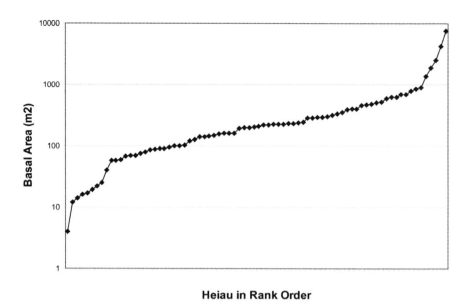

FIGURE 3.3. Rank-order plot of Kahikinui and Kaupō *heiau* in ascending order of size; the *y*-axis is on a logarithmic scale.

Architectural Features of *Heiau*

Heiau in Kahikinui and Kaupō exhibit a significant diversity of architectural features. The functions of some of these features can be inferred from descriptions in the ethnohistorical literature (see chapter 2), while others are more ambiguous or obscure.

Internal divisions, courts. With the exception of some of the square enclosure and notched forms, a majority of the *heiau* in our survey have some kind of internal division, so that the space within the enclosed perimeter consists of two distinct rooms, or courts. This division may consist of a true wall, as is the case in the notched *heiau* KAU-333A, where two distinct rooms are separated in this manner. More often, this division is marked by a change in elevation, with one room, or court, higher than the other and with a low retaining wall (typically one or two stone courses high) delineating the step up to the higher court. In notched *heiau*, the higher court is typically the smaller room, which often also contains branch coral or coral heads or a large waterworn (*'alā*) stone. Throughout Polynesia, height is associated with higher status and rank; we therefore infer that these elevated spaces were considered more sacred (*kapu*) than the adjacent lower rooms. In some of the smaller *heiau*, these elevated spaces may have been roofed over and contained ritual paraphernalia. In other, larger *heiau* (such as KIP-1 and KIP-728), the upper courts were clearly open to the sky but, in such cases, often exhibit altars or secondary platforms at one end.

Secondary platforms. Small platforms, ranging from 2 to 5 m on a side and elevated 25–50 cm above the adjacent level of the court, are found within some *heiau*, usually abutting a wall or situated in a corner. A good example is at site KIP-75, which has a low stone platform measuring 5 by 5 m in the southwestern corner of the enclosure. Another example is seen in KIP-1156, where a slightly smaller platform is nestled into that enclosure's northwestern corner. KIP-1010 also has such a platform abutting the eastern wall of the larger notched enclosure. Site KIP-1 exhibits a platform that is centrally positioned near the eastern end of the upper court but detached from the eastern wall; this platform is likely to have formed the base for an ʻanuʻu, or lana-nuʻu-mamao, structure (sometimes called an oracle tower). In other sites, these platforms may have served as foundations for small thatched houses.

Kou Heiau (KAU-995) is unique in having a group of four abutting platforms near the eastern end of the court, just below the imposing terraced altar. These form a distinctive cruciform pattern in plan view; two of them are covered in large numbers of branch coral fingers. The central platform of this group may well have supported an ʻanuʻu tower. (There are also other low platforms lining both sides of the court at Kou Heiau, but these appear to be burial platforms added in the early postcontact period.)

Benches or altars. A number of sites exhibit raised benches or terraces, typically faced with one or two courses of stone, running along the inner faces of enclosure walls; these may have served as "altars," or places where offerings were placed. They may also have served as benches, upon which were set small portable images or other representations of deities or ʻaumākua. A good example can be seen at LUA-4, where a low bench about 80 cm wide runs along the inside of the western wall. Another example is at LUA-3, where a low, stone-filled terrace almost 2 m wide runs along the inner base of the northern wall. A different kind of altar is present at KIP-405, where there is a low terrace in front of a curved wall, 1 m high, in the northeastern corner of the structure; placed on this terrace were several branch corals and flaked ʻiliʻili cobbles. At Pōpōiwi—a massive luakini heiau associated with Maui king Kekaulike—a small terrace or altar, carefully faced with stone on three sides, sits at the northern end of the room labeled B in the plan (see fig. 8.16).

Uprights. Throughout Eastern Polynesia, upright stones are known to have functioned as representations of, or as "vessels" containing the essence of, deities and ancestors. Kirch and Green (2001, 254) argue that such uprights had their origins in the posts (pou) of sacred god houses in Ancestral Polynesia. In Hawaiʻi the best-known examples of such upright stones are the shrines called pōhaku o Kāne (Kamakau 1976, 130). Site KIP-1317 is probably such a pōhaku o Kāne, consisting of a large, natural ʻaʻā outcrop to which a small enclosure has been attached on the western side.

Upright stones occur on at least 14 heiau. Sometimes these consist of waterworn cobbles (ʻalā stones) that have been transported from one of the boulder beaches, as in the case of AUW-11, a koʻa with an elongate, waterworn upright probably representing Kūʻula, the god of fishermen (fig. 3.4). Distinctive, elongate waterworn uprights also occur on the Loʻaloʻa luakini temple platform in Kaupō. More often, uprights consist of naturally elongated or slablike ʻaʻā boulders. At NAK-30—a heiau in the Nakaaha uplands of Kahikinui—two such ʻaʻā uprights were placed in the northeastern and southeastern corners of the temple platform. Uprights also occur in rows: at site KIP-1, three large ʻaʻā boulders (about 1 m high) stand together near the southwestern corner of the lower court. Another example is at KIP-77, where a row of seven upright slabs forms an alignment defining one edge of a room.

Pits. Pits or depressions are a relatively common feature at Kahikinui and Kaupō temples, occurring at 17 sites. Most of these are circular in plan view and range between 1 and 2 m in diameter, although a few are rectangular. Some, such as a circular pit in the platform at KIP-567 or the pit next to a large natural boulder at KIP-1151, were carefully constructed with stone facings, while other pits are less formal. In a few cases (such as KIP-567), these pits con-

FIGURE 3.4. Interior of site AUW-11, a *ko'a* in Auwahi, Kahikinui, with an upright, waterworn Kū'ula stone. (Photo by P. V. Kirch)

tained branch coral. Based on ethnohistorical descriptions, the most likely function of such pits was for the disposal of offerings.

Cists.   Small, rectangular spaces formed by stone slabs set on edge to form cists are found at a few sites. A particularly good example is at WF-AUW-403, where *pāhoehoe* slabs define a small cist set into the wall separating the two rooms. A similar rectangular cist was constructed into the floor of KIP-77. Such cists were probably covered over and may have served as places to store ritual paraphernalia or other sacred objects when not in use.

Pavings and house foundations.   As described in chapter 2, various kinds of thatched houses are known to have existed within the enclosing walls of *heiau*. In some sites, the former presence of such houses is indicated either by foundation curbstones or by areas of dense *'ili'ili* paving (typically rectangular in area). The best example of a large house within a major *heiau* is at Kou (KAU-995), where a foundation, about 20 m long and 6 m wide, is indicated by a low wall of *'a'ā* cobbles and by a thick *'ili'ili* paving. This particular house is likely to have been a *hale mana*, a unique feature of *luakini heiau* (Valeri 1985, 239). On a number of other sites, areas of *'ili'ili* pavement are suggestive of the former presence of thatched structures.

Pathways.   Hawaiian temple ritual involved highly proscribed movements and actions within the sacred precincts; hence, it is not surprising to find some structural evidence for pathways that may have marked the repeated movements of priests or other officiants. The most formal example is at site ALE-1, where two parallel rows of upright *pāhoehoe* slabs define such a pathway connecting the larger, lower court to two higher, presumably increasingly sacred, courts. At ALE-4, a single row of upright *pāhoehoe* slabs (a parallel row may previously have existed) also seems to define one edge of a pathway leading toward the northern wall. At KIP-

1010, a carefully positioned set of flat *pāhoehoe* slabs leads from a small, walled enclosure just outside the southern temple enclosure wall, across the wall, and into the main court. The small exterior enclosure at KIP-1010 may have been the foundation for a thatched house in which the priest prepared himself before entering the temple via the slab-lined pathway.

Niches. Small niches, usually less than 50 cm across and formed by a kind of lintel slab construction, are a feature seen in the walls of a few sites. Such niches occur in the interior faces of the northern walls of both the smaller (western) and larger (eastern) notched *heiau* at KIP-1010. A similar niche was noted in the interior face of the western enclosure wall at KIP-77. The function of such niches is not certain, although they could have served as places for depositing offerings.

Seating slabs. Two structures, both notched *heiau*, have unusual, large flat slabs of *pāhoehoe* lava that have been carefully positioned within the centers of their larger courts, apparently as seats for *kāhuna* officiating at these temples. At WF-AUW-403 the *pāhoehoe* slab is nearly 1.5 m across and positioned so that a priest seated on it would have a superb view of the eastern horizon, with Pu'u Hōkūkano and the Luala'ilua Hills. A similar situation exists at site NAK-29.

Branch coral offerings. Branch pieces or, more rarely, entire heads of the branching coral species *Pocillopora meandrina* occur on temple sites in Kahikinui and Kaupō both in surface contexts, such as on altars or benches, and incorporated within stone walls (fig. 3.5), terraces, or platforms (we refer to the latter as "architecturally integral" coral offerings). We believe that these corals were placed either as dedicatory offerings during temple construction or as ritual offerings during temple use. Whereas the surface corals could have been placed on the altars or pavements anytime during temple use, the architecturally integral corals were sealed in situ at

FIGURE 3.5. A branch coral head in the wall of site ALE-140, exposed by wall collapse. (Photo by P. V. Kirch)

the time that the walls or platforms were constructed. Close examination of the corals under a low-power microscope reveals that their fine structures (verrucae) are typically fully intact and unweathered. This is strong evidence that the corals were collected while living and growing on basalt substrates of their habitat in the shallow waters immediately offshore, and not from beach contexts, where they would have quickly become rolled and rounded.

Branch corals were observed at 39 of the 78 *heiau* structures we studied. They are most abundant at coastal sites, especially on *ko'a*, although in some cases they are also abundant on larger *heiau*, such as Kou (site KAU-995). Entire coral heads are found only on these coastal sites. On *heiau* situated in the uplands, one usually finds only a few coral branches. While coral heads and branches are ubiquitous at coastal *ko'a* sites, coral branches also occur on all of the main architectural types of *heiau*.

The ideological or ritual significance of coral heads or branches broken off of coral heads is not well understood in Hawaiian ethnohistory, although it is known that corals had ritual significance in a number of Polynesian societies (Rowland 2007). Coral blocks, for example, were incorporated into the facades of temple (*marae*) altars in the Society Islands (Sharp et al. 2010). For Hawai'i, Malo (1951, 175) describes a rite associated with the *luakini*, or king's war temple, during which the participants purified themselves by bathing in the ocean, then "carried with them pieces of coral, which they piled up outside of the *heiau*" (see also Handy 1927, 281; Valeri 1985, 319). The Hawaiian word *ko'a* is polysemic, with glosses of "coral," "fishing grounds," and "fishing shrine" (Pukui and Elbert 1986, 156). There are hints in Hawaiian mythology that coral was associated with the creator god Kāne (Beckwith 1940, 59; 1951, 55, 58). Sun-bleached coral, turning pure white, may also have been associated with Lono, among whose ritual symbols were white barkcloth streamers and white bird skins (Valeri 1985, 15). Why branch corals were incorporated into or placed on some *heiau* and not on others is unclear; if corals were a symbol of a particular deity (Kāne or Lono, for example), then their presence might be an indication that the temple was dedicated to that deity.

Types of *Heiau* in Kahikinui and Kaupō

Several previous attempts at developing classifications of *heiau* architecture have been made (Bennett 1931; McAllister 1933b; Kolb 1991), none of them particularly successful (see chapter 2). All of these classificatory schemes, however, have begun with the fundamental distinctions among (1) walled enclosures, (2) platforms, and (3) terraced forms of *heiau* stone foundations. These same distinctions are also present in the *heiau* of Kahikinui and Kaupō. Rather than attempt yet another formal taxonomy, we have opted for a simpler categorization of these temples into five main groups, with a sixth "catchall" group of unusual or unique structures. This is, we admit, an ad hoc approach, an exercise in "grouping" rather than formal "classification" in the strict sense (Dunnell 1971). That is to say, our groups (or "types") have been created by combining sets of structures that display certain common traits or similarities; they are not "classes" formed by the intersection of ideational criteria. As such, these groups are useful as a heuristic device for describing the range of variation in southeastern Maui *heiau*; they are not necessarily applicable beyond our study region, nor would we expect them to encompass all of the variation to be found in temple foundations in other regions of the archipelago.

NOTCHED ENCLOSURES

The term "notched *heiau*" appears to have been coined by Winslow Walker ([1930]) to describe one of the most frequently encountered kinds of temple enclosure on Maui Island. The "notch" is a square or rectangle from which one corner has been removed, resulting in a structure with six wall segments rather than four (fig. 3.6). Presumably, it was not the presence of the notch itself that was important to the temple architects but, rather, the presence of a small chamber created by this offset. The floors of these smaller rooms are typically elevated above that of the larger adjoining spaces, often demarcated by a terrace step or alignment or sometimes by a wall. The smaller rooms often also contain branch coral offerings. It is likely that these smaller, elevated spaces served as the sanctum sanctorum of the temples, where offerings (and possibly also images) were placed by priests or other officiants who were seated in the larger, adjoining spaces.

There are 24 notched enclosures in our corpus (31 percent of all *heiau* sites), the most abundant of any of our *heiau* groups (see table 3.1). The position of the notch varies, with 12 of the *heiau* having the notched corner situated in the northwest quadrant, 7 with the notch in the northeast quadrant, 3 in the southeast quadrant, and just 2 in the southwest quadrant (the positioning of notches and their possible significance is treated in greater detail in chapter 4). Where the notch is in the northwest or northeast, the smaller room is therefore on the north side of the temple (19 cases); where the notch is in the southeast or southwest, the smaller room is therefore on the south (5 cases).

FIGURE 3.6. Aerial view of site KIP-77, a notched *heiau* in the Kīpapa-Nakaohu uplands of Kahikinui. (Photo by P. V. Kirch)

In size (as measured by basal area), notched *heiau* range from 57 to 660 m², with an average (mean) size of 224 m². In the case of site KIP-1010, which consists of two notched *heiau* constructed adjacent to each other, the combined structure totals 1,385 m² in area. All notched *heiau* have core-filled walls.

Although notched *heiau* are by definition enclosures, they do not in all cases have walls that extend entirely around their peripheries. While some are fully enclosed, others are open on one side, as with WF-AUW-100, which is open to the south. In other cases, one or more sides are formed by a terrace with an elevated facade, as with WF-AUW-403. The L-shaped space defined by the enclosing walls or other perimeter facings is often internally differentiated into two discrete rooms, either by an internal dividing wall (as in KAU-333A) or by a change in elevation marked by a step (as in KIP-1010).

Upright stones are present at two notched *heiau*: WF-AUW-343 has a *pāhoehoe* upright set into the eastern wall, while KIP-77 has a row of seven upright *'a'ā* stones. In addition, site ALE-140 has two waterworn cobbles that may have represented deities. Pits or depressions are rare, with only two sites (KIP-77 and NAK-29) exhibiting these features. Cists, however, are present at three notched structures (WF-AUW-403, KIP-77, and NAK-29).

Branch coral (either individual pieces or in some cases entire coral heads) is present on 13 of the notched *heiau* and is abundant on three of the sites (LUA-4, LUA-29, and KIP-273A). Branch coral appears to be absent on 11 notched *heiau*, although in some cases obscuring vegetation makes it impossible to be absolutely certain that coral was absent.

Square or U-Shaped Enclosures

Thirteen structures are included in the group of square or U-shaped enclosures (see table 3.1). Of these, nine are fully enclosed, and four are open on one side and designated "U-shaped" (although the term "boxed U" might be more apt, as the corners are right angles, not rounded). These sites range in size from fairly small structures (such as KIP-273B at 70 m² and WF-AUW-391 at 86 m²) up to Kou Heiau (KAU-995), which has a basal area of 2,500 m². The mean size for square and U-shaped enclosures is 451 m², but this is skewed somewhat by the inclusion of Kou Heiau in the sample. Excluding Kou, the mean size is 280 m², just slightly larger than that for notched *heiau*. With one exception (WF-AUW-574), all of the *heiau* in this group have core-filled walls.

A number of square enclosures exhibit internal features or divisions of the space into discrete rooms or courts. At KIP-75, a stone platform is situated in the southeastern corner, perhaps as an altar or the foundation for an *'anu'u* tower. At KIP-1156, low walls subdivide the interior into three rooms, two of which have earthen floors while the third space (in the northwestern corner) is slightly elevated and paved with *'a'ā* rubble (fig. 3.7). Site LUA-3 is similarly subdivided into three rooms, the largest of which has a kind of bench or altar running along the face of the northern enclosing wall.

The U-shaped enclosures are similar to square enclosures except that one side has been left open. A particularly impressive example is KAU-411, situated atop a prominent ridge spur in Kaupō, oriented so that the open side exposes anyone within the temple precincts to a superb view across the core of the extensive Kaupō dryland field system. The largest U-shaped temple of all is Kou Heiau (KAU-995), where the U is open to the west, with a high, terraced altar making up the eastern side. Kou is unusual in also having an extension of the northern wall, which turns abruptly and runs for some distance directly north; this wall extension also incorporates a formal entryway.

FIGURE 3.7. Aerial view of site KIP-1156, a square enclosure in the Kīpapa uplands of Kahikinui. Note the internal dividing walls. (Photo by P. V. Kirch)

Only two upright stones were observed at square or U-shaped enclosures: a waterworn upright on WF-AUW-574 and a possible fallen upright on the wall of KIP-75. Pits are also less common at these *heiau*, with two small pits (or possibly image holes) in the walls of site LUA-3 and a pit in the thick northern wall of WF-AUW-574. Branch coral was present at six of the square or U-shaped enclosures, being abundant at two of those.

Elongated Double-Court Enclosures

The 10 structures within the group of elongated double-court enclosures all have ground plans that are considerably longer than they are wide (hence, "elongated") and have their interiors divided into two main courts, one elevated higher than the other (see table 3.1). In some cases, such as ALE-1 and NUU-81, one of the main enclosure walls has a right-angle jog or offset, giving the structure something of a "notched" appearance. In the case of KIP-1, the two courts (each of which is nearly square in plan) are offset such that the structure has a "double-notched" plan (fig. 3.8). The temples in this group are all quite large, ranging from 225 m² for ALE-211 up to 1,870 m² for NUU-101; the mean basal area is 735 m². In all cases, the walls were constructed with the core-filled technique.

A number of the elongated double-court temples exhibit complex internal features. Site ALE-1, for example, not only has a division separating the two main courts but also has two other low terrace alignments, a slab-lined pathway, a bench or altar running along one wall, a small house foundation, and an area of *'ili'ili* gravel paving that probably marks the position of another

FIGURE 3.8. Aerial view of site KIP-1, an elongated double-court *heiau* in the Kīpapa uplands of Kahikinui. (Photo by P. V. Kirch)

thatched superstructure. KIP-1 has several pits, a raised platform that may be the base for an *ʻanuʻu* tower, a paved area that is probably a house foundation, and a row of three *ʻaʻā* uprights. Sites KIP-1, NUU-100, and NUU-101 all have buttressed exterior wall faces. Pits appear on four of the sites, and uprights on two. *ʻIliʻili* pavements indicating the presence of thatched houses occur on three sites.

Branch coral was noted on seven of the elongated double-court *heiau*, although it was not particularly abundant on any of them. In two cases, obscuring vegetation makes it uncertain whether or not branch coral was present.

Platform or Terraced *Heiau*

The distinction between a platform and a terrace is simply that platforms are elevated above the surrounding terrain on all sides, whereas a terrace has one or more sides level with, or sometimes cut into, the terrain. Platform and terraced *heiau* are distinct from the preceding groups in lacking enclosing walls, although secondary walls or retaining walls are present in some cases. There are 14 structures in this group (see table 3.1). The smallest (KIP-275) has an area of just 80 m², while the largest, Loʻaloʻa Heiau (KAU-994), covers an impressive 4,200 m². Excluding Loʻaloʻa, the other 13 platform or terraced structures have a mean basal area of 250 m², consistent with the mean values for most notched *heiau* and for square and U-shaped enclosures.

The majority of platform and terraced *heiau* have maximum elevations on the order of about 1 m on their downslope sides, although a few exceed this height. In most cases, the platform or terrace is retained by a well-constructed, multicourse facade with large *ʻaʻā* boulders at the base and proceeding to smaller boulders and cobbles in the higher courses. In a few cases, these retaining walls are buttressed with secondary terracing. Such is the case at site KIP-80 (which we interpret as a *kahua* for *hula* performance, rather than a *heiau* in the strict sense), which has a succession of three descending terraces supporting the main platform. Loʻaloʻa Heiau (KAU-994) is a truly massive platform, constructed with an estimated 10,400 m³ of stone incorporated into the structure (Kirch, Holson, and Baer 2009); its stepped eastern facade rises to a height of 7 m above ground level.

Pits are fairly common, occurring on six of the platform or terraced temple sites. Uprights occur on three of the sites. Branch coral was observed on just three of these sites, although in some cases obscuring vegetation made it difficult to be certain that branch corals are absent.

Fishing Shrines (*Koʻa*) and Agricultural Shrines

This group of smaller shrines includes six that are clearly *koʻa* (fishing shrines), two agricultural shrines, and a *pōhaku o Kāne* (see table 3.1). The *koʻa* have an average basal area of just 25 m², and all are situated close to the shoreline. They typically are low enclosures, sometimes with an internal division into two courts, the inland court elevated slightly higher, with a low stone terrace step separating them. One *koʻa* at Auwahi (AUW-11) contains a classic upright, waterworn Kūʻula stone (see fig. 3.4); another site (LUA-39) has a large, waterworn basalt cobble that probably had the same function. Site LUA-36 incorporates a large slab of conglomerate beach rock in its eastern facade (see fig. 3.2); this may also have been a representation of Kūʻula. All of the *koʻa* shrines have abundant branch coral pieces and often entire coral heads placed on them.

The six *koʻa* described above are all free-standing structures. However, a seventh structure that may also have been a *koʻa* is a small enclosure attached on its northern side to a larger rectangular enclosure at Auwahi; the larger enclosure appears to have been a habitation site. This small, attached enclosure is covered in substantial quantities of branch coral, which suggests that it functioned as a *koʻa*.

Two sites in the Kīpapa area of Kahikinui, both directly associated with extensive dryland agricultural complexes, are interpreted as agricultural shrines. Site KIP-1306 is a small platform with a central pit and a niche (possibly for offerings) at the base of its primary facade; the shrine is situated in the northeastern corner of an extensive dryland planting swale. The second agricultural shrine (KIP-1398) is a terrace constructed against a large ʻaʻā outcrop whose vertical slabs form natural uprights; pieces of branch coral are scattered over the terrace surface. The shrine is located along the eastern edge of a large swale containing dozens of planting mounds.

The final site in the shrine category is a large, natural ʻaʻā outcrop forming a distinctive upright (KIP-1317). A small stacked-stone enclosure was constructed at the base of this upright on the western side. Carbonized material from a test excavation in the enclosure yielded a surprisingly early radiocarbon date (see the section "Chronology of *Heiau* in Kahikinui and Kaupō"). We interpret this shrine as having functioned as a classic *pōhaku o Kāne*.

Unusual or Unique *Heiau*

Seven structures in our sample do not fit readily within the five main groups already discussed (see table 3.1). Site KIP-115 combines aspects of a notched *heiau* with those of a low terrace type, but the structure may have been unfinished. The same is true of site KIP-1025. KIP-405 is primarily a large terrace, but with several walls and a small enclosure, as well as a unique altar feature. KIP-414 is an irregularly shaped enclosure, notable for its stacked rather than core-filled walls; however, the presence of branch coral strongly suggests that it was a *heiau*. Site KIP-1146 consists of a natural ridge with stone terraces on one side and a long, stacked wall ending in a large, upright slab; the site does not appear to have a residential function, and the presence of the large upright suggests a ritual function. Also unusual is site KIP-1307, a combination of terrace and partially walled enclosure, situated to the east of a major agricultural swale complex; we interpret this site as a likely men's house (*mua*) that also functioned as a small *heiau*.

The largest site in the entire corpus of Kahikinui and Kaupō temples, Pōpōiwi Heiau (KAU-324), could arguably be placed in the elongated double-court group of *heiau*, as its main structures form two massive courts surrounded by a thick wall. However, Pōpōiwi also displays a complex array of subsidiary terraces, walls, and enclosures that make it unique.

Heiau with Multiple Construction Phases

Prior excavation of *heiau* on Hawai'i, Maui, and O'ahu Islands has shown that temples were often constructed over time in a series of phases, with later construction covering over or incorporating older structures or features (e.g., Kolb 1991, 1992, 1999; Ladd 1969, 1973; Thurman 2015). It is probable that many of the *heiau* in Kahikinui and Kaupō were similarly rebuilt in two or more successive stages over time. Indeed, Kolb's excavations at both Lo'alo'a and Pōpōiwi Heiau in Kaupō demonstrated multiple construction phases at those large and complex *luakini* temples (Kolb 1991). At five sites, however, it is possible to infer at least two construction phases from the surface structural evidence, even without the aid of excavation. In addition, three *heiau* exhibit evidence of postcontact structural modifications.

Site AUW-9, a coastal temple in Auwahi, appears today as a high platform attached on the south to a rectangular walled enclosure that is partly open to the west. Upon close inspection, however, the large base foundation stones of a core-filled wall can be traced along the western side of the structure (where it is currently open). Thus, the configuration of the original enclosure wall was clearly of the notched type. The western wall and part of the original southern wall were evidently robbed of all but their base courses in order to construct the high platform that now dominates the temple's architecture.

At first glance, site LUA-29 in Luala'ilua appears to be a classic notched *heiau*. Detailed mapping, however, revealed that the structure consists of two separate enclosures: the smaller western enclosure was originally a small *ko'a*, to which the larger eastern enclosure was later added with its walls abutting the shrine, resulting in the notched *heiau* configuration. ^{230}Th dating of branch corals from each of the enclosures confirmed that the smaller western enclosure was constructed between AD 1612–1618, about 45 years before the eastern enclosure was added, between circa AD 1656–1662 (Kirch, Mertz-Kraus, and Sharp 2015).

Another example of a *heiau* that was added onto and expanded in size is ALE-4, Wailapa Heiau in Alena. Here the clue is provided by the distinct construction styles employed. The original structure was an impressive U-shaped enclosure with high, well-built, core-filled walls. To this a second enclosure was later added on the southern side, with crudely constructed, stacked-stone walls that abut and are not well integrated with the original structure. This secondary construction altered the temple from one of a U-shaped type to an elongated double-court type.

Site KIP-1010 in the Kīpapa uplands is the largest temple in Kahikinui District, consisting of two notched enclosures constructed immediately adjacent to each other (fig. 3.9). The western

FIGURE 3.9. Aerial view of site KIP-1010, with smaller and larger twin notched enclosures. Note the acute-angled "canoe prow" wall corner in the near foreground. (Photo by P. V. Kirch)

enclosure is considerably smaller, but exhibits several distinctive features (the canoe-prow form of the southeastern corner, an internal terrace division, a niche in the northern wall) that are precisely replicated in the larger, eastern notched enclosure. We believe that the western enclosure is the original temple and that the eastern enclosure was added later, precisely duplicating all of the features of the older temple at an enlarged scale. It may be that the original temple design was seen to have been especially efficacious and hence was replicated at a later date when a larger temple was desired.

The last example of precontact enlargement of a temple is KAU-411, a large U-shaped enclosure that sits astride a high ridge in Kaupō District. The western arm of the large enclosure abuts a small structure that, despite its walls being in considerable disrepair, appears originally to have been a small notched *heiau*. Thus, it seems that the original temple here was a notched *heiau*, which was abandoned after the larger U-shaped enclosure was constructed on the eastern side of the original *heiau*.

Three temples exhibit modifications that were made in the postcontact period, following the demise of the *kapu* system and the adoption of Christianity (after 1820). The first of these is KIP-728 in the Kīpapa uplands, originally a large, elongated double-court *heiau*. Kirch (2014, 229–241) describes in detail the evidence, both from the site's configuration and from excavations, for the transformation of this temple into a residential site, sometime between 1830 and 1850. This modification involved the removal of parts of the original *heiau* wall and the construction of a small, high-walled enclosure with a doorway on the west, within the bounds of the original temple. That this smaller enclosure was occupied as a house site in the mid-nineteenth century was confirmed by our excavations, which yielded various household items of Euro-American origin (nails, ceramic sherds, glass, buttons, etc.). An underlying stratigraphic horizon, however, revealed a disturbed pavement with coral and pig bone, dating to the period of *heiau* use. The transformation of this structure from a temple to a house is thought to have been undertaken by either Helio Kaiwiloa or Simeon Kaoao, two fervent Catholic converts of the mid-nineteenth century in Kahikinui (see Kirch 2014, 229–241).

A similar scenario is evidenced at site KIP-307, again possibly the work of either Kaiwiloa or Kaoao. Here the original temple was of the notched type, with the western parts of the enclosing wall being removed to build two smaller, high-walled rectangular enclosures with doorways, typical of postcontact houses. One of the houses lies within the original temple precincts, while the other one sits just to the west of the temple.

In Kaupō, site NUU-79 provides another example of a rectangular postcontact house enclosure constructed at the western end of a large temple of the elongated double-court variety. In this case the temple's western walls were robbed of stones in order to build the house enclosure. Also in Kaupō, the *luakini* temple of Pōpōiwi has a similar, high-walled rectangular house enclosure situated on top of a terrace near the southwestern corner of the complex; this too, has the appearance of a postcontact addition.

Distribution and Topographic Settings of *Heiau*

Of the 78 structures in our sample of *heiau*, 58 are situated in Kahikinui District and 20 in Kaupō District (fig. 3.10). In elevation above sea level, the sites range from some that are near the

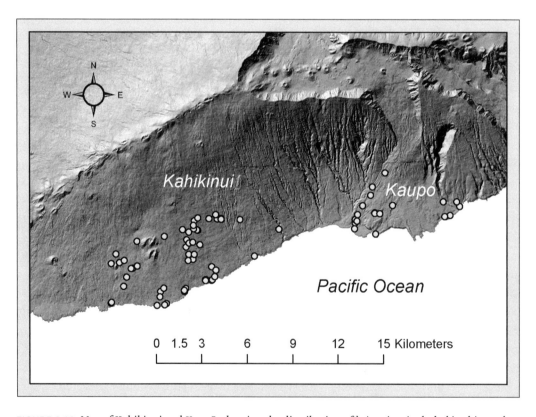

FIGURE 3.10. Map of Kahikinui and Kaupō, showing the distribution of *heiau* sites included in this study.

shoreline at an elevation of around 10 m, up to a maximum of 725 m in the case of site KIP-414 (see table 3.1). The frequency distribution of *heiau* by 100 m increments of elevation above sea level (fig. 3.11) demonstrates that the greatest concentration of temples (including *ko'a*) is in the immediate coastal zone up to 99 m elevation. Only 5 sites are found between 200–299 m, but site density then increases in the upland agricultural zone, with a total of 35 *heiau* between 300 and 800 m elevation. Most of these upland structures probably functioned as *heiau ho'oulu 'ai*, or agricultural temples.

The most common topographic setting for *heiau* in Kahikinui and Kaupō is on gently sloping terrain consisting of older, weathered flow slopes of either *'a'ā* or *pāhoehoe* lava. Forty sites are situated on such gentle flow slopes, usually where the terrain afforded good views in one or more directions. Another favored setting is on lava ridges, promontories, or knolls; 19 sites occupy such positions. It is less common for temples to be situated in depressions or swales, which generally have poor viewsheds, although 10 sites do occupy such settings. Three temples in the Nu'u area of Kaupō are located at the inland heads of long swales that were extensively terraced for dryland cultivation; thus, the temples overlooked the cultivated fields. Finally, a few sites—5 in all—were built either into the edge of or on top of younger, massive *'a'ā* flows.

At least 22 sites are situated close to intermittent stream channels, sometimes just a few meters away and overlooking the channel. In 20 of these cases the *heiau* lies to the west of the stream channel, but in two cases the *heiau* is to the east of the channel. Most of these drainage channels flow only after heavy rains deposited by *kona* storms. The resulting runoff was frequently artificially directed into swales that were terraced for cultivation; hence, the intermittent flow was important in the subsistence economy of this arid landscape.

Chronology of *Heiau* in Kahikinui and Kaupō

The archaeological landscape of Kahikinui and Kaupō—including agricultural terraces and walls, house sites, shrines, and temples—can be likened to a palimpsest, the term used by

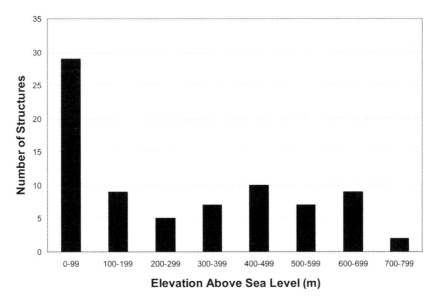

FIGURE 3.11. Frequency distribution of *heiau* by elevation above sea level.

paleographers to describe a parchment or manuscript that was written on, scraped off, and rewritten on, sometimes repeatedly. The vast leeward slopes of southeastern Maui first began to be permanently settled by Polynesians in the fifteenth century; by the time of European contact in the late eighteenth century they were home to thousands of people. Over the course of roughly four centuries of increasingly intensive occupation and land use, thousands of stone structures were constructed, used and sometimes reused or their stones robbed to build new structures. The final set of structures was ultimately abandoned when the region's population declined precipitously in the nineteenth century.

Such an archaeological landscape at first appears to be temporally static, timeless. Even though we can surmise that individual structures were probably constructed at different periods, there is no reliable way to determine their respective ages based on surface surveys. Establishing a chronology can be accomplished only through excavation, obtaining materials from stratified contexts within the sites, most often charcoal or other organic remains that are put into chronological context through radiocarbon or ^{230}Th dating. Through the careful excavation of a series of such sites, combined with the dating of many samples, the archaeological palimpsest can painstakingly be deconstructed and deciphered.

Of the 78 *heiau* structures in our study, 40 have had some excavation undertaken within them, and all of these have yielded samples that have been dated either by the radiocarbon or ^{230}Th methods (see table 3.1). These excavations have mostly been small test pits, or soundings (typically 1 m² in size), dug in order to sample the subsurface contents of *heiau* and to help unravel construction sequences. *Heiau* excavations on Maui were initially undertaken by Michael Kolb as part of his doctoral research (Kolb 1991) and later with his National Science Foundation–funded project on Kahikinui temple sites (Kolb and Radewagen 1997; Kolb 2006). In all, Kolb excavated in 19 sites in Kahikinui and Kaupō. Kirch and his Berkeley students test-excavated at 14 *heiau* in Kahikinui District and at another 5 sites in Nuʻu *ahupuaʻa* in Kaupō. Most recently, Alex Baer test-excavated in 4 other temples in Kaupō as part of his Berkeley doctoral research project (Baer 2015, 2016). Summaries of all these excavations are included in the site descriptions in part II.

Two dating methods have been applied to the Kahikinui and Kaupō *heiau* sites. The first is the well-known radiocarbon method, usually dating carbonized wood or other plant material recovered from excavated contexts. For Kahikinui and Kaupō, there are 59 radiocarbon dates from *heiau* contexts. The second method, ^{230}Th dating (also called U/Th or U-series dating), also depends on a radiometric "clock," in this case the gradual decay of uranium to thorium, but the method is limited in application to specimens of coral. Observing that branch corals of the species *Pocillopora meandrina* were often found in Kahikinui *heiau* contexts (where they presumably had been placed as offerings), Kirch and Sharp (2005) pioneered this method of dating Hawaiian temples, later expanding their sampling program (Kirch, Mertz-Kraus, and Sharp 2015). Forty-six ^{230}Th dates are now available from Kahikinui and Kaupō temples or shrines. The advantage of ^{230}Th over radiocarbon dating lies in the much higher precision of the ^{230}Th method, along with the fact that ^{230}Th dates do not present the same complexities of "calibration" required for radiocarbon dates.

In part II of this book we provide individual radiocarbon and ^{230}Th dates within the context of the sites from which they were obtained. Here we summarize the overall dating corpus for what this tells us regarding the larger temporal patterns of *heiau* construction and use in Kahikinui and Kaupō.

Radiocarbon Dates from Kahikinui and Kaupō Heiau

Due to differences in the ways that the Kolb and Kirch research teams selected and processed samples for radiocarbon dating, it is instructive to treat the two sets of samples separately. Kolb's radiocarbon dates were obtained from excavated samples of wood charcoal that were not botanically identified to genus or species. The lack of botanical identification leaves open the possibility that some of Kolb's samples were of older hardwoods, which could result in an "inbuilt age" effect. This is to say, while the radiocarbon ages themselves provide accurate estimates of the ages of the dated wood charcoal, in some cases the age of the wood could be considerably older than the "target event" of when the wood was burnt (see Dean 1978 regarding "target" versus "dated" events). Prior to submission for radiocarbon dating, all of Kirch and Baer's excavated samples were sorted and botanically identified using a reference collection of carbonized Hawaiian woody plants (these identifications were undertaken by James Coil or by Marjeta Jeraj). Only short-lived taxa, preferably shrubby plants such as *Chenopodium* spp., *Chamaesyce* spp., and the endocarps (hard seed cases) of candlenut (*Aleurites moluccana*) were then selected for dating. This procedure minimized the potential for an inbuilt age effect.

Additionally, Kolb's radiocarbon dates were not corrected for isotopic fractionation through measurement of their $\delta^{13}C$ values; at least, such values are not reported. The issue in this case is that many species of native Hawaiian dryland forest shrubs and trees have $\delta^{13}C$ values significantly different from the norm of approximately –25 per mil, resulting in corrected ages that can be as much as one to two centuries older than the uncorrected ages. If a measured radiocarbon age has not been corrected for variation in $\delta^{13}C$, it is not possible to know whether the reported age may be significantly younger than the true age of the sample. Samples from Kirch's Kahikinui and Baer's Kaupō excavations were all measured for $\delta^{13}C$ values and appropriately corrected; these samples showed that fully 34 percent of radiocarbon dates from Kahikinui required correction for isotopic fractionation of ^{13}C in order to obtain an accurate age. We presume that a similar percentage of Kolb's samples are likely to have been affected in this manner.

A final difference in the two sets of radiocarbon dates concerns their stratigraphic contexts. Of Kolb's 24 Kahikinui radiocarbon dates, fully half were obtained from contexts described as "basal," meaning that they were taken from the base of an excavation, typically at the interface between the cultural deposits and the underlying subsoil (see Kolb 2006, table 1). Such basal contexts are unlikely to date the actual construction or subsequent use of the *heiau*; rather, they most likely date some prior event such as initial land clearance or agricultural use of the area. In some cases the interval between this initial land use (the basal date) and temple construction may have been relatively short, but in other cases it could have been quite long, even a century or more. In our view, basal samples are best considered as providing *terminus post quem* dates—that is, dates after which the temple was constructed, recognizing that a considerable time gap could be involved. In the case of Kirch's *heiau* dates, all samples were recovered from cultural contexts directly related to temple construction or use. This is also true of Baer's samples from four Kaupō *heiau*, with one exception in which a basal context was dated—an exception that proved instructive, because the date for that basal sample is significantly older than two other dates from use contexts within the same site (see site KAU-999 in part II for details).

Keeping these qualifiers in mind, we first consider the set of 24 radiocarbon dates obtained by Kolb from 19 temple sites in Kahikinui District (all dates are reported in Kolb 2006, table 1). Figure 3.12 is an OxCal plot of the probability distributions for these 24 radiocarbon ages, with all dates calibrated using the IntCal13 calibration curve (Reimer et al. 2013); the dates are plotted in chronological order, with the oldest at the bottom of the graph. We emphasize that the 11 oldest dates are all from contexts described by Kolb as "basal." In at least some cases, these basal samples no doubt significantly predate the time of actual *heiau* construction and use. Given that these dates were also obtained on unidentified wood charcoal (and are thus potentially susceptible to an inbuilt age effect), it is prudent to reject the three oldest dates (Beta-95909, -122898, and -122891) as being unreliable indicators of *heiau* chronology.

If the three oldest basal dates are rejected as unreliable, the Kolb radiocarbon corpus indicates that temple construction in Kahikinui commenced sometime between the fifteenth and sixteenth centuries. The relatively large standard errors associated with most of these dates, and

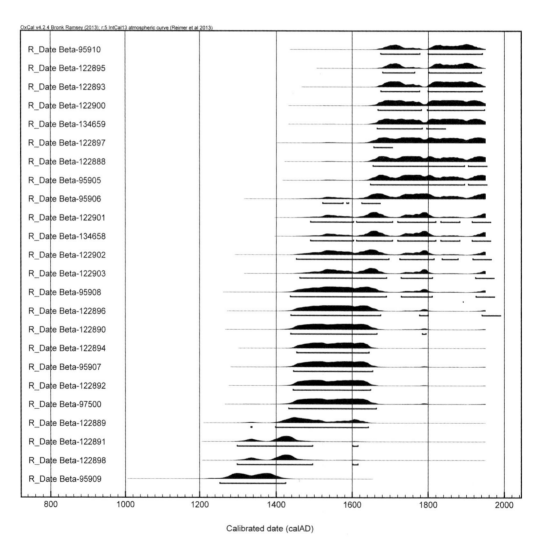

FIGURE 3.12. OxCal plot of 24 radiocarbon dates obtained by M. Kolb from excavations in 19 *heiau* sites in Kahikinui District.

the correspondingly wide probability distributions seen in figure 3.12, constrain the degree of precision regarding the onset of temple construction.

The second set of radiocarbon dates, the 35 dates obtained by Kirch and Baer from 23 sites in both Kahikinui and Kaupō, is plotted in figure 3.13, again after calibration with the IntCal13 curve. The oldest date (AA-38689) clearly stands apart, with an unusually early age of cal AD 996–1161, a date that is close to the estimated time of initial Polynesian colonization of the Hawaiian

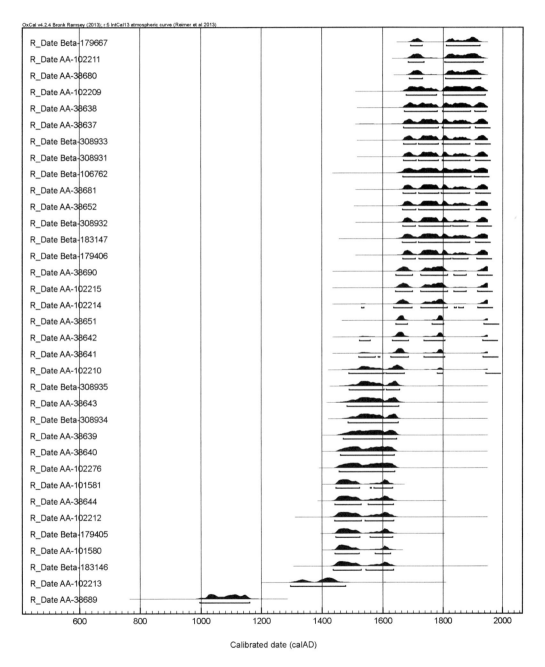

FIGURE 3.13. OxCal plot of 35 radiocarbon dates obtained by P. V. Kirch and A. Baer from 23 temple sites in Kahikinui and Kaupō Districts.

Islands (Athens, Rieth, and Dye 2014; Kirch 2011b). This date comes from the unique *pōhaku o Kāne* shrine in the Kīpapa uplands (site KIP-1317). As discussed by Kirch (2014, 82–83), the simple enclosure at the base of this natural outcrop may then date to an initial brief phase of exploration of Kahikinui soon after the arrival of people in the islands. Numerous radiocarbon dates from habitation sites throughout Kahikinui indicate that permanent occupation of the district did not commence until the early fifteenth century (Kirch 2014, 80–82, fig. 18).

The second-oldest radiocarbon date in figure 3.13 (AA-102213) comes from Baer's excavation at site KAU-999 in Kaupō, the only sample in this series from a basal context. This date is considerably older than two other radiocarbon dates from the same structure (AA-102214, -102215); we therefore reject it as a valid date for the *heiau* itself.

If those two earliest dates are set aside, the remaining 33 radiocarbon dates plotted in figure 3.13 display a consistent temporal pattern, with an initial set of six dates indicative of *heiau* construction beginning in the mid-to-late fifteenth century, followed by another seven dates spanning the late fifteenth through the sixteenth centuries. The youngest 20 dates all have probability distributions spanning the mid-seventeenth century through the nineteenth, although on ethnohistoric grounds we know that *heiau* use did not continue more than a few years beyond the overthrow of the *kapu* system in 1819.

The Kirch-Baer series of radiocarbon dates is more precise than the Kolb series in part due to the use of the accelerator mass-spectrometry (AMS) technique, with standard errors in the range of 30–40 years, as opposed to standard errors of 50–70 years for the Kolb series. This accounts for the wider probability distributions seen in figure 3.12 as opposed to the shorter distributions in figure 3.13. Nonetheless, when we discount the three oldest dates in figure 3.12 and the two oldest dates in figure 3.13, for reasons explained above, the two series of radiocarbon dates exhibit remarkably consistent temporal patterns. Taken together, the two series indicate that temple construction in Kahikinui and Kaupō began no earlier than the late fifteenth century, with construction and use continuing until the abolishment of the temple system in late 1819.

In the foregoing discussion, we have left aside two other sets of radiocarbon dates obtained by Kolb (1991) from his initial work at the major *heiau* sites of Loʻaloʻa (KAU-994) and Pōpōiwi (KAU-324) in Kaupō. These monumental sites are both much grander than anything else in our sample, and both are ethnohistorically documented *luakini* temples associated with Maui king Kekaulike (Kamakau 1961, 66); for these reasons they deserve separate consideration. Furthermore, these two sites have complex architecture and were evidently constructed in a series of phases over considerable time periods.

Kolb (1991) obtained six radiocarbon dates from Loʻaloʻa, a massive terraced platform in the eastern part of Kaupō District. These dates, plotted in figure 3.14 after calibration with IntCal13, suggest that the earliest stages at Loʻaloʻa date to between the late fifteenth and early seventeenth centuries, with later construction phases between the late seventeenth and early nineteenth centuries. Although not highly precise, this chronology is consistent with the broader pattern for Kahikinui and Kaupō based on the larger Kolb and Kirch-Baer date series already discussed.

Kolb (1991) obtained five radiocarbon dates for Pōpōiwi Heiau, plotted in chronological order in figure 3.15. These dates are significantly older than the Loʻaloʻa series; at face value, they suggest that Pōpōiwi is a much older temple than anything else in southeastern Maui. Indeed, the two oldest dates (Beta-40369, -40640) would place initial construction at Pōpōiwi between the late twelfth and early fourteenth centuries. Both of these dates, however, are from basal con-

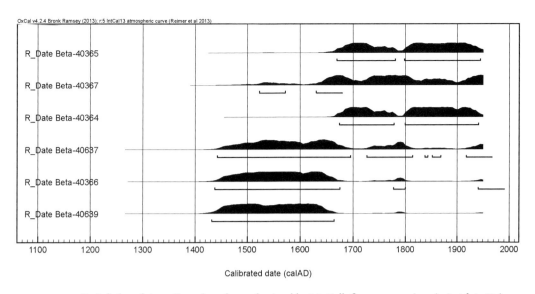

FIGURE 3.14. OxCal plot of six radiocarbon dates obtained by M. Kolb from excavations in Loʻaloʻa Heiau (KAU-994), Kaupō District.

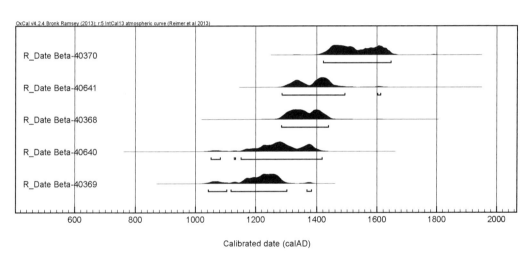

FIGURE 3.15. OxCal plot of five radiocarbon dates obtained by M. Kolb from excavations in Pōpōiwi Heiau (KAU-324), Kaupō District.

texts that are likely to predate actual *heiau* construction; while they may accurately date cultural activity on the site, they must be rejected as indications of the temple's age. Kolb (1991) reports the middle date (Beta-40368) as having been recovered from within the massive *makai* (seaward) terrace; this would put an early stage of Pōpōiwi within the time span from the fourteenth to the early fifteenth century, which is certainly possible. The two youngest dates were both obtained from fire pit features within a major terrace (PP12 on Kolb's plan, 1991, fig. 5.9). From the limited number of reliable dates, Pōpōiwi appears to have had a long history of construction and rebuilding; it was, of course, still in use at the time of Kekaulike's reign in the early eighteenth century. Future excavations, combined with extensive dating, will be required to refine our understanding of the chronology of this important site.

^{230}Th Dating of Corals from Kahikinui and Kaupō *Heiau*

Branch corals (either full coral heads or, more commonly, branches broken off from coral heads) are present at most *heiau* in the coastal zone and more selectively at temple sites in the uplands of Kahikinui and Kaupō. Corals, which take up uranium present in seawater while they are living, can be precisely dated using the ^{230}Th method. ^{230}Th dating is more precise than radiocarbon dating, with standard errors of 2–10 years at two standard deviations (95% confidence level); the method also does not require calibration as radiocarbon dating does. Kirch and Sharp (2005) initially applied ^{230}Th dating to branch corals from surface contexts (probably placed as dedicatory offerings) from seven *heiau* in Kahikinui. Their results contrasted with Kolb's (1994) radiocarbon chronology for Maui temples, suggesting that the Kahikinui temples had been constructed over a short period of about 60 years between circa AD 1580 and 1640.

Kirch and Sharp (2005) pointed out that the compressed chronology indicated by the initial ^{230}Th coral dating results corresponded with the period during which, according to Hawaiian oral traditions, Maui king Piʻilani and his immediate successors, Kiha-a-Piʻilani and Kamalālā-walu, consolidated the island into a single polity (Fornander [1878–1885] 1996, 2:87, 97–98, 121–123; Kamakau 1961, 22–33; Kirch 2010, 99–102). Royal genealogies indicate that these rulers controlled Maui between about AD 1570 and 1630 (Kirch 2010, table 3.1; Hommon 2013, 267). Kirch and Sharp (2005) advanced the hypothesis that the construction of a "ritual control hierarchy" of temples was an integral part of the Piʻilani dynasty's strategy for consolidation of power and extraction of surplus from the Kahikinui region.

Kolb (2006) and Weisler et al. (2006), reacting to Kirch and Sharp (2005), argued that because the initial set of dated corals from Kahikinui *heiau* were all from surface contexts, they might have been placed on the temples some time after the structures were constructed. While Kolb agreed that the 60-year temple-building interval suggested by the ^{230}Th dating of corals was "intriguing," he argued that "linking these ^{230}Th dates to temple construction is problematic because near-surface coral depositions lack the stratigraphic precision of ^{14}C samples collected from basal architecture" (Kolb 2006, 663). To address these valid critiques, Kirch and Sharp extended their investigation and dating of corals at Kahikinui temples, adding 39 new ^{230}Th dates (plus four replicates). The enlarged series of coral dates included not only corals from surface contexts but also a significant number of samples extracted from architecturally integral contexts (such as wall fill) unambiguously associated with temple construction. Including the original set of seven dates, 46 ^{230}Th dates on corals from 26 temple sites are now available. All but one of the ^{230}Th dated temples or shrines are situated in Kahikinui District; Kou Heiau in Kaupō (KAU-995) was also dated. Full analytical results of these ^{230}Th dates were published by Kirch, Mertz-Kraus, and Sharp (2015). Here we briefly summarize the results.

Figure 3.16 plots the temporal frequency of 26 ^{230}Th dated temples by 50-year age intervals, combined with a plot of cumulative probability distribution of the coral dates. This graph uses only the oldest date for each temple (i.e., that most closely indicating initial construction), so that each temple site is indicated only once on the chart. Figure 3.16 can therefore be considered a plot of temple construction dates, rather than a plot of temple use spans. Three sites are significantly earlier than the others, one falling into the period AD 1050–1099 and two into the period AD 1300–1349. The other 23 sites fall between AD 1500 and 1699. There is a spike between AD 1550 and 1599, when 9 temples were constructed, followed by the building of 8 more temples between

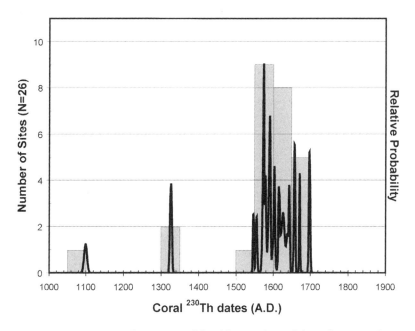

FIGURE 3.16. Frequency histogram of the oldest ^{230}Th coral dates from temple sites (N = 26), plotted by 50-year age intervals, combined with a plot of the cumulative probability distribution (after Kirch, Mertz-Kraus, and Sharp 2015, fig. 4).

AD 1600 and 1649, and another 5 temples between AD 1650–1699. In all, 22 of the 26 temples, or 85 percent, were constructed during the 150-year period between AD 1550 and 1699.

The 46 coral dates reported by Kirch, Mertz-Kraus, and Sharp (2015) include 24 from surface contexts and 22 from architecturally integral contexts such as wall fill. Figure 3.17 shows the distribution of corals from these two kinds of context using the same 50-year intervals as in figure 3.16. The two subsets display nearly identical patterns, with both the surface and the architecturally integral subsets evidencing the same major phase of temple construction and use between AD 1550 and 1699. Of the four early dates, two are from surface contexts, and two from architecturally integral contexts. Both subsets also show continued temple use into the period from AD 1750 to 1799. These data indicate that no significant bias is introduced when corals from surface contexts are used to infer the timing of Kahikinui temple construction. They also reinforce the interpretation that most surface corals in Kahikinui *heiau* are dedicatory offerings, as originally suggested by Kirch and Sharp (2005).

Four of the 46 ^{230}Th coral dates significantly predate the main cluster between AD 1500 and 1799 (see fig. 3.17). Two of these earlier dates come from the large coastal *heiau* of Kou in Kaupō (site KAU-995). Kirch, Mertz-Kraus, and Sharp (2015, table 1) dated five coral samples from Kou, three from surface contexts and two from architecturally integral contexts. The two early dates both came from collapsed segments of the temple's northern wall, where they were part of wall fill. KOU-CS-5A dates to AD 1099 ± 8, while KOU-CS-4 dates to AD 1396 ± 7. Both samples were derived from the same architectural context, but Kirch, Mertz-Kraus, and Sharp (2015) regard it as unlikely that either sample actually dates the construction of the massive northern wall. Oral traditions (Kamakau 1991, 112) refer to the arrival at Kou of Kāne and Kanaloa from Kahiki (Tahiti),

Archaeology of Heiau • 69

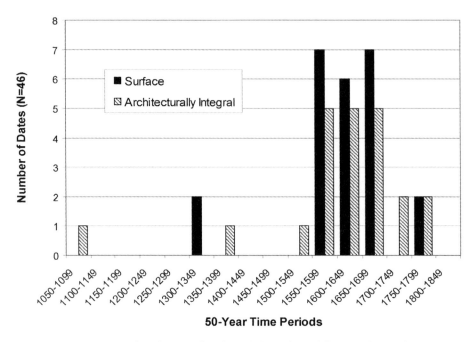

FIGURE 3.17. Frequency distribution of ^{230}Th coral dates derived from surface and architecturally integral contexts, plotted by 50-year age intervals.

where they discovered a freshwater source (probably the nearby spring of Waiū). It seems probable that Kou has been a locus of ritual activity from a very early time, during which coral offerings were made. When the northern wall was constructed, a large quantity of lava rubble was collected and heaped up as wall fill, presumably incorporating these and other pieces of coral that were already present on the site. The actual age of the extant massive structure at Kou is, in our view, more accurately indicated by the three other samples from surface contexts on the standing architecture: sample KOU-CS-3, dating to AD 1634 ± 9, comes from the main altar platform; samples KOU-CS-1 and KOU-CS-2 date to AD 1779 ± 5 and 1794 ± 4, respectively. These dates suggest that the massive architecture may have been built during the reigns of either Kamalālāwalu or Kauhi-a-Kama, two Maui rulers preceding Kekaulike, and that the temple continued to be used even into the period following initial European contact (AD 1778).

The other two early ^{230}Th dates come from two *heiau* in the Kahikinui uplands, both from surface contexts. Site KIP-1, an elongated double-court temple in Kīpapa, has a coral date of AD 1325 ± 3, while site WF-AUW-338, a notched *heiau* in the Auwahi uplands, has a coral date of AD 1328 ± 3. Because both of these early dates are from coral branches in surface contexts, we must be cautious in interpreting them as indicative of actual temple construction; it is possible that these coral branches were taken from an important, older temple (such as Kou) and placed on the upland temples at a later time.

Chronology of *Heiau* Construction and Use in Southeastern Maui

The 46 ^{230}Th dated corals from 26 *heiau* in Kahikinui (plus Kou in Kaupō) support the interpretation of a major phase of temple construction beginning in the second half of the sixteenth century and continuing through the seventeenth century. Moreover, the coral dating chronology is consistent with that provided by the radiocarbon date corpus, which shows temples beginning to be built no earlier than the late fifteenth century and continuing up through the eighteenth century, with continued use as late as 1819. Given the higher precision of the ^{230}Th dates, however, the coral dating chronology is tighter, placing initial temple construction as commencing about a century later than the time frame indicated by the radiocarbon dates. Since the corals are believed to have been placed in and on the sites primarily at the time of construction (as dedicatory offerings), their dates may be more indicative of the true onset of temple building in southeastern Maui.

The chronologies provided by both the ^{230}Th coral dates and the radiocarbon dates are consistent with the hypothesis presented by Kirch and Sharp (2005) that a rapid phase of temple building was initiated by the successive Maui Island rulers Piʻilani, Kiha-a-Piʻilani, and Kamalālāwalu, whose reigns have been calculated on the basis of royal genealogies to have spanned the period from approximately AD 1570 to 1610 (Kirch 2010, table 3.1). According to Hawaiian traditions (Fornander [1878–1885] 1996; Kamakau 1961), these rulers consolidated several formerly independent Maui chiefdoms into a single, island-wide polity as part of a process of archaic state formation (Hommon 2013; Kirch 2010, 2012). Elaboration and expansion of a hierarchical system of temples would have provided a means for control over agricultural production (through a network of *heiau hoʻoulu ʻai*, or fertility temples), as well for the regular collection of surplus in the form of tribute (*hoʻokupu*) provided by the inhabitants of each *ahupuaʻa* during the annual Makahiki rites dedicated to Lono.

One final aspect of temple chronology remaining to be addressed is that of possible temporal patterns in *heiau* morphology. Based on his initial work on several Maui *heiau*, Kolb (1991, 1992) argued that there was a progression from early terraced forms to later platform-type structures. Kolb also suggested that temples with walled enclosures were constructed in both early and later periods. Kirch, Mertz-Kraus, and Sharp (2015, table 2) summarized the evidence from the ^{230}Th dates for the chronology of different morphological kinds of *heiau*. The most common form of Kahikinui temple, the notched enclosure, is found throughout the sequence; the same is true of square enclosures. The two examples of elongated double-court *heiau* that were dated with ^{230}Th samples are both relatively early; however, other examples of the elongated double-court type in Nuʻu were dated by radiocarbon to a much later period, indicating that this form also has a long history. In sum, it appears that all of the main types of *heiau* were constructed and in use from at least the sixteenth through the eighteenth century and that there is no clear-cut temporal succession of *heiau* architectural forms.

Heiau Function: Evidence from Excavations

Not only have the excavations in Kahikinui and Kaupō *heiau* lent important insights into the chronology of temple construction in the region, but they have also provided new evidence on the activities once undertaken within these structures. Obviously, many of the ritual acts and performances that are described in Hawaiian ethnohistoric sources regarding *heiau* would have left few, if any, material traces. Recitation of sacred chants and the offering of prayers did not leave material traces, although the cooking, offering, and discard within temple precincts of such items as pigs, dogs, fish, or other foodstuffs should leave recognizable faunal or botanical remains. In addition to such predictable traces, however, excavations have produced evidence of other kinds of activity that were not initially anticipated. The results of test excavations conducted by Kirch at 18 *heiau* in Kahikinui and Kaupō are described in detail in part II; here we briefly summarize the results. Details of excavations conducted by Kolb at 15 *heiau* in Kahikinui (Kolb and Radewagen 1997) in 1995–1996 remain unpublished.

Test excavations in 16 *heiau* in Kahikinui and Kaupō produced both artifacts and faunal remains. Given that *heiau* were regarded primarily as places of ritual activity, the presence of a significant range of artifacts was initially surprising (table 3.2). Basalt flakes are virtually ubiquitous at *heiau*, occurring in high density in several sites (e.g., AUW-9, KIP-1307, NUU-100, and NUU-153). Some of the flakes may simply have been expediently produced for use in butchering or preparing sacrificial offerings (such as pig and dog), but in the sites where flake density is high, it is apparent that adzes were also being manufactured and/or rejuvenated. Indeed, unfinished adze preforms as well as finished adzes were recovered at several sites. A hammerstone was also found at site NUU-79. The lithic assemblages at sites KIP-1307 and NUU-153, in particular, clearly demonstrate adze production.

Bone working—and in particular the production of bone fishhooks—is another kind of activity evidenced at several sites. Bone fishhooks were found at AUW-9, KIP-77, and NUU-100, with one unfinished hook at the last site (fig. 3.18). Coral abraders, used to manufacture hooks, were found at AUW-9 and KIP-77, while cut bone was found at KIP-77 and NUU-100.

Both adze production and the manufacture of fishing gear are regarded as male activities, based on ethnohistorical sources. The evidence for both kinds of activities at a number of *heiau*

TABLE 3.2. Artifacts recovered from *heiau* excavations.

SITE	BASALT FLAKES/M²	VOLCANIC GLASS/M²	ADZE	ADZE PREFORM	ABRADER	FISHHOOK	OTHER ARTIFACT(S)
AUW-9	242	—	—	—	1	1	—
AUW-11	128	—	—	—	—	—	—
KIP-405	5	2	—	—	—	—	—
KIP-75	32.5	5	—	—	—	—	—
KIP-115	45.6	57	—	—	—	—	—
KIP-77	128	210	—	—	2	1	Cut bone, basalt core
KIP-1156	32.6	—	—	—	—	—	—
KIP-1307	254	9	1	—	—	—	—
MAH-231	29	1	—	—	—	—	—
MAW-2	84	—	1	—	—	—	—
NUU-79	70	1	—	—	—	—	Hammerstone
NUU-81	95	1	—	1	—	—	—
NUU-100	286	2	—	1	—	2	Cut bone
NUU-153	405	4	1	2	—	—	—
NUU-175	67	—	—	—	—	—	Basalt core

Note: Values for basalt flakes and volcanic glass flakes are normalized per m².

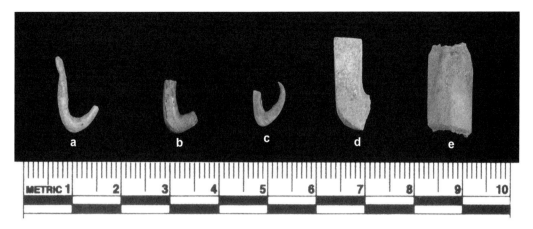

FIGURE 3.18. Fishhooks and worked bone from Kahikinui and Kaupō *heiau* sites: *a*, one-piece fishhook (AUW-9-TP1-2-3); *b*, one-piece fishhook bend (KIP-77-P19-5-1); *c*, one-piece fishhook point (NUU-100-TU1-3-1); *d*, unfinished fishhook (NUU-100-TU1-4-2); *e*, cut bone (NUU-100-TU1-4-1).

suggests that Hawaiian men regularly carried out such craft activities within temple precincts. Both Malo (1951, 32–33) and Kamakau (1976, 18) refer to four *kapu* periods within each lunar month, totaling nine days, during which men were required to sequester themselves in *heiau*, apart from the womenfolk. It seems likely that during these periods the men worked at specialist craft activities within the temple enclosures.

Kirch et al. (2012) used energy-dispersive X-ray fluorescence analysis to geochemically characterize a sample of 328 lithic artifacts from Kahikinui, including basalt flakes from several excavated *heiau* sites. Much of this material proved to derive from local Maui sources, but materials from nonlocal sources were also present. Strikingly, however, much of the imported, nonlocal basalt occurred in *heiau* contexts:

> It is notable that, of 10 specimens in group I sourced to the Mauna Kea adze quarry on Hawai'i Island, 7 come from sites 76, 77, and 115, the Naka'ohu temple complex. When we examine the distribution of local versus nonlocal sources in commoner residential versus ritual contexts, whether in terms of absolute numbers or in terms of relative frequencies, it is evident that nonlocal sources are much more heavily represented in the ritual [*heiau*] contexts. Whereas in residential contexts local sources on average comprise 76% of the basalt artifacts, in ritual contexts local sources comprise only 56%, with extralocal sources making up the other 44% (Kirch et al. 2012, 1060).

The discovery of higher frequencies of imported basalt in *heiau* contexts suggests that control over the distribution of such materials may have been facilitated by elites, including priests.

The *heiau* test excavations also yielded a diversity of faunal remains, summarized in table 3.3. Pigs are reported to have been one of the most important sacrificial offerings at *heiau*. Kamakau remarks that the dedication of certain kinds of *heiau* "called for pig eating." He adds, "When it was heard that some prominent person had a heiau to dedicate, *ho'okapu*, and that many pigs had been baked, the *po'e mua akua* came from all about to the pig-eating" (1976, 132). Pig bones

Archaeology of Heiau • 73

TABLE 3.3. Faunal material from *heiau* excavations.

SITE	PIG	DOG	MEDIUM MAMMAL	RAT	BIRD	FISH	TURTLE	MARINE MOLLUSK
AUW-9	—	1	3	—	1	183	—	146
AUW-11	—	12	8	—	4	580	4	56
KIP-75	1	0.5	—	45.5	13	61	1	33
KIP-77	10	1	6.3	21.6	5.6	26.3	—	—
KIP-115	—	—	—	—	—	—	—	4.6
KIP-405	—	—	—	—	—	38	—	15
KIP-728	3.2	0.6	—	—	—	—	—	—
KIP-1156	—	—	—	—	—	—	—	56.3
KIP-1307	1	0.5	9.5	2.5	1	36.5	—	784.5
MAH-231	—	0.5	18.5	—	—	X	—	—
MAW-2	—	—	—	—	—	—	—	X
NUU-79	4	—	—	—	—	—	—	28
NUU-81	1	—	1	—	—	—	—	89
NUU-100	—	2	—	—	—	2	—	70
NUU-153	—	—	1	—	0.5	6	—	103.5
NUU-175	1	—	—	—	—	—	—	—

Note: Values are NISP/m^2; X = present but quantity not determined.

or teeth were present at 7 of the sites tested. However, dog bones or teeth were present at 8 sites, and "medium mammal" bones that could be from either pig or dog were present at 7 sites.

Bones of the small Pacific rat (*Rattus exulans*) were found at 3 sites; while some of the rat bones may simply represent natural deposition, the relatively high frequency of bones in sites KIP-75 and KIP-77 raises the question of whether rats might have been consumed at those sites. Although the Hawaiian ethnohistorical literature does not mention the eating of rats, they were known to have been consumed in some other Eastern Polynesian societies (such as Mangaia and New Zealand). Bird bones were present at 6 sites and included both chicken (*Gallus gallus*) bones and those of several wild species. By far the most frequent vertebrate faunal material present was fish bone, found in 9 of the sites. Most frequent of all, however, were the shells of marine mollusks, present in 12 sites, sometimes in fairly dense concentrations. Both the fish bones and the marine mollusks are likely to represent the remains of meals consumed by the men while passing the *kapu* days within the temple precincts, rather than sacrificial offerings. Finally, 2 sites contained turtle bone, again most likely the remains of meals consumed within the *heiau*.

4

Heiau, Landscape, and Sky

Viewsheds and Orientations

The *heiau* in Kahikinui and Kaupō, in common with all Hawaiian temples and indeed temples throughout Polynesia, generally display a clear dominant axis of orientation, with a variety of architectural elaborations marking the sacred (*kapu*) direction (see chapter 1) and the entrance typically on the opposite side. A number of factors could have influenced the "direction faced." The most obvious, perhaps, is that they were constrained by the local topography. We shall address this possibility in the systematic analysis that follows, but to anticipate the principal conclusion, *heiau* orientations in Kahikinui and Kaupō cannot be attributed entirely, or indeed in any great measure, to the local lay of the land. Another important possibility is that some Hawaiian *luakini heiau* were oriented toward enemy territory, such as a neighboring island, during a time of war (Valeri 1985, 256).

The ethnohistorical sources are unclear about the extent to which conspicuous features in the surrounding landscape, or the appearance of celestial bodies, might have influenced temple placement and orientation. For example, while there are many references to observations of the Pleiades for calendrical purposes, in the Hawaiian Islands as throughout Polynesia (Kirch and Green 2001, 261–264), there are rarely specific indications of the place of observation. Two accounts of sunrise observations that were used to mark the passage of the seasons, at Honouliuli on Oʻahu (Kamakau 1976, 14) and at Cape Kumukahi on Hawaiʻi Island (Emerson 1909, 197), are linked to natural features (a hill called Puʻu o Kapolei and lava pillars, respectively) but not to human constructions or to temple sites. The same is true of sunset observations made from the west coast of Kauaʻi (Kamakau 1976, 14). Finally, it has been suggested that Ka-Ulu-a-Pāʻoa Heiau on Kauaʻi Island incorporated a significant alignment upon the June solstice sun along the foot of the Nā Pali cliffs forming the island's northwestern coastline, in the direction of Niʻihau Island (see Ruggles 1999b, 55–64). The latter case is only indirectly supported in the ethnography, in that this alignment, in the reverse direction, appears to be described in a sacred chant (Emerson 1909, 114) said to have been performed on Niʻihau (Ruggles 1999b, 46).

Yet the existence in Mangareva of a large flat platform from which solstice observations are known to have been undertaken in order to keep the lunar calendar on track with the seasonal year—attested in the ethnohistory (noted by the Catholic missionary Honoré Laval) and verified archaeologically and archaeoastronomically (Kirch 2004a)—shows that Polynesians did use constructions such as a platform, or flat stone ("priest's stone"), at astronomical observation places; it also shows that they sometimes placed markers such as small upright stones as artificial foresights where no suitable natural marker or prominent topographic feature was available. Nonetheless, the platform in Mangareva is not a temple (*marae*) as such (2004a, 14).

Despite the lack of ethnohistorical evidence, the possibility that conspicuous features in the surrounding landscape or the appearance of celestial bodies might have influenced the design and orientation of Hawaiian temples remains a strong one, worthy of investigation. In particular, as we argued in chapter 2, astronomical observations by priests could have been crucial for regulating the calendar and ensuring the orderly functioning of society. In this chapter we explore this issue further by analyzing the spatial relationships revealed by archaeological, landscape, and archaeoastronomical field data.

Terminology

We follow established convention (e.g., Ruggles 1999a, ix) in using the term **precision** to refer to the degree of refinement of a measurement and the term **accuracy** to refer to how well a measurement reflects the true value of the entity being measured. Thus, an instrument could record the length of a wall down to the millimeter or an angle down to the second of arc, but owing to various errors, the recorded values could be as much as several centimeters or minutes of arc different from the true value. It is important not to overstate the precision, for doing so leads to a false sense of accuracy.

We also use the term **precision** in relation to the activities or conceptions of the builders, as opposed to our determination of the current state of the material remains. Thus, on the basis of measurements precise (in the first sense) to 0.1° and considered accurate to within 0.25°, we might conclude that a wall was deliberately aligned upon an astronomical event to a precision (in the second sense) of around 1°.

A (true) **azimuth** is a clockwise bearing from true north. This horizontal angle specifies the direction of an oriented structure or of an observed point in the landscape or sky or on the horizon. An **altitude** is the vertical angle of an observed point, such as a point on the horizon, above the level of the observer. We use **elevation** exclusively to mean the vertical height of a location above mean sea level. (Confusingly, the meanings of "elevation" and "altitude" are often switched round by different authors.)

The astronomical **declination,** which is explained in this chapter's section "Appearance of the Sky," is quite distinct from magnetic declination, which is the angular difference between magnetic north and true north.

Landscape Setting

As we described them in chapter 1, the temples of Kahikinui and Kaupō lie on the lower reaches of the southern slopes of Haleakalā Volcano, above a coastline running broadly south-southwest to north-northeast. A typical Kahikinui viewshed is characterized by ground sweeping up to the Haleakalā ridgeline (west of Red Hill, the site of Haleakalā Observatory, at UTM 786100 2292420; 3,050 masl) in the northwest and north; sloping lava ridgelines visible to the southwest and south and to the northeast and east; and views of the open sea to the southeast and south, with Hawai'i Island visible beyond, to the southeast and south-southeast, in good weather conditions. Depending on the distance of the site from the coast and its elevation, and also on its local situation varying from atop a lava ridge to within a dip or swale, views in some of these directions may be more or less restricted. From the westernmost *ahupua'a*—Auwahi and Luala'ilua—apart from sites along the coast, part or all of Kaho'olawe Island is visible to the west. From the Kaupō sites, the view to the northwest and north is dominated instead by the southern rim of the Haleakalā Crater, running eastward from Red Hill to the lower summit of Haleakalā itself (UTM 794440 2291680; 2,500 masl), and by the Kaupō Gap, a break in this rim between the peaks of Haleakalā and Kūiki (UTM 798240 2292670; 2,300 masl).

A number of cinder cones within the Kahikinui and Kaupō Districts are conspicuous from sites in their vicinity. These include the hills listed in table 4.1. There are also a number of cinder cones on the ridgeline leading up to the summit of Red Hill from the west, whose prominence depends on the position of the *heiau* on the slopes below. One of these, Kanahau (UTM 781700 2290100; 2,640 masl), stands out particularly as viewed from several of the upland Kīpapa *heiau*, including KIP-275, KIP-307, KIP-330, and KIP-405, because of its conspicuous red color.

A number of coastal promontories also form striking visual landmarks, including Wekea Point (UTM 781110 2278320) in Kahikinui and Ka Lae o Ka 'Ilio (Kailio Point) (UTM 798080 2282530) and Moku'ia Point (UTM 802840 2285080) in Kaupō.

Appearance of the Sky

All the celestial bodies appear to circulate once each day affixed to the celestial vault, which extends both above and below the visible horizon. Declination is a synonym for latitude on the celestial sphere, so that any given star moves, effectively, around a line of constant declination. By convention, the declination of the north celestial pole is taken to be +90° and that of

TABLE 4.1. Prominent cinder cones within the Kahikinui-Kaupō landscape.

CINDER CONE	UTM E	UTM N	ELEVATION (MASL)
Kaho'olawe summit	753000	2275420	450
Pu'u Pīmoe	774000	2281250	540
Pu'u Hōkūkano	778350	2280850	450
Luala'ilua northwestern peak	780320	2282460	730
Luala'ilua northeastern peak	780860	2282560	730
Luala'ilua southeastern peak	780600	2281670	600
Manukani	782070	2284900	1,100
Kahua	784000	2288800	2,170
Pu'u Pane	786540	2286620	1,230
Pu'u Māneoneo	795070	2283380	110
Pu'u Ahulili	800620	2290000	1,420

Note: UTM coordinates (zone 4) and elevations (meters above sea level) are quoted to the nearest 10 m.

the south celestial pole −90°; the celestial equator has a declination of 0°. At the latitude of the Kahikinui and Kaupō temples (20.6° N), the declinations of horizon points vary from around −71° for sea horizons around due south, up to about +(69 + h)° for horizons around due north whose altitude is h°. From the *heiau* in our sample the northern horizon, the Haleakalā Ridge, has an altitude of between about 12° and 17°, so the northerly declination is generally above +80° and can be as high as +86°.

Table 4.2 lists, in order of their declination, the principal Hawaiian *hōkū* (stars and other asterisms such as constellations, clusters, etc.) documented in historical sources from the early nineteenth century onward, according to the compilation by Johnson, Mahelona, and Ruggles (2015). Planets are excluded (see the section "Planets" below). Frequently, different historical sources may cite one or several alternative names for a star, or several stars for a name, in which case the Hawaiian name indicated in table 4.2 is that considered the most strongly attested in the historical literature. In the case of Sirius, exceptionally, two names ('A'ā and Hōkūho'okelewa'a) are included.

A star has a particular declination, while a constellation or cluster covers a range. The declination or declination range of any *hōkū* appears unchanging over a few years but gradually changes over a timescale of decades and centuries, owing to precession (short for "the precession of the equinoxes") (Ruggles 2005, 345–347; 2014d, 473–479). Thus, in table 4.2 the declinations are quoted for the years AD 1500, 1600, 1700, and 1800.

STARS

Viewed from a fixed location, any given asterism will rise or set in the same place night after night, but progressively earlier, by around four minutes each night. In general, the annual cycle of the stars is marked by a distinctive sequence of events. At some time of year, typically for a few weeks, the star will not be visible at all because it rises and sets at approximately the same time as the sun and passes through the sky during daylight. Then, once the star's rising precedes sunrise by a sufficient interval, it will begin to be visible, albeit very briefly at first, in the predawn sky. This is known as the "heliacal rise." Over subsequent months the star will rise progressively earlier in the night until it is seen to rise just as the sky gets sufficiently dark in the evening following sunset (after this it will appear in the darkening evening sky already risen). This culturally significant event is generally referred to as the "acronychal [or acronical] rise," although it is technically the "apparent acronychal rise," the true acronychal rise occurring when the star rises just as the sun sets (it is not visible at this time).

At around the same time as the acronychal rise, there occurs the "cosmical set," the cultural (apparent) event being when the star is first seen to set before the predawn sky lightens too much. (The "true" event occurs when the star sets at sunrise.) Thereafter, it sets progressively earlier each night until the time comes when it is last seen in the evening sky after sunset. This event is known as the "heliacal set" and heralds the next annual period when the star is not visible at all. For further details, see Ruggles (2014b, 463–465). (Note that in the literature, there is some confusion over terminology. The "heliacal set" is sometimes referred to as the "acronychal set," since "acronychal" means "happening at sunset," but the true acronychal set occurs when the star and the sun set at the same time. The "cosmical set" is also sometimes referred to as the "acronychal set," giving a symmetry to the heliacal and acronychal events, but this runs counter to the classical meaning of "acronychal".)

TABLE 4.2. Known Hawaiian asterisms and their declinations between AD 1500 and 1800.

		HAWAIIAN STARS			DECLINATION (IN DATE AD)			
STAR	MAGNITUDE	HAWAIIAN NAME	MEANING OF NAME	NIH PAGE	1500	1600	1700	1800
Polaris (α UMi)	2.0	Hōkūpaʻa	Fixed star	157	+86.6°	+87.1°	+87.7°	+88.2°
Dubhe (α UMa)	2.0	Hiku Kahi, part of Nā Hiku	First of the seven	150	+64.4°	+63.8°	+63.3°	+62.8°
Megrez (δ UMa)	3.3	Hiku Ahā, part of Nā Hiku	Fourth of the seven	150	+59.8°	+59.3°	+58.7°	+58.1°
Merak (β UMa)	2.3	Hiku Alua, part of Nā Hiku	Second of the seven	150	+59.0°	+58.5°	+58.0°	+57.4°
Alioth (ε UMa)	1.8	Hiku Lima, part of Nā Hiku	Fifth of the seven	150	+58.7°	+58.1°	+57.6°	+57.0°
Mizar (ζ UMa)	2.2	Hiku Ono, part of Nā Hiku	Sixth of the seven	150	+57.6°	+57.0°	+56.5°	+56.0°
Phecda (γ UMa)	2.4	Hiku Kolu, part of Nā Hiku	Third of the seven	150	+56.5°	+55.9°	+55.4°	+54.8°
Alkaid (η UMa)	1.9	Hiku Pau, part of Nā Hiku	Last of the seven	151	+51.9°	+51.3°	+50.8°	+50.3°
Capella (α Aur)	0.1	Hōkūlei	Wreath star	156	+45.3°	+45.5°	+45.6°	+45.8°
Deneb (α Cyg)	1.3	Unknown	—	—	+43.6°	+43.9°	+44.2°	+44.6°
Vega (α Lyr)	0.0	Keoe	—	182	+38.4°	+38.5°	+38.5°	+38.6°
Castor (α Gem)	1.6	Nānāmua	Look forward	199	+32.9°	+32.7°	+32.5°	+32.3°
Pollux (β Gem)	1.2	Nānāhope	Look behind	199	+29.1°	+28.9°	+28.7°	+28.5°
Arcturus (α Boo)	−0.1	Hōkūleʻa	Clear star; star of gladness	156	+21.9°	+21.3°	+20.8°	+20.2°
Aldebaran (α Tau)	0.9	Hōkūʻula; part of Kaomaʻaikū	Red star	158	+15.4°	+15.6°	+15.9°	+16.1°
Regulus (α Leo)	1.4	Kau-ʻōpae [possible]	—	177	+14.3°	+13.9°	+13.4°	+12.9°
Altair (α Aql)	0.8	Humu	Triggerfish	161	+7.7°	+7.9°	+8.1°	+8.4°
Betelgeuse (α Ori)	0.5	ʻAua, part of Nā Kao	—	146	+7.2°	+7.3°	+7.3°	+7.4°
Procyon (α CMi)	0.4	Ka ʻŌnohi Aliʻi	The chief's eyeball	172	+6.4°	+6.2°	+6.0°	+5.7°
Rigel (β Ori)	0.2	Puanakau, part of Nā Kao	—	205	−8.9°	−8.7°	−8.6°	−8.4°

(continued)

TABLE 4.2. Known Hawaiian asterisms and their declinations between AD 1500 and 1800. *(continued)*

HAWAIIAN STARS

STAR	MAGNITUDE	HAWAIIAN NAME	MEANING OF NAME	NIH PAGE	DECLINATION (IN DATE AD)			
					1500	1600	1700	1800
Spica (α Vir)	1.0	Hikianalia	—	149	−8.5°	−9.0°	−9.6°	−10.1°
Sirius (α CMa)	−1.4	ʻAʻā *or* Hōkūhoʻokelewaʻa	Burning bright Canoe-guiding star	145 154	−16.1°	−16.2°	−16.3°	−16.4°
Antares (α Sco)	1.1	Lehua-kona [possible]; part of Ka Makau Nui o Maui	Southern *lehua* flower	187	−25.2°	−25.5°	−25.7°	−26.0°
Adhara (ε CMa)	1.5	Unknown	—	—	−28.4°	−28.5°	−28.6°	−28.7°
Fomalhaut (α PsA)	1.2	Unknown	—	—	−32.2°	−31.7°	−31.2°	−30.7°
Shaula (λ Sco)	1.6	Part of Ka Makau Nui o Maui	—	169	−36.6°	−36.7°	−36.8°	−36.9°
Canopus (α Car)	−0.6	Ke Aliʻi o Kona i Ka Lewa	Chief of the southern heavens	180	−52.5°	−52.5°	−52.6°	−52.6°
Gacrux (γ Cru)	1.6	Part of Newe	—	200	−54.3°	−54.9°	−55.4°	−56.0°
Mimosa (β Cru)	1.3	Part of Newe	—	200	−56.9°	−57.5°	−58.0°	−58.6°
Hadar (β Cen)	0.6	Ka-maile-mua [possible]	The first/earlier *maile* (vine)	169	−57.9°	−58.4°	−58.9°	−59.4°
Achernar (α Eri)	0.5	Unknown	—	—	−59.8°	−59.3°	−58.8°	−58.2°
Rigil Kentaurus (α Cen)	−0.3	Ka-maile-hope [possible]	The last/later *maile* (vine)	168	−59.1°	−59.4°	−59.8°	−60.2°
Acrux (α Cru)	0.8	Kaulia [possible]; part of Newe	—	175	−60.3°	−60.9°	−61.4°	−62.0°

OTHER HAWAIIAN ASTERISMS (CONSTELLATIONS, CLUSTERS)

HAWAIIAN ASTERISM	MEANING OF NAME	WESTERN EQUIVALENT	NIH PAGE	DECLINATION (MINIMUM AND MAXIMUM IN DATE AD)							
				1500		1600		1700		1800	
				MIN.	MAX.	MIN.	MAX.	MIN.	MAX.	MIN.	MAX.
Nā Hiku	The seven	Ursa Major	196	+51.9°	+64.4°	+51.3°	+63.8°	+50.8°	+63.3°	+50.3°	+62.8°
Mūlehu	—	Schedar, Caph, and Tsih (α, β, and γ Cas) [possible]	195	+53.8°	+58.0°	+54.3°	+58.5°	+54.9°	+59.1°	+55.4°	+59.6°
Makaliʻi	Little eyes	The Pleiades	191	+22.3°	+22.8°	+22.6°	+23.1°	+23.0°	+23.5°	+23.3°	+23.8°
Ka Nuku o Ka Puahi	The opening of the fireplace	The Hyades plus Aldebaran	171	+14.3°	+17.9°	+14.6°	+18.2°	+14.9°	+18.5°	+15.1°	+18.7°
Nā Kao	The darts	Orion *or* Orion's Belt	198	−10.0°	+7.2°	−9.9°	+7.3°	−9.8°	+7.3°	−9.7°	+7.4°
			198	−2.3°	−0.8°	−2.2°	−0.7°	−2.1°	−0.6°	−2.1°	−0.5°
Ka Makau Nui o Maui	Maui's big fishhook	Tail of Scorpius	169	−42.5°	−18.3°	−42.6°	−18.6°	−42.8°	−18.9°	−43.0°	−19.3°
Newe	—	The Southern Cross (Crux)	200	−60.3°	−54.3°	−60.9°	−54.9°	−61.4°	−55.4°	−62.0°	−56.0°

Note: The asterisms are sourced from *Nā Inoa Hōkū* (Johnson, Mahelona, and Ruggles 2015; "NIH"), and the entries are ordered by declination (in AD 1600). Stellar (apparent) magnitudes down to (numerically, up to) 1.6 are taken from Ridpath (2004, 111, table 48). Declinations and other magnitudes are taken from Stellarium (www.stellarium.org) v. 0.15.0 (2015). Stars whose Hawaiian identification is uncertain and other bright stars (magnitude [lower for brighter stars] at least 1.6) whose Hawaiian names are unknown are also included. Planets are excluded.

Exceptions to this are circumpolar stars, which circulate in the sky close to one of the celestial poles and never set. As seen from the Kahikinui-Kaupō temples, the north celestial pole, due north, has an altitude of 20.6° (equal to the latitude of the *heiau* sites). Since the north point on the horizon typically has an altitude of between 12° and 17°, the north celestial pole is never more than about 9° above the horizon, meaning that only stars with a declination greater than about +81° will be circumpolar. The only bright Hawaiian star in this category is Hōkūpa'a (Polaris).

SUN

The sun's rising and setting positions and path across the sky vary depending on the time of year. The sun is at its most northerly at the time of the June solstice (June 21 in the Gregorian calendar), when its declination is about +23.5°; at this time it rises in the east-northeast and sets in the west-northwest. It reaches its most southerly position at the December solstice (December 21), when its declination is about −23.5°; at this time it rises in the east-southeast and sets in the west-southwest. The more exact figures (quoted now to a precision of 0.05°) for the upper limb, center, and lower limb of the sun in around AD 1600 are as follows:

SOLSTICE	UPPER LIMB	CENTER	LOWER LIMB
June solstice	+23.75°	+23.5°	+23.25°
December solstice	−23.25°	−23.5°	−23.75°

Thus, a horizon point with a declination between +23.25° and +23.75° represents a point behind which the sun would have risen or set at the time of the June solstice around AD 1600; one with a declination between −23.75° and −23.25° aligns on the rising or setting point of the December solstitial sun. The change in these figures over a timescale of a few centuries is negligible: precession does not affect the motions of the sun, and the factor that does—the changing obliquity of the ecliptic (Ruggles 2014d, 479–481)—has a much smaller effect, altering the limiting (solstitial) declinations by only approximately 0.05° in 500 years.

Given that the ethnographic record contains direct references to observations of both the rising sun and the Pleiades for calendrical purposes, it is ironic that the declination of the Pleiades is very close to that of the June solstice sun. Gradually shifting because of precession, the declination range covered by the Pleiades started to overlap with that of the June solstice sun from about AD 1650, and by AD 1800 they coincided, rendering the Pleiades and June solstice sun alignments indistinguishable.

The solar rising arc extends between azimuths of approximately 66° and 115°, depending on the altitude of the horizon. Likewise, the setting arc extends between approximately 245° and 293°. Day by day, the rising or setting position of the sun shifts steadily along the respective arc, slowing down as it approaches, and speeding up again as it moves away from, its limiting positions at the solstices. For any position within the solar rising or setting arc, we can deduce from the declination the equivalent Gregorian date on which the sun would have risen or set at that point (this varies slightly from year to year because of the leap-year cycle). Table 4.3 gives the data for AD 1600; these do not change significantly over the period in question.

Broadly speaking, the equinoxes represent the approximate halfway point along this arc.

TABLE 4.3. Mean declination of the center of the sun at approximately 23-day (1/16-year) intervals through the year in AD 1600.

NO. OF DAYS FROM JUNE SOLSTICE	GREGORIAN DATE	MEAN DECLINATION OF CENTER OF SUN	NO. OF DAYS FROM JUNE SOLSTICE	GREGORIAN DATE	MEAN DECLINATION OF CENTER OF SUN
0	Jun 21	+23.5°	183	Dec 21	−23.5°
23	Jul 14	+21.7°	205	Jan 12	−21.6°
46	Aug 6	+16.7°	228	Feb 4	−16.1°
68	Aug 28	+9.6°	251	Feb 27	−8.1°
91	Sep 20	+0.9°	274	Mar 22	+0.8°
114	Oct 13	−8.0°	297	Apr 14	+9.6°
137	Nov 5	−15.9°	320	May 7	+16.9°
160	Nov 28	−21.5°	342	May 29	+21.7°
183	Dec 21	−23.5°	365	Jun 21	+23.5°

However, this "spatial equinox" is distinct from the "true" (astronomical) equinox, which corresponds to declination 0°. This in turn is distinct from the "temporal equinox" (the halfway point in time between the solstices), as is evident from the nonzero declinations on days 91 and 274 in table 4.3, and various other ways in which the "equinox" could be defined (Ruggles 1997). Although equinoctial sunrise or sunset is widely assumed to be a "target" of evident significance across cultures, this assumption is highly questionable, since it is based on a Western-style conception of space and time as abstract reference axes (Ruggles 1997, 2017). Unlike the solstitial directions, which represent the physical limits of the sunrise and sunset arcs, the "halfway" points (either in space or time) are not readily distinguishable. They are unexceptional in the sky and are not generally marked by any tangible reference points on the land. Nonetheless, we do refer to the ("true," astronomical) equinoxes for reference purposes in the text and include the sun's path at the equinoxes, again as points of reference, in the visualizations of the horizon and sky.

MOON

The moon's declination varies significantly from day to day and its longer-term cycles are more complex than the annual cycle of the sun, varying both over a monthly cycle and a longer, 18.6-year cycle (lunar node cycle). This results in eight limiting declinations known as the "lunar standstill limits," much discussed by archaeoastronomers as potential alignment targets even though setting up such alignments would present considerable challenges in practice (Ruggles 2014d, 466–469; González-García 2014, 494–495). For the record, we list them in table 4.4. As with the sun, these are unaffected by precession and change only negligibly over a period of a few centuries.

While the phase cycle of the moon forms the basis of the Polynesian lunar calendar and was well known to Hawaiians, there is no evidence to suggest, and no reason to believe, that the rising and setting position of the moon, and particularly the 18.6-year cycle and the lunar standstill limits, were of any interest or significance in Polynesian and Hawaiian culture. Nonetheless, we consider this possibility where it arises at some of the *heiau* sites.

TABLE 4.4. Declination of the center of the moon at the "lunar standstill limits" around AD 1600.

LUNAR "TARGET"	DECLINATION
Northern major standstill limit	+28.3°
Northern minor standstill limit	+18.0°
Southern minor standstill limit	−18.7°
Southern major standstill limit	−29.0°

Note: The figures quoted are the mean apparent declinations, which take account of mean lunar horizontal parallax (Ruggles 1999a, 23, 36–37), about 0.35° at latitude 20.6° N for the targets in question.

Planets

The Hawaiians recognized the five planets visible to the naked eye, and several names are recorded for each of them, including distinct names for Venus as evening or morning star (Johnson, Mahelona, and Ruggles 2015, 210). The motions of the planets among the fixed stars are complex and their declinations vary over ranges similar to those of the sun and moon. For this reason, it is almost impossible to distinguish a planetary alignment on the basis of alignment evidence alone, and planetary alignments are generally only taken seriously by archaeoastronomers when they are substantiated by clear historical or ethnographic evidence (Šprajc 2014).

Data Selection Methodology and General Approach

All oriented structures must point somewhere, and the mere existence of an alignment of apparent astronomical significance does not prove that it was intentional or meaningful. For this reason, data selection methodology is a central issue in archaeoastronomy. Furthermore, deliberate alignment upon astronomical objects or events or, equally, upon conspicuous features in the visible landscape are just two among many factors that could have influenced orientation. The most obvious methodological challenge is to demonstrate that putative astronomical or topographic alignments have not simply been selected from innumerable possibilities (implying the need for some sort of statistical verification of the intentionality of alignments) while recognizing that people, unlike laws of the physical universe, rarely if ever behave with absolute consistency, however powerful and restrictive the protocols governing their behavior (Ruggles 2011, 8–15; 2014c). This affects both our evaluation of alignment target possibilities and the selection of potential aligned structures.

As regards astronomical targets, if we determine the declination of a horizon point, we can deduce which astronomical bodies rose or set there at a given epoch in the past. In this sense, the "data speak for themselves," but the problem is the sheer number of possibilities. There are a great many stars in the sky, and their positions change over the centuries owing to precession. Even given just the 15 brightest stars and a 500-year time span, we would be able to find an "explanation" for about one in three of all alignments. Fortunately, we have good knowledge of some of the stars and asterisms that were most significant to Hawaiians, and we have accurate ^{230}Th dates for many of the sites. If it were not for this, there would be too many possibilities to have confidence that any stellar alignments identified were in fact deliberate (Ruggles 2014f). The sun's motions are simple and can be easily considered; but those of the moon and planets are more complex and problematic, as already explained. This is not to say that lunar and planetary alignments did not exist; it merely means that we cannot hope to identify them with any great confidence from the alignment evidence unless they are directly attested or strongly backed up by independent (ethnohistorical) data. As regards putative topographic alignments, we avoid undue proliferation of possibilities by confining our attentions to the conspicuous features already identified in the "Landscape Setting" section above.

The Selection of Potential Aligned Structures

Some archaeoastronomers favor "data-driven" approaches in which an attempt is made to identify and measure the main orientation of all, or a representative sample, of the monuments of a given group, ignoring other alignments of possible significance at the sites in question. Such an approach has had remarkable success in western Mediterranean Europe, where it has demonstrated that orientation was a significant factor in the design of many local groups of later prehistoric tombs and temples, and that the orientations were generally determined astronomically and probably in relation to the sun (Hoskin 2001). Others emphasize the need to develop interpretations that are viable from the perspective of anthropology and consistent with the broader cultural context, but with a much more flexible, and ultimately subjective, approach to the structural alignments that are considered. The debate has much in common with debates over interpretive approaches in landscape archaeology (e.g., Fleming 2006). For a full discussion, see Ruggles (2011, 2014e).

In the course of our work in Kahikinui and Kaupō we have merged aspects of these two approaches, by attempting a systematic analysis first, in order to identify general trends, and then undertaking a detailed analysis that helps us develop more specific and nuanced interpretations. On-site, in addition to identifying the primary direction of orientation, an attempt was made to determine the orientation of all extant walls or platform edges, where these were sufficiently straight and in a good enough state of repair. We also investigated any additional features of interest, such as possible sighting-stones or other observational devices, and canoe-prow corners (whose diagonal orientation may indicate the direction in which they were intended to "point"). Through the documentation presented here and in supplementary online materials, including site plans and computer-generated horizon profiles, our selection decisions and conclusions are open to reassessment and to be reworked independently from the basic archaeological and archaeoastronomical data.

Methods: Determining Viewshed and Orientation Data at Each *Heiau*

Survey Procedure

The aim of our surveys was to determine whether each *heiau* site as a whole, or specific components such as walls or platform edges, might have been deliberately placed and oriented in relation to visually prominent features in the surrounding landscape or in relation to the appearance of certain astronomical bodies. In order to do this, we sought to establish the azimuth of each oriented structure of possible interest (the two directions yield azimuths 180° apart) and the **altitude** of the horizon point upon which it aligns. From these parameters, together with the latitude of the site, we can deduce the astronomical **declination** of the horizon point, and from this—as explained earlier—we can ascertain the astronomical objects that would have risen or set there at any given epoch in the past (Ruggles 1999a, 22–23; 2014b, 460–463). (An interactive version of the GETDEC program used to calculate declinations [Ruggles 1999a, 169] is available at www.cliveruggles.com.)

Total station surveys. Wherever possible, surveys were undertaken using a Leica TCR-705 Total Station (theodolite with electronic distance-measurement [EDM] device). This was set up at a suitable position within, or in the immediate vicinity of, each *heiau* site. Theodolite (horizontal and vertical angle) readings were taken of points along relevant segments of the visible horizon, while angle-plus-distance readings were taken at intervals along each oriented component (wall, platform edge, etc.), focusing on points considered to give the most reliable indication of its intended orientation, such as on intact segments of wall facing. This effectively fixed these points in space, enabling the azimuth from one to another (or the best-fit azimuth along a line of points) to be deduced.

Sun-azimuth observations. A critical aspect of the use of a theodolite for archaeoastronomical work is the need to determine the direction of true north to sufficient accuracy in order to calibrate horizontal circle readings and convert them to azimuths (the "plate bearing to azimuth" [PB–Az] correction). For our surveys we took observations of the plate bearing of the sun, timed to the nearest second, and then compared this with the azimuth of the sun at that moment as determined using the MICA (Multiyear Interactive Computer Almanac) program from the US Naval Observatory (http://aa.usno.navy.mil/software; see also Prendergast 2014, 395–396). This was possible thanks to the use of a sun filter and a corner eyepiece that permitted sun observations to be made by direct sighting (rather than projection) at any time of the day, including when the sun was close to the zenith.

Except where weather conditions intervened, a total of 12 sun-azimuth readings were taken at each site: three on each limb of the sun (preceding and following), necessary because the center of the sun's disc could not be determined accurately, with all six readings being repeated on each face of the theodolite. This is standard procedure in order both to eliminate gross errors (mistakes) and to provide an estimate of random errors (Ruggles 1999a, 164–171). The latter were generally well within 1 minute of arc, with a couple of exceptions where we were surveying in unfavorable conditions such as high winds.

Time determination. Possible systematic errors are more dangerous because they can affect all readings in a similar way and not be readily detectable. A potential source of systematic error comes from the need to determine the difference between watch time (WT) and universal time (UT)—in other words, to calibrate the watches being used on site. During periods of fieldwork, WT was normally compared with UT on a daily basis using the telephone time signal provided by Hawaiian Telcom. Where this was not possible, WT and the time given by one (2002, 2003, 2008) or two (2011) further digital watches were compared with each other on a daily basis in order to check consistency. In addition, WT-UT was determined for some days immediately before and after each period of fieldwork, in order to establish a linear variation with time.

During the 2011 trip, bracketing WT-UT determinations were made in the normal way using the Hawaiian Telcom time signal, but these were bracketed in turn using further determinations against British Telecom's TIM (123) telephone time-signal in the United Kingdom before and after the fieldwork period. This comparison revealed a difference of up to 27 seconds between the estimates of UT obtained using the Hawaiian Telcom and British Telecom calibrations (varying linearly with time), which translates into a difference of some 7 arc minutes between the deduced azimuth readings. For the 2011 data, this discrepancy was resolved thanks to two cases where two sets of sun-azimuth readings were taken at the same site some 2.5 and 4 hours apart, respectively (the sites concerned were actually on Hawai'i Island, surveyed by Ruggles during the following week). In each case, only the British Telecom calibration provided consistency between both sets of readings, which clearly showed the British Telecom calibration to be correct. However, this raises the question of the reliability of the Hawaiian Telcom calibrations used in earlier years. Fortunately, the implied systematic error is never greater than about 0.1°.

Compass-clinometer surveys. Where a lack of time or a particularly inaccessible location made it impracticable to carry heavy equipment to the site, azimuths and altitudes were determined using handheld prismatic compasses and compass-clinometers, sighting directly on horizon points or along relevant oriented features such as walls. Wherever possible, compass readings of azimuths were corroborated in one or more of several ways: taking repeated measurements, taking measurements along the same structure in opposite directions, using two different instruments, and/or using readings by different people. The magnetic azimuths thus determined were converted into true azimuths by adding 10 degrees—the mean magnetic declination in this area varied from 10.0° in 2002 down to 9.8° in 2011 (data from www.ngdc.noaa.gov)—but uncertainties remain owing to the possible existence of local anomalies that, if averaging to a nonzero value over the extent of the site, would introduce a systematic error. Where feasible, this possibility was eliminated by taking readings of reference targets that could be identified on maps, thus providing an independent determination of their azimuth.

Precision. Despite the much greater measurement precision that is possible when using a total station, the azimuths of structures are quoted only to the nearest 0.1°, which represents a lateral uncertainty of about 1 cm along a length of 5 m. This is the greatest precision that can generally be justified on archaeological grounds, and in some cases (e.g., short walls) even this may represent false accuracy. By comparison, the apparent angular diameter of the sun or moon is 0.5°.

Handheld compass measurements are precise only to 1° (or at best 0.5°) in the first place and, as already noted, are also subject to uncertainties owing to local variations in the magnetic field. Where we were able to check compass readings against total station readings made at the same site, the results indicated that the compass readings were generally accurate only to within about 2°. One difficulty is that the presence of magnetite in underlying 'a'ā formations, particularly where there may be concentrations of dense dikestone, can result in local magnetic anomalies (as demonstrated, e.g., by Palmer [1927] in the case of Nihoa and Necker Islands). In a few cases (e.g., WF-AUW-359), we were able to identify and correct for such errors by means of the digitally generated horizon profiles (see the section "Visualizations of Landscape and Sky").

Where azimuths have been measured directly, they are marked (either to a precision of 0.1°, if measured by total station, or to a precision of 1°, if measured only by magnetic compass) on plane-table plans included in part II. In some cases, the azimuths may not appear consistent with the orientations of the structures concerned (wall faces, platform edges, etc.) as they appear in the plan relative to one another or to the north arrow. In these cases the measured and quoted azimuths should be taken as definitive; discrepancies may be due to the surveyed azimuths' being based only on segments of intact wall facing, as opposed to the extent of rubble collapse, or to inaccuracies inherent in the process of plane-table survey, which is not well suited to determining structural orientations.

Altitude measurements taken by total station or handheld clinometer are at least as accurate as the equivalent azimuth readings, which means that the precision to which it is meaningful to quote declinations is limited to the precision of the corresponding azimuth: 0.1° where surveyed by total station or 1° where measured by magnetic compass.

In order to convert azimuths and altitudes to declinations, it is also necessary to know the latitude of the place (Ruggles 1999a, 22). However, given that the greatest justifiable precision in the azimuths and declinations is 0.1°, there is no need to determine the latitude more accurately than this. Thus, while the location of each *heiau* has been accurately determined by GPS, it is sufficient for the purposes of the archaeoastronomical analysis to know that all the Kahikinui and Kaupō *heiau* have a latitude of 20.6° N (+20.6°).

Horizon Visibility

In the past, as now, a distant horizon might have been covered by forest or obscured by intervening trees. Although the ecological and archaeological evidence can give a broad indication of the likely vegetation cover in different areas, we can never know the details precisely enough to know the extent to which any particular putative astronomical sight line might have been affected. In order to adopt a consistent approach, we have generally considered the appearance of each horizon bare of vegetation, making allowance (as and where necessary) at the interpretative stage for the possibility that the horizon concerned was in fact forested.

In several cases, our measurements of the horizon have been curtailed by obstruction by large invasive trees or, more often, simply by the weather. From most of the locations studied, the slopes of Haleakalā are largely obscured by cloud during the daylight hours except in the early morning and late afternoon. The peaks of Mauna Kea and Mauna Loa on Hawai'i Island to the southeast, although prominent when visible, are rarely seen.

Visualizations of Landscape and Sky

For the reasons just given, as well as to have a convenient means of displaying horizon profiles that also accurately depict the astronomical data, we have collaborated with Andrew Smith of the Physics Department at the University of Adelaide, Australia. His Horizon program (www.agksmith.net/horizon/) produces a graphic representation of the complete horizon from any given location, generated from digital terrain model (DTM) data—in this case, the USGS National Elevation Dataset (NED), a 1/3-arc second DTM with a horizontal resolution of approximately 10 m (Gesch et al. 2002; Gesch 2007). An example is shown in figure 4.1, in which shading is used to give a broad indication of the relative distances of different landscape features and to add a degree of visual realism. The visualizations are overlain with azimuth, altitude, and declination grids and have also been configured to show the rising and setting tracks of the solstitial and equinoctial sun and of the Pleiades at a date appropriate to the dating of the

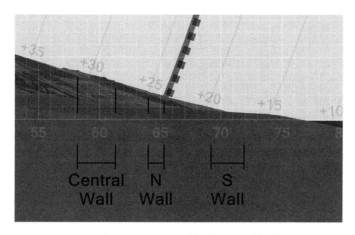

FIGURE 4.1. Part of a DTM-generated horizon profile (site LUA-4, to the ENE) overlaid with azimuth, altitude, and declination grids and showing the rising tracks of the June solstice sun (*solid shaded line*) and of the Pleiades in AD 1700 (*dotted line*). The horizontal bars indicate the direction of orientation of the three WSW–ENE-oriented wall segments. (Graphic by Andrew Smith)

site concerned. Sections of horizon on which structures such as walls and platform edges align are clearly marked.

While the Horizon program is an extremely useful tool for reconstructing parts of (or, in a few cases, complete) distant horizons that could not be photographed or surveyed during our site visits, it is much less reliable for nearby horizons (say, only a few hundred meters distant) because of the lack of adequate resolution in the DTM at small distances. Local horizons are also those affected most critically by vegetation and, in addition, are most likely to have been altered significantly by changes in the landscape itself.

Wherever possible, annotated profile photographs were used for "ground-truthing"—in other words, to provide an additional consistency check between the horizon data generated from the DTM and the surveyed data. At some sites where only compass-clinometer surveys were possible, such as WF-AUW-359, systematic differences between DTM-generated horizon profiles and surveyed points enabled us to identify, and correct for, errors due to local magnetic anomalies.

It is also possible to visualize the configurations of stars in particular directions by overlaying photographs upon sky projections created using the program Stellarium (www.stellarium.org) v. 0.15.0 (2015), as is done in figure 8.10 and figure 8.26 in Part II of this book. Another example can be seen in Kirch, Ruggles, and Sharp (2013, 61), where the profile to the west-northwest at LUA-29 is produced in this way.

Determining the Four Directions of Orientation

Despite the variations in their design, all of the *heiau* in Kahikinui and Kaupō display a clear "four directionality" in the sense that they consist of, or are elaborations of, one or more broadly rectangular structures (platforms, enclosures, or courts). For the purposes of the systematic analysis, we therefore attempt to define the two roughly perpendicular axes that will provide our best estimate of the intended orientation, if any, in each of the four directions. Clearly, the precision to which it is meaningful to express the azimuths in question depends on the length, straightness, and degree of parallelism of the walls, platform edges, or other structures aligned in the requisite directions, and their state of preservation.

The assortment of configurations and states of preservation with which we are faced means that we have no consistent way of estimating "error bars" in any formal sense, so instead we determine, for each axis at each site, a mean azimuth A together with a difference ΔA such that $(A - \Delta A)$ to $(A + \Delta A)$ defines an azimuth range best expressing the range of possibilities for the intended orientation (and a similar range differing by 180° in the opposite direction).

In an ideal case where there are two long, straight structures (e.g., opposite walls or platform edges) reasonably well preserved, then we would obtain a best estimate of the orientation of each (for a wall, as opposed to a platform edge, we might have orientation estimates from intact facing on either the interior or exterior side or on both sides) and take their average to obtain the mean azimuth value (in both directions). Generally, however, the situation is more complex and we make informed, if subjective, choices. Details are given on a case-by-case basis in the catalog so that others can examine the basic data and our decisions and rework our analyses if they wish.

Determining the Main Direction of Orientation

In many cases, it is evident which of the four directions was the sacred (*kapu*) direction of orientation, owing to one or more of the following architectural indications:

- *Raised altar platform.* Examples include LUA-4, ALE-4, ALE-121, KIP-330, NUU-81, and KAU-995 (Kou).
- *More substantial enclosing wall.* Examples include WF-AUW-359, WF-AUW-493, KIP-77, KIP-275, KIP-366, KIP-1025, and KAU-994 (Loʻaloʻa).
- *Raised terrace(s) and/or lower court in opposite direction.* Examples include WF-AUW-338, ALE-1, KIP-275, KIP-410, KIP-1010 (W and E), KIP-1151, and KAU-324 (Pōpōiwi).
- *Entrance or open side in opposite direction.* Examples include AUW-11, WF-AUW-100, WF-AUW-403, WF-AUW-574, LUA-3, LUA-36, KIP-1, KIP-273 (A and B), and NAK-30.
- *Other.* Other strong indicators of the *kapu* direction include a wall elaborated by buttressing (WF-AUW-391, KIP-1, KAU-333B), upright stones placed on a wall (KIP-75, KIP-1146) or on the side of a platform (NAK-30), a cobble pavement (KIP-567), a niche for offerings built into a wall (KIP-1010 [W and E]), and the quality of construction of an enclosure relative to an adjacent terrace (NUU-175).

Such indications are sometimes ambiguous. Thus, at KIP-405 the raised altar platform is placed on a corner, and at KIP-1151 and NUU-81 there are more substantial walls on two adjacent sides. In these three cases, another factor clarifies the situation. At some other sites, however, different factors seem to indicate different *kapu* orientations. Thus, WF-AUW-176A has a raised altar platform on the northern side but an entrance on the western side, which would suggest an eastern orientation; LUA-1 has a "bedrock altar" at the eastern end but a more substantial enclosing wall on the north. Again, other factors clarify the situation in both these cases.

Where these factors do not provide a clear indication of the *kapu* direction, we can sometimes take into account one or more secondary indicators:

- *Concentration of branch coral or other artifacts.* At AUW-6 this is the best indication of the intended direction of orientation, but generally it reinforces other factors, as at LUA-29 (later stage of construction), KIP-330, and KIP-567. At WF-AUW-359, however, the visible branch coral is concentrated to the north while the more prominent wall is to the east.
- *Longer along primary axis.* This in itself identifies only two opposite directions but can help narrow down the possibilities. Thus, at KIP-414 the longer axis is east–west, and artifacts are concentrated to the north and east, indicating that east is the *kapu* direction. Sites where the longer axis is useful in helping to identify the main direction of orientation include WF-AUW-176A, LUA-1, NAK-29, NUU-79, and NUU-175. However, at a few sites (WF-AUW-403, KIP-307, and KIP-366) the primary axis appears to be the shorter one.
- *Broadening out in direction of indication.* At MAH-231 this was a factor in determining the likely direction of orientation. However, it is not a universally reliable factor. At some sites, such as WF-AUW-100 and ALE-140, the structure does indeed widen out in the direction of orientation, but at others (e.g., NAK-30 and KAU-994 [Loʻaloʻa]) the opposite is the case, and at yet others (AUW-11, KIP-188, NUU-175) it (also) broadens out in one of the directions perpendicular to the *kapu* direction as indicated by other factors.
- *Open view in direction of indication.* At WF-AUW-391 the fact that the only open view is in one of the four directions seems to provide a clear confirmation of the *kapu* direction, and at MAH-231 it helps narrow down the possibilities. However, in other cases, such as ALE-140 and KIP-1151, the *heiau* appears to be oriented in the one of the four directions that does *not* have an open view, although at ALE-140 we have discovered an observation device in the opposite direction.

A summary of all the data regarding the direction of orientation is given in table 4.5 (primary and secondary indicators) and table 4.6 (notched corners and canoe-prow corners).

We appreciate that there is a degree of subjectivity in identifying the principal direction of orientation and for this reason give some consideration to all four directions in the analysis that follows. We have also examined other factors that might be relevant in determining the *kapu* direction. One is the orientation of notched corners—that is, corners of notched *heiau* (see table 4.6).

In 14 of 30 cases (including possible notched corners and variants) where the direction of orientation seems clear, the notch is placed at the front left (taking the wall in the direction of orientation as the front); in 10 cases, the notch is at the front right; and in 6 cases, it is at the back left. None is at the back right. This implies that there is some correlation between the placement of the notch and the *kapu* direction, with the great majority of notched corners at the front. Furthermore, all 6 cases with the notch at the back left are east-facing, raising the possibility that if their orientation has been misidentified and was in fact northerly rather than easterly, then notched corners might always have been placed at the front. The sites in question are WF-AUW-403, LUA-1 (possible notch), KIP-410, KIP-414 (variant design), NAK-29, and NUU-175. As can be seen in table 4.5, some factors do indicate a northerly rather than an easterly orientation at nearly all of these sites. This would also imply that the orientation of MAW-2 is likely to be northerly rather than easterly and that NAK-27 may have been oriented to the west.

There are fewer (obvious or possible) canoe-prow corners, mostly to the southeast (8 of 12 cases). Excluding NUU-78, where this feature is less certain, 5 of 11 cases have the canoe-prow

Landscape and Sky • 91

TABLE 4.5. Primary and secondary indicators of the *kapu* direction of Kahikinui and Kaupō *heiau*.

	PRIMARY INDICATORS						SECONDARY INDICATORS			
SITE	RAISED ALTAR PLATFORM	MORE SUBSTANTIAL ENCLOSING WALL	RAISED TERRACE(S) AND/OR LOWER COURT IN OPPOSITE DIRECTION	ENTRANCE OR OPEN SIDE IN OPPOSITE DIRECTION	OTHER (SEE TEXT)	CONCENTRATIONS OF BRANCH CORAL OR OTHER ARTIFACTS	LONGER AXIS	BROADENS OUT IN THIS DIRECTION	OPEN VIEW	MAIN DIRECTION OF ORIENTATION, AS IDENTIFIED (SEE PART II FOR DETAILS)
AUW-6	—	—	—	—	—	**N**	—	—	N, S	N
AUW-9 (first phase)	—	N, E	—	—	—	—	—	S	N, S	N
AUW-9 (second phase)	**S**	—	—	—	—	—	—	S	N, S	S
AUW-11	—	—	—	**N**	—	—	—	W	N, S	N
WF-AUW-100	—	N	—	**N**	—	—	—	N	N, S	N
WF-AUW-176A	**N**	N, W	—	E	—	—	N, S	—	N, S, W	N
WF-AUW-338	—	—	**N**	—	—	—	N	—	N, S, W	N
WF-AUW-343	—	E	—	**N, E**	—	—	N, S	—	N, S, W	E or N
WF-AUW-359	—	**E**	—	—	—	N	—	—	N, E, S, W	E
WF-AUW-391	—	E, S	—	—	**E**	—	—	—	E	E
WF-AUW-403	—	—	N	**E**	—	—	N, S	—	N, E, S, W	E
WF-AUW-493	—	**E**	—	—	—	E?	—	—	N, E, S, W	E
WF-AUW-574	—	N	—	**N**	—	—	—	—	N, E, S	N
LUA-1	**E**	N	—	N? E?	—	—	E, W	—	N, E, S, W	E
LUA-3	N	—	N	**N**	—	—	—	—	N, S	N
LUA-4	**N**	—	N	—	—	N?	—	—	N, E, S	N

Site									
LUA-29 (koʻa)	**N**	—	**N**	—	S	—	—	N, S, W	N
LUA-29 (notched enclosure)	—	—	**N**	—	N	—	—	N, S, W	N
LUA-36	—	E	**E**	—	—	—	—	N, E, S, W	E
LUA-39	N	—	**N**	—	N	—	—	N, S	N
ALE-1	N	N, E, W	**N**	—	—	N, S	—	N, S	N
ALE-4	**N**	—	**N**	—	—	N, S	—	N, S	N
ALE-121	N	—	**N**	—	N?	—	—	N, E, S	N
ALE-140	**W**	W	**W**	—	—	—	N, W	N, E, S	W
ALE-211	—	—	**N**	—	—	N, S	—	N, E, S, W	N
KIP-1	E	E	E	E	—	E, W	—	N, S, W	E
KIP-75	E, S	N	—	**E**	—	—	—	N, S	E
KIP-77	—	E	E	—	—	E, W	—	N, E	E
KIP-188	—	E	—	—	—	—	N	N, S, W	E
KIP-273A	—	—	**N**	—	—	—	—	N, S	N
KIP-273B	—	—	**E**	—	—	E, W	—	N, E, S	E
KIP-275	—	**E**	—	—	—	—	—	N, S	E
KIP-306	—	N	E?	—	N	—	—	N, S	N
KIP-307	—	N, E	N, W?	—	—	E, W	—	N, E, S, W	N
KIP-330	**E**	—	**E**	—	E	—	—	N, E, S	E
KIP-366	—	**E**	—	—	—	N, S	—	N, E, S, W	E
KIP-405	N, E	**E**	E	—	N, E	—	—	N, E, S, W	E
KIP-410	—	—	**E**	—	—	—	—	N, E, S, W	E
KIP-414	—	—	—	—	N, E	**E, W**	—	N, E, S, W	E
KIP-567	—	N, E	—	**E**	E	—	—	N, E, S	E

(continued)

TABLE 4.5. Primary and secondary indicators of the *kapu* direction of Kahikinui and Kaupō *heiau*. (*continued*)

	PRIMARY INDICATORS					SECONDARY INDICATORS				
SITE	RAISED ALTAR PLATFORM	MORE SUBSTANTIAL ENCLOSING WALL	RAISED TERRACE(S) AND/OR LOWER COURT IN OPPOSITE DIRECTION	ENTRANCE OR OPEN SIDE IN OPPOSITE DIRECTION	OTHER (SEE TEXT)	CONCENTRATIONS OF BRANCH CORAL OR OTHER ARTIFACTS	LONGER AXIS	BROADENS OUT IN THIS DIRECTION	OPEN VIEW	MAIN DIRECTION OF ORIENTATION, AS IDENTIFIED (SEE PART II FOR DETAILS)
KIP-728	—	—	N, E	—	—	—	N, S	—	—	N
KIP-1010 (W)	—	—	**N**	E	**N**	—	N, S	E	N, E, S	N
KIP-1010 (E)	—	N, E	**N**	N	**N**	—	N, S	E	N, E, S	N
KIP-1025	—	**E**	—	—	—	—	E, W	—	N, E, S, W	E
KIP-1146	—	—	N	—	E	—	E, W	—	N, E, S, W	E
KIP-1151	—	N, E	**E**	E?	—	—	—	—	N, S, W	E
KIP-1156	N	—	N, E	N	—	—	—	N	N, E, S	E
NAK-27	—	—	—	—	—	—	**E, W**	—	N, S, W	E?
NAK-29	—	**E**	N, E	E?	—	—	E, W	—	N, E, S	E
NAK-30	—	N	N	**E**	**E**	—	—	W	N, (E), S	E
NAK-34	—	**N**	—	—	—	—	—	—	N, S	N
MAH-231	—	—	N?	S	—	—	—	**N, W**	N, S, W	N
MAW-2	—	—	—	—	—	—	—	—	N, E, S	E or N
NUU-78	N?	—	—	—	—	—	N, S	S	?	N?
NUU-79	E?	**E**	N	E	—	—	E, W	—	N, (E), S, W	E
NUU-81	**W**	N, W	W	—	—	—	E, W	—	N, S, W	W
NUU-100	W	**W**	**W**	—	—	—	E, W	—	N, S, W	W

Site									
NUU-175	—	N	—	—	E, W	—	N	S	E
NUU-188	—	—	—	—	E, W	—	E	S	E?
KAU-324 (Pōpōiwi)	N	—	**N**	—	N, S	—	—	N, E, S, W	N
KAU-333A	—	N, E	—	**N**	E, W	—	—	N, S, W	N
KAU-333B	—	—	E	—	—	—	—	N, S, W	E
KAU-536	—	—	**E**	—	E, W	—	—	—	E
KAU-994 (Loʻaloʻa)	—	—	E	—	E, W	—	W	N, E, S	E
KAU-995 (Kou)	**E**	E	—	—	E, W	—	—	N, (E), S, W	E

Note: The principal architectural features or other factors considered to indicate the *kapu* direction at each *heiau* site are marked in bold. For clarity, we have simplified the four directions of orientation as N, E, S, or W. Thus, for example, the directions of orientation identified as ENE, E, and ESE in the catalog (part II) are all included here simply as E. The actual orientations may deviate from the cardinal or intercardinal directions in question by up to 30°.

TABLE 4.6. Directionality of notched corners and canoe-prow corners in relation to the identified *kapu* direction.

SITE	MAIN DIRECTION OF ORIENTATION*	NOTCHED CORNER	NOTCHED CORNER RELATIVE TO MAIN DIRECTION	CANOE-PROW CORNER	CANOE-PROW CORNER RELATIVE TO MAIN DIRECTION
AUW-9 (first phase)	N	NW	Front left	—	—
WF-AUW-100	N	NE	Front right	—	—
WF-AUW-176A	N	NE	Front right	SE	Back right
WF-AUW-338	N	NW	Front left	—	—
WF-AUW-343	E or N	NE	Front left or front right	—	—
WF-AUW-359	E	NE	Front left	—	—
WF-AUW-391	E	NE	*Front left*	—	—
WF-AUW-403	E	NW	Back left	—	—
WF-AUW-493	E	—	—	SW	*Back right*
LUA-1	E	NW	*Back left*	—	—
LUA-4	N	NW	Front left	—	—
LUA-29 (notched enclosure)	N	NW	Front left	NW	Front left
ALE-1	N	NW	Front left	—	—
ALE-4	N	NW	Front left	—	—
ALE-140	W	SW	Front left	—	—
ALE-211	N	NE	Front right	—	—
KIP-75	E	—	—	SE	*Front right*
KIP-77	E	SE	Front right	—	—
KIP-188	E	NE	Front left	—	—
KIP-273A	N	NW	Front left	—	—
KIP-275	E	NE	*Front left*	—	—
KIP-307	N	NW	Front left	—	—
KIP-330	E	—	—	SE	*Front right*
KIP-410	E	NW	Back left	—	—
KIP-414	E	NW	*Back left*	SE	*Front right*
KIP-1010 (W)	N	NE	Front right	SE	Back right
KIP-1010 (E)	N	NE	Front right	SE	Back right
KIP-1156	E	—	—	NE	Front left
NAK-27	E?	SW	Back right?	—	—
NAK-29	E	NW	Back left	—	—

SITE	MAIN DIRECTION OF ORIENTATION*	NOTCHED CORNER	NOTCHED CORNER RELATIVE TO MAIN DIRECTION	CANOE-PROW CORNER	CANOE-PROW CORNER RELATIVE TO MAIN DIRECTION
NAK-34	N	NE	Front right	—	—
MAW-2	E or N	NW	Front left or back left	—	—
NUU-78	N?	NE	*Front right?*	SE	*Back right?*
NUU-81	W	NW	Front right	—	—
NUU-100	W	NW	Front right	—	—
NUU-175	E	NW	Back left	—	—
NUU-188	E?	SE	Front right?	—	—
KAU-324 (Pōpōiwi)	N	—	—	SE	Back right
KAU-333A	N	NW	Front left	—	—
KAU-536	E	SE	Front right	—	—

Note: Possible or variant cases are shown in italics.

* See table 4.5 on simplifications made for clarity.

corner at the back right with respect to the assumed *kapu* direction, with 4 at the front right, 2 at the front left, and none at the back left. This appears to be a simple consequence of the preference for placing canoe-prow corners in the southeast (in the broad direction of Hawaiʻi Island) and of the northerly or easterly orientation of the *heiau*. However, with only two exceptions it may be that there was a preference for situating the canoe-prow corners on the right as opposed to the left. One of the two exceptions is KIP-1156, where several indicators suggest that the orientation might have been to the north rather than to the east (see table 4.5), a modification that would remove the anomaly. The other is LUA-29, where, despite several factors indicating that the *kapu* direction was to the north, the anomaly would be removed only by postulating a westerly (or southerly) orientation.

The seven *koʻa* sites do not display any obvious consistency as regards the main direction of orientation. Five of them (AUW-11, LUA-29 [earlier phase], LUA-39, ALE-121, and KIP-306) are oriented to the north, while two (LUA-36 and KIP-330) appear to have been oriented to the east. One (KIP-330) has a possible canoe-prow corner.

Systematic Analysis

The systematic analysis that follows is based on orientation data from 54 *heiau* sites. The following sites appear in the catalog but do not form part of the systematic analysis for the reasons stated:

SITE(S)	REASON FOR EXCLUSION FROM SYSTEMATIC ANALYSIS
KIP-76, -80, -394, -424; NUU-1, -172	Not a *heiau* (KIP-394, NUU-1, and NUU-172 are thought to be elite residence sites; KIP-424 is thought to be a possible men's house within a residential complex; KIP-76, an oven house; and KIP-80, a *hula* platform)
KIP-1146	Irregular, with no identifiable principal orientation (but site has suspected ritual and calendrical function)
AUW-6; ALE-211; KIP-410, -728, -1306, -1307, -1317, -1398; MAH-363; NUU-78, -151, -153, -188, -424; KAU-333, -411, -536, -996, -999, -1001	Not surveyed
KIP-115	Not surveyed; irregular, possibly unfinished

Of the sites analyzed, all but three where the principal direction could not be reliably determined—WF-AUW-343, MAW-2, and NUU-101—also appear in the "principal direction" analysis.

THE DISTRIBUTION OF AZIMUTHS

The azimuth data for the direction of orientation are summarized in table 4.7, which also includes for comparison the values quoted in Kirch (2004b) and Ruggles (2007). A supplementary version of this table available in the online materials (table S4.1) includes additional columns

TABLE 4.7. Azimuths of Kahikinui and Kaupō temples in the direction of orientation.

SITE	DIRECTION THE SITE FACES	MEAN AZIMUTH (A)	SEMIWIDTH OF AZIMUTH RANGE (ΔA)	CF. KIRCH (2004A)	CF. RUGGLES (2007)
AUW-9	N	0.6	4.2	—	176
LUA-39	N	1.0	1.0	—	—
MAH-231	N	3.0	1.0	—	—
WF-AUW-176A	N	8.1	1.7	—	—
AUW-11	NNE	17.0	1.0	—	—
WF-AUW-100	NNE	18.9	0.2	—	—
LUA-29	NNE	33.6	1.0	—	299
NAK-27	ENE	58.8	0.4	—	—
KIP-75	ENE	60.5	2.5	69	59
KIP-77	ENE	60.9	0.6	71	60
KIP-567	ENE	61.0	0.1	64	61
NAK-29	ENE	64.3	0.8	—	—
KAU-994 (Loʻaloʻa)	ENE	66.3	0.1	—	—
KIP-1156	ENE	67.0	0.3	73	—
NAK-30	ENE	67.1	0.3	—	—

SITE	DIRECTION THE SITE FACES	MEAN AZIMUTH (A)	SEMIWIDTH OF AZIMUTH RANGE (ΔA)	CF. KIRCH (2004A)	CF. RUGGLES (2007)
KIP-273B	ENE	70.4	2.0	83	—
KIP-330	ENE	73.0	2.0	—	—
KIP-275	ENE	77.4	9.1	—	—
KIP-1151	ENE	78.7	2.4	82	—
NUU-175	E	79.7	2.2	—	—
LUA-36	E	81.0	1.0	—	—
KIP-366	E	82.0	2.0	—	—
KIP-1025	E	85.0	3.6	87	—
WF-AUW-391	E	86.0	4.0	—	—
WF-AUW-493	E	87.0	4.0	—	—
KIP-1	E	87.3	0.6	89	87
WF-AUW-359	E	92.0	5.0	—	—
KIP-188	E	94.0	5.0	—	96
KIP-405	E	94.0	4.0	88	87
WF-AUW-403	E	94.7	0.4	—	—
KIP-414	E	95.0	2.0	92	94
LUA-1	E	97.0	4.0	—	97
NUU-79	E	97.8	2.5	—	—
KAU-995 (Kou)	ESE	107.8	1.3	—	—
ALE-140	WSW	244.5	1.3	241	—
NUU-100	W	278.1	0.3	—	—
NUU-81	WNW	286.3	1.2	—	—
ALE-4	NNW	327.5	0.4	333	—
LUA-4	NNW	335.5	2.4	340	337
KAU-324 (Pōpōiwi)	NNW	341.9	2.6	—	—
KIP-307	NNW	344.8	1.0	345	—
LUA-3	N	350.0	2.1	351	351
NAK-34	N	351.0	1.0	—	350
ALE-1	N	351.9	0.5	350	352
KIP-273A	N	351.9	1.8	93	—
KIP-1010 (E)	N	353.4	2.4	—	353
WF-AUW-338	N	355.0	0.5	—	—
ALE-121	N	355.0	5.0	—	—
KIP-1010 (W)	N	357.3	2.7	0	354
KIP-306	N	358.0	1.0	—	—
WF-AUW-574	N	359.0	3.0	—	—

Note: A version of this table that includes columns summarizing the elevation of the *heiau* and, where known, the chronology, is available in the supplementary online materials (table S4.1).

summarizing the elevation of the *heiau* and, where known, the chronology. The corresponding table for all four directions (table S4.2) is also available in the supplementary online materials.

The azimuth data are visualized as radar plots for all four directions in figure 4.2 and for the main direction of orientation in figure 4.3. The most obvious trend conveyed by these illustrations is that all the *heiau* in our sample, without exception, are broadly cardinally oriented to within about 30°. There is no example that is oriented even approximately along the intercardinal directions. (The shorter axis at NAK-27 [mean az. 143.4°/323.4°] comes closest, being within 9° of the intercardinal direction. The closest azimuth in the direction faced is 33.6°, at LUA-29, 11.4° from the intercardinal direction.) Indeed, these two figures clearly demonstrate, first, that the principal orientations are significantly clustered and, second, that this cannot be attributed entirely, or indeed in any great measure, to the local lay of the land, with its contours running east-northeast to west-southwest (az. ~65°–245°), roughly parallel to the coastline.

The clustering of the principal orientations is even more marked among upland *heiau* at an elevation of ~400 m or more (fig. 4.4). This plot reveals three clear groupings: one around north, centered slightly to the west of north (az. 351° to 3°); one to the east-northeast (az. 59° to 67°); and one to the east, centered slightly to the south of east (az. 87° to 97°). Both Kirch (2004b) and Ruggles (2007) previously noted three orientation clusters broadly to the north, east-northeast, and east, although the new data set revises some of their provisional azimuth values by several degrees; in three instances the main direction of orientation was differently identified (see table 4.7). The east-northeast cluster comprises four values between 59° and 61°, one at 64°, and two at 67°, which in itself strongly suggests that this group is not simply reflecting a propensity to orient *heiau* along the topographic contour lines, which run at around 63°–65°.

When the *heiau* at lower elevations are included, the northerly orientation group extends between azimuths of 328° and 8°, with outliers at 17°, 19°, and 34°, while the east-northeasterly and easterly groups merge and spread from azimuth 59° to 98°, with an outlier at 108°. An additional three sites are oriented westward, with azimuths of 245°, 278°, and 286°.

There are no clear correlations between the *heiau* orientations and the known chronology of specific temples. However, four of the six sites with orientations that stand out from the main groups to the east and north and for which dating information is available—namely, AUW-11 (az. 17°), KAU-995 (Kou) (az. 108°), NUU-100 (az. 278°), and NUU-81 (az. 286°)—all yield relatively late dates of around AD 1650 or later.

The systematic analysis of azimuths is insufficient in itself for identifying specific topographic referents because the azimuths of those referents vary from site to site. Discussion of these is left to the "Prominent Topographic Features" section below. However, in chapter 3 we mentioned the possibility that canoe-prow corners may have been oriented toward significant landmarks; this suggests an additional set of directions of possible significance that should be examined at this point. The directions concerned are best estimated by the orientation of the line bisecting the corner. We present the relevant data in table 4.8 and figure 4.5.

The majority of canoe-prow features are found on the southeastern corners of *heiau* enclosures (see also table 4.6), with estimated azimuths between ~110° and ~160°. This correlates, broadly speaking, with the direction of Hawai'i Island, which spans an approximate azimuth range of 125°–155° from the most westerly sites in Kahikinui and of 135°–170° from the most easterly in Kaupō. This supports the idea that these corners tended to be associated with topographic targets and were specifically oriented toward enemy territory. In particular, the canoe-prow

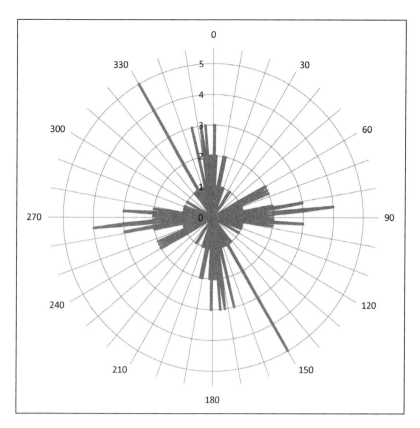

FIGURE 4.2. Radar plot showing the mean azimuths of Kahikinui and Kaupō temples in all four directions.

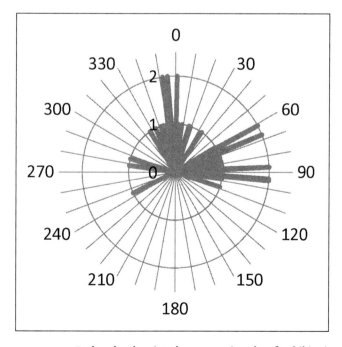

FIGURE 4.3. Radar plot showing the mean azimuths of Kahikinui and Kaupō temples in the main direction of orientation.

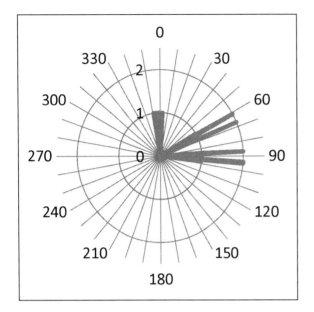

FIGURE 4.4. Radar plot showing the mean azimuths of upland Kahikinui and Kaupō temples (elevation ~400 m or above) in the main direction of orientation.

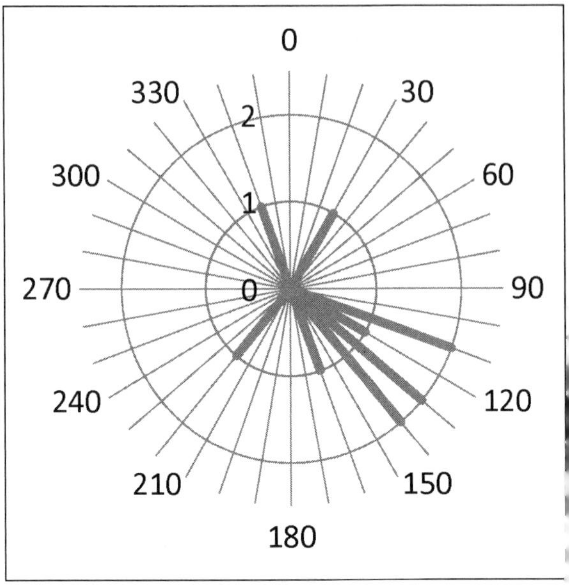

FIGURE 4.5. Radar plot showing the orientations of "canoe-prow corners," estimated to the nearest 5° or 10°.

TABLE 4.8. Orientation of canoe-prow corners at Kahikinui and Kaupō temples.

SITE	STATUS	AZIMUTH OF CANOE-PROW CORNER
WF-AUW-176A	—	~130°
WF-AUW-493	Possible	~220°
LUA-29	—	~340°
KIP-75	Possible	~110°
KIP-330	Possible	~110°
KIP-414	Variant	~140°
KIP-1010 (W)	—	140°
KIP-1010 (E)	—	~130°
KIP-1156	—	30°
NUU-78	Possible	~160°
KAU-324 (Pōpōiwi)	—	120°

Note: Estimated by the direction of the line bisecting the angle of the corner. The azimuths are quoted to the nearest 5° or, when marked with a tilde (~), to the nearest 10°. For further information, see part II.

corners at the two large, north-facing KIP-1010 enclosures, as well as that at the sizeable and irregular KIP-414, may well have been intended to align upon the summit of Mauna Kea, while the impressive canoe-prow corner at Pōpōiwi (az. ~120°) was intended to face Waipiʻo, an important seat of the Hawaiʻi Island ruling chiefs, even though the alignment is not very accurate.

What is apparently an anomalously placed canoe-prow corner at KIP-1156 may be better explained as a result of the skewing of the east-northeastern wall so that it was aligned directly upon the red cinder cone Kanahau.

Azimuth ranges. Returning to the main four directions of orientation and the principal, *kapu* direction, we can elaborate on our initial results if we introduce an indication of our uncertainty about the intended direction in each case. Uncertainties can arise for three reasons:

1. Limitations in the original precision, e.g., where a wall segment is curved rather than straight, or "parallel" walls are not strictly parallel. This presents us with a range of possibilities: it might be that the intended orientation was within that range, or that the exact orientation was of no relevance to the builders.
2. Limitations due to the state of preservation.
3. Limitations in the accuracy of our determination of the orientation (in particular, where a compass was used rather than a total station).

Only the last of these can be treated in a statistically rigorous way; limitation types 1 and 2 are inevitably subjective. For each alignment, we have determined an azimuth range, taking into account the various uncertainty factors, and have documented our choices in the catalog of sites (part II). In table 4.7 the azimuth range is expressed in terms of a semiwidth (ΔA), so that, for example, at KAU-995 (mean az. 107.8°, semiwidth 1.3°), the azimuth range is from 107.8° − 1.3° to 107.8° + 1.3°—that is, 106.5° to 109.1°. These azimuth ranges are shown in the various profile diagrams.

Using the azimuth range, we can express our knowledge about the alignment as a spread of probability over azimuth. A simple way to do this is to use a normal curve (Gaussian hump) with its mean at the mean azimuth and its standard deviation equal to the semiwidth of the azimuth range. This weights the probability in favor of azimuths closer to the mean azimuths and places 68% of the total probability within the azimuth range.

Finally, we can produce a cumulative probability histogram, or curvigram (Ruggles 1999a, 50–52), of all of the alignments by adding together the component probabilities at each azimuth value. However, it is important to emphasize that, given the various subjective judgments involved in this procedure, the resulting curvigram should be regarded not as a tool suitable for formal statistical analysis but as an aid to visualization.

The curvigram thus obtained for the azimuths of Kahikinui and Kaupō temples in the direction of orientation is shown in figure 4.6. The data for all four directions are also shown, in lighter shading. These data underline the absolute preference for broadly cardinal orientation and the total avoidance of the intercardinal directions (45°, 135°, 225°, 315°). For the directions of orientation, they also show a relatively even spread of probability within the two main clusters around north (az. ~330° to ~20°) and east (az. ~60° to ~110°), although when all the data in all directions are included, there is an evident concentration toward the cardinal directions at the

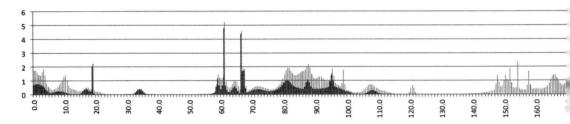

FIGURE 4.6. Cumulative probability histogram of the azimuths of Kahikinui and Kaupō temples in the direction of orientation (*darker shading*) and in all four directions (*lighter shading*).

centers of the ranges, specifically between 350° and 5° (also 170° and 185°) and between 80° and 95° (also 260° and 275°).

The curvigram shows probability spikes around 61° and 67° (actually centered at 61.0° and 66.3°), arising because the east-northeasterly alignments of KIP-567 and KAU-994 (Loʻaloʻa) are very precisely defined. This reinforces the conclusion that *heiau* with orientations to the east-northeast are not simply oriented along the topographic contours, and it invites an astronomical explanation, since the horizon in this direction is generally a featureless uniform slope.

The Distribution of Declinations

Table 4.9 presents the declination data for the direction of orientation, corresponding to the azimuth data in table 4.7. A supplementary version of this table available in the online materials, table S4.3, includes additional columns summarizing the elevation of the *heiau* and, where known, the chronology. The corresponding supplementary table for all four directions, table S4.4, is also available in the online materials.

The tables indicate the declination range for each alignment by listing, as for the azimuths, a mean value together with the semiwidth ($\Delta\delta$). Generally speaking, this is acceptable because there is an approximately linear relationship between azimuth and declination: the rate of increase/decrease of declination with azimuth is effectively uniform, and variations in altitude along the horizon are of little consequence because the astronomical bodies rise and set nearly vertically (actually, at 20.6° to the vertical, the angle corresponding to the latitude). However, for alignments within a few degrees of north or south, the rate of increase/decrease of declination with azimuth changes significantly, decreasing to zero at due north and south, and astronomical bodies are no longer steeply rising or setting. Here, the maximum and minimum declinations do not necessarily occur at the ends of the azimuth range, and the mean declination does not necessarily represent the "most likely" value. At AUW-9, for example, the declination varies from +79.9° at the left-hand end of the azimuth range (az. 356.4°) up to +80.8° in the middle (az. 0.6°) and then down again to +79.7° at the right-hand end (az. 4.8°). The mean and semiwidth values quoted in table 4.9 are chosen so as to span the declinations in the range. Thus, AUW-9 is listed with a mean declination of +80.3° and a semiwidth ($\Delta\delta$) of 0.5°.

With the declination data, as with the azimuth data, it is possible to create a curvigram, again using the semiwidth for the standard deviation of each Gaussian hump. However, as with the azimuth curvigram, and for the same reasons, the declination curvigram should be seen as a visualization aid rather than as a formal statistical tool.

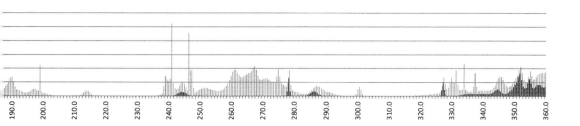

TABLE 4.9. Declinations of Kahikinui and Kaupō temples in the direction of orientation.

SITE	DIRECTION FACED	MEAN DECLINATION (δ)	SEMIWIDTH OF DECLINATION RANGE (Δδ)	RISE/SET
MAH-231	N	85.3	0.6	Rise
KIP-1010 (W)	N	84.7	1.0	Set
KIP-306	N	83.0	0.2	Set
KIP-1010 (E)	N	82.5	2.0	Set
LUA-39	N	81.7	0.1	Rise
WF-AUW-574	N	81.6	0.5	—
NAK-34	N	80.9	0.8	Set
WF-AUW-338	N	80.3	0.3	Set
AUW-9	N	80.3	0.5	—
ALE-121	N	79.9	2.2	Set
KIP-273A	N	79.6	1.3	Set
WF-AUW-176A	N	78.8	1.0	Rise
ALE-1	N	78.7	0.4	Set
LUA-3	N	77.1	1.6	Set
KIP-307	NNW	73.9	0.9	Set
KAU-324 (Pōpōiwi)	NNW	72.0	2.4	Set
AUW-11	NNE	71.5	0.8	Rise
WF-AUW-100	NNE	70.0	0.2	Rise
LUA-4	NNW	64.1	2.3	Set
ALE-4	NNW	57.1	0.4	Set
LUA-29	NNE	55.9	1.0	Rise
KIP-75	ENE	29.2	2.6	Rise
KIP-77	ENE	28.8	0.6	Rise
KIP-567	ENE	28.5	0.1	Rise

(continued)

TABLE 4.9. Declinations of Kahikinui and Kaupō temples in the direction of orientation. (*continued*)

SITE	DIRECTION FACED	MEAN DECLINATION (δ)	SEMIWIDTH OF DECLINATION RANGE ($\Delta\delta$)	RISE/SET
NAK-29	ENE	25.4	0.9	Rise
KAU-994 (Loʻaloʻa)	ENE	23.4	0.1	Rise
NAK-30	ENE	22.5	0.3	Rise
KIP-1156	ENE	22.3	0.3	Rise
KIP-273B	ENE	18.6	2.0	Rise
NUU-81	WNW	17.8	1.2	Set
KIP-330	ENE	16.7	2.0	Rise
NUU-100	W	9.5	0.3	Set
LUA-36	E	8.2	0.9	Rise
KIP-366	E	7.2	1.9	Rise
WF-AUW-391	E	4.5	4.2	Rise
KIP-1025	E	4.3	3.3	Rise
WF-AUW-493	E	3.6	3.5	Rise
WF-AUW-359	E	−1.9	5.1	Rise
LUA-1	E	−2.3	3.9	Rise
WF-AUW-403	E	−3.6	0.3	Rise
KIP-188	E	−4.2	4.7	Rise
KIP-405	E	−4.3	3.7	Rise
NUU-79	E	−5.1	2.3	Rise
KIP-414	E	−5.2	1.8	Rise
KAU-995 (Kou)	ESE	−16.9	1.2	Rise
ALE-140*	WSW	—	—	—
KIP-1*	E	—	—	—
KIP-1151*	ENE	—	—	—
KIP-275*	ENE	—	—	—
NAK-27*	ENE	—	—	—
NUU-175*	E	—	—	—

Note: A version of this table, including columns summarizing the elevation of the *heiau* and, where known, the chronology, is available in the supplementary online materials (table S4.3).

* Site whose horizons in the direction of orientation are too close for the declination to be accurately determinable.

A declination curvigram has the advantage that peaks of probability can be compared directly against the declinations corresponding to the rising and setting positions of the sun, moon, and stars, as detailed earlier in this chapter. The main limitation concerns alignments close to north or south, where a Gaussian hump produced in this way can be a poor representation of the actual spread of probability. For this reason, the relevant figures exclude declinations above +70° and below −60°, which would represent alignments close to north and south. Fortunately, the only bright Hawaiian star outside this declination range (see table 4.2) is Hōkūpa'a (Polaris), which is so close to the celestial north pole that it never rises or sets anyway but rather, as seen from the Kahikinui and Kaupō sites, hangs in the sky not far above the high northern horizon.

A cumulative probability histogram of the declination data for the main direction of orientation is presented in figure 4.7, including, for reference, the declinations of the sun at the solstices and equinoxes and of principal Hawaiian stars at different times during the chronological period of interest. The corresponding data for all four directions are shown in figure 4.8.

In figure 4.7, the two tallest spikes are centered at +28.5° and +23.4° and correspond to the azimuth spikes at 61.0° (KIP-567) and 66.3° (KAU-994 [Lo'alo'a]) already noted, but the latter is now seen to be reinforced by an adjacent spike around +22.4°, to which two sites (NAK-30 and KIP-1156) contribute. (Two indications in nonfacing directions, at ALE-140 and LUA-4, also spread some probability around declination +23°, but too thinly to show up clearly in figure 4.8.) This

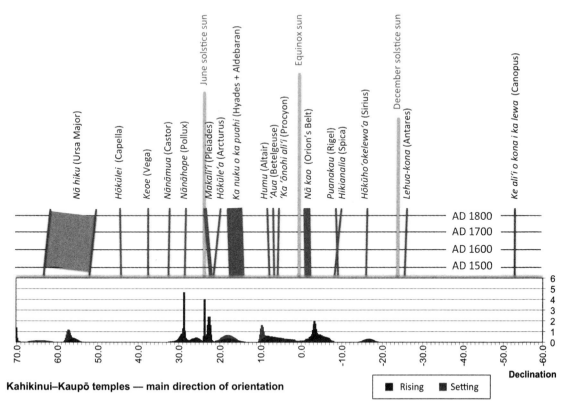

FIGURE 4.7. Cumulative probability histogram of the declinations of Kahikinui and Kaupō temples in the main direction of orientation. Darker shading indicates the easterly (rising) direction; lighter shading, the westerly (setting) direction. The declinations of the solstitial and equinoctial sun, and of principal Hawaiian stars during the period of interest, are also shown for comparison.

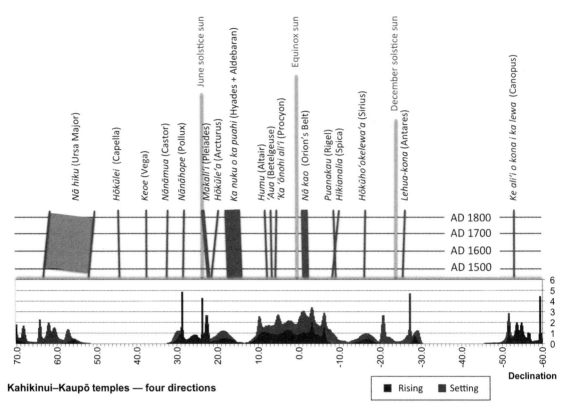

FIGURE 4.8. Cumulative probability histogram of the declinations of Kahikinui and Kaupō temples in all four directions. Darker shading indicates the easterly (rising) direction; lighter shading the westerly (setting) direction. The declinations of the solstitial and equinoctial sun, and of principal Hawaiian stars during the period of interest, are also shown for comparison.

cluster strongly suggests an explanation in terms of the Pleiades (Makaliʻi), which themselves span a declination range 0.5° wide, increasing from +22.3° to +22.8° around AD 1500 to +23.3° to +23.8° around AD 1800. Unfortunately, the two sites where we have the most precise chronological indication, ALE-140 and LUA-4, do not yield precise estimates of the declinations (see table S4.4), while at NAK-30 the reverse is true. For KIP-1156, the calibrated date range from a single radiocarbon determination (see part II) implies that the site is unlikely to be older than about AD 1650, when the declination of the Pleiades was between +22.8° and +23.3°, while the most likely declination range for the alignment (see table 4.9) is between +22.0° and +22.6°. While these data do not provide strong support for the Pleiades association, they do not refute it. On the other hand, the date fit at KAU-994 [Loʻaloʻa]) is very good. Assuming, as argued in part II, that the massive east platform and terraces date to around AD 1650 to 1700, implying that the Pleiades were seen somewhere between declinations of +22.8° and +23.5°, this is completely consistent with the declination range for the alignment, +23.3° to +23.5°.

The spike at +28.5° correlates with the Hawaiian star Nānāhope (Pollux). A ^{230}Th date at KIP-567 indicates a construction date close to AD 1574, when Nānāhope's declination was +28.9° (see table 4.2). The Heiau Ridge sites KIP-77 and KIP-75—which face most likely declinations of +28.8° and +29.2°, respectively—also reinforce this spike, although the spread is thin and wide in the latter case. The archaeological dating evidence suggests that both sites were constructed during the late seventeenth or eighteenth century, when Nānāhope's declination was around +28.7°, so it is plausible that this star was a target. Also, all three are high-elevation sites, above 400 m. On the other hand, NAK-29 (also at high elevation, and constructed close to AD 1600) appears to face neither Makaliʻi or Nānāhope but a range of declinations between the two; no site is evidently aligned upon Nānāhope's companion star, Nānāmua (Castor).

The accumulation of probability around +17.5° is due to three coastal *heiau*, KIP-273B, NUU-81, and KIP-330, which face declination ranges centered on +18.6°, +17.8°, and +16.7°, respectively. Although none of these is sharply defined, they could well relate to Ka Nuku o Ka Puahi (the V shape of the Hyades together with Aldebaran [α Tau]), which stretched from declination +14.6° to +18.2° around AD 1600, rising to +15.0° to +18.6° around AD 1750. All three of the *heiau* appear to have been constructed within this relatively late date range, around AD 1620 in the case of KIP-273B and later for the others. Two of the large Kaupō temples, KAU-995 (Kou) and KAU-324 (Pōpōiwi), also yield similar declinations but not in the direction faced.

A scatter of probability humps between +10° and −10° is evident among the directions faced (see fig. 4.7), and a more substantial accumulation of probability in this range becomes evident when the data in all four directions are included (see fig. 4.8). However, there is no obvious correlation between the maxima in the probability distributions and the declinations of stars known to have been significant to the Hawaiians. As all these alignments fall within the solar arc, an alternative explanation is that they are associated with sunrise and sunset. However, they do not fill the range: declinations from +10° down to −10° represent the path of the sun over a time span of only about four weeks on either side of the equinoxes (see table 4.3). This may indicate no more than a broadly cardinal orientation and perhaps that the orientation was fixed by reference to sunrise or sunset at a particular time in the spring or autumn as determined by the lunar calendar (whose months and days vary with respect to the solar year according to the phase cycle of the moon). On the other hand, 5 of the 13 sites that broadly face sunrise (or, in one case, sunset) within a month of the equinox also face Ka Lae o Ka ʻIlio (Kailio Point) (see the section "Prominent Topographic Features" below); these are LUA-36, KIP-366, KIP-405, KIP-414, and KIP-1025. This could imply either that the topographic referent was more important than the astronomical one or that both were used in conjunction, perhaps for tracking the changing position of sunrise.

Certainly, there is no evidence of a particular interest in the equinox itself—which is unsurprising (see the section "Sun" earlier in this chapter)—nor in the December solstice. The sharp spike at the June solstice is due to Loʻaloʻa and could be also attributable to the Pleiades, as already discussed: the two possibilities overlap in the late seventeenth and eighteenth centuries.

The most likely declination range for the direction faced by KAU-995 (Kou) is between −18.1° and −15.7° (see table 4.9), which appears as an isolated probability hump in figure 4.7. This coastal *heiau* of late date thus faces the rising point of ʻAʻā/Hōkūhoʻokelewaʻa (Sirius), whose declination was −16.2° around AD 1600, decreasing slightly to −16.4° around AD 1800. In figure 4.8 this probability hump is seen to be reinforced by some nonfacing directions of orientation; these derive from KIP-330, WF-AUW-100, KAU-324 (Pōpōiwi), and, marginally, KIP-273B, with declination ranges centered on −15.6°, −16.0°, −16.6°, and −18.1°, respectively. On the other hand, KIP-330 and KIP-273B face the rising of Ka Nuku o Ka Puahi to the east-northeast, so the alignment in the opposite direction (west-southwest), which yields the declinations quoted here, is likely to be coincidental. All of these *heiau* are coastal, apart from WF-AUW-100, whose elevation is a little above 200 m.

The accumulation of probability humps extending from −20° down to −30° seen in figure 4.8 is largely if not exclusively composed of "opposite directions" for *heiau* facing east-northeast and is unlikely to be of significance.

Whether some of the probability humps appearing on the left of the graphs in figures 4.7 and 4.8, around declination +55° and above, are related to Nā Hiku (Ursa Major), is an open question, but we draw attention to the evidently topographic nature of the northerly group of alignments as a whole (see the section "Prominent Topographic Features" below).

In sum, the systematic analysis of declinations indicates that some of the Kahikinui and Kaupō *heiau* may have been oriented with respect to Makaliʻi (the Pleiades), Nānāhope (Pollux), Ka Nuku o Ka Puahi (Hyades-Aldebaran), and ʻAʻā/Hōkūhoʻokelewaʻa (Sirius), and others may have been oriented toward the rising and setting of the sun, broadly around (but not exactly at) the equinoxes. Some of these generic conclusions will be subjected to more detailed examination in what follows.

Prominent Topographic Features

There is no straightforward way to undertake a systematic analysis of prominent topographic features, since prominence is a subjective concept and the conspicuity of visible landmarks varies considerably from place to place and according to the lighting (depending on the time of day and atmospheric conditions). For example, the summit of Haleakalā, which appears from Loʻaloʻa and Pōpōiwi as a conspicuous pointed peak, appears as a rounded prominence only at the end of a long, flat plateau from sites farther west in Nuʻu. However, we can attempt an analysis of major features in the visual topography of the Kahikinui-Kaupō area. Earlier in this chapter we identified a number of striking points within the visual topography, including prominent cinder cones (see table 4.1) and coastal promontories. In what follows we examine possible correlations between the orientations of the *heiau* and the directions of these landmarks, where they are visible. The basic data are shown in table 4.10. Additional topographic referents considered of possible significance at specific sites or groups of sites are discussed in the section "Specific Structural Orientations" below.

Two main trends stand out from table 4.10. First, 12 *heiau* stretching from Auwahi to Nakaaha—namely, AUW-11, WF-AUW-100, LUA-39, ALE-121, KIP-273A, KIP-275, KIP-307, KIP-330, KIP-405, KIP-1151, KIP-1156, and NAK-29—are aligned northward at least approximately upon the conspicuous red cinder cone of Kanahau on the Haleakalā Ridge, whose azimuth varies from 331° at NAK-29 to 18° at AUW-11. At KIP-1156, although the secondary axis is poorly defined owing to a particularly skewed east-northeastern wall, the skewed wall is itself aligned directly upon Kanahau, which could provide an explanation for its anomalous orientation. Of these 12 temples, 6 (AUW-11, WF-AUW-100, LUA-39, ALE-121, KIP-273A, and KIP-307) appear (from other criteria; see "Determining the Main Direction of Orientation" above) to face this northerly direction, while the remaining 6 face eastward. KIP-405, KIP-1156, and NAK-29 are upland sites; KIP-1151 is at ~300 m, WF-AUW-100 is at ~200 m, and the remainder are coastal. The upland sites ALE-211, KIP-410, and KIP-424, although not included in the systematic analysis, also appear to fit this pattern.

The second main trend is that eight of the *heiau* in the same general area—namely, LUA-36, ALE-121, KIP-307, KIP-366, KIP-405, KIP-414, KIP-1010 (W), and KIP-1025—are aligned eastward at least approximately upon Ka Lae o Ka ʻIlio (Kailio Point). Of these, five (LUA-36, KIP-366, KIP-

TABLE 4.10. Azimuths of prominent cinder cones (see table 4.1), other peaks, and coastal promontories, where visible, in comparison with the orientations of the Kahikinui and Kaupō temples.

SITE	KAHO'OLAWE (HIGHEST POINT)	PU'U PĪMOE	PU'U HŌKŪKANO	LUA'ILUA HILLS, NW PEAK	LUA'ILUA HILLS, NE PEAK	LUA'ILUA HILLS, SE PEAK	MANUKANI	KAHUA	PU'U PANE	PU'U AHULILI	KANAHAU	HALEAKALĀ (SUMMIT)	ALENA POINT	KA LAE O KA 'ILIO (KAILIO POINT)	MAUNA KEA (HAWAI'I ISLAND)	MAUNA LOA (HAWAI'I ISLAND)
AUW-9	—	305	8	—	—	—	—	—	—	—	18	—	—	—	133	148
AUW-11	—	305	8	—	—	—	—	—	—	—	18	—	—	—	133	148
WF-AUW-100	262	—	340	28	35	40	—	—	—	—	16	—	—	—	134	149
WF-AUW-176A	261	283	315	29	38	48	—	—	—	—	16	—	—	—	134	149
WF-AUW-338	260	—	278	27	38	51	—	—	—	—	14	—	—	—	134	149
WF-AUW-343	260	—	280	30	40	53	—	—	—	—	15	—	—	—	134	149
WF-AUW-359	—	273	~230	52	57	74	43	—	—	—	20	—	—	—	134	149
WF-AUW-391	—	—	~240	49	58	76	42	—	—	—	19	—	—	—	134	149
WF-AUW-403	258	274	~117	59	64	77	47	—	—	—	23	—	—	—	134	149
WF-AUW-493	257	266	182	68	70	88	50	—	—	—	23	—	—	—	134	149
WF-AUW-574	—	—	—	26	38	53	—	—	—	—	14	—	—	—	134	149
LUA-1	256	—	223	~94	—	~133	44	—	—	—	15	—	—	—	135	150
LUA-3	—	—	—	343	351	344	8	—	—	—	3	—	—	—	134	149
LUA-4	—	291	306	350	349	342	4	—	—	—	0	—	—	—	134	149
LUA-29	265	293	313	—	358	352	9	—	—	—	4	—	—	—	134	149
LUA-36	265	292	308	—	351	344	5	—	—	—	1	—	—	77	134	149
LUA-39	—	—	—	—	354	347	8	—	—	—	3	—	—	—	134	149

(continued)

TABLE 4.10. Azimuths of prominent cinder cones (see table 4.1), other peaks, and coastal promontories, where visible, in comparison with the orientations of the Kahikinui and Kaupō temples. *(continued)*

SITE	KAHO'OLAWE (HIGHEST POINT)	PU'U PĪMOE	PU'U HŌKŪKANO	LUA/ILUA HILLS, NW PEAK	LUA/ILUA HILLS, NE PEAK	LUA/ILUA HILLS, SE PEAK	MANUKANI	KAHUA	PU'U PANE	PU'U AHULILI	KANAHAU	HALEAKALĀ (SUMMIT)	ALENA POINT	KA LAE O KA 'ILIO (KAILIO POINT)	MAUNA KEA (HAWAI'I ISLAND)	MAUNA LOA (HAWAI'I ISLAND)
ALE-1	—	—	—	339	346	338	6	—	—	—	2	—	—	—	134	150
ALE-4	—	—	—	318	**327**	314	351	—	—	—	354	—	—	—	135	150
ALE-121	—	—	—	322	329	317	**352**	—	—	—	**355**	—	—	79	135	150
ALE-140	—	—	—	—	**329**	—	352	—	—	—	355	—	—	—	135	150
KIP-1	—	—	—	—	253	236	338	—	49	—	352	—	—	—	136	151
KIP-75	—	—	—	—	—	—	317	—	41	—	345	—	—	—	136	151
KIP-77	—	—	—	—	—	—	316	—	39	—	344	—	—	93	136	151
KIP-188	—	—	—	—	260	244	330	—	44	—	349	—	—	—	136	151
KIP-273A	—	—	—	—	308	—	**336**	—	19	—	**346**	—	—	—	136	151
KIP-273B	—	—	—	—	308	—	336	—	19	—	346	—	—	81	136	151
KIP-275	—	—	—	—	308	—	336	—	19	—	346	—	—	—	136	151
KIP-306	—	—	—	—	305	294	333	—	16	—	344	—	—	—	136	151
KIP-307	—	—	—	—	304	293	332	—	—	—	**344**	—	—	80	136	151
KIP-330	—	—	—	—	301	290	330	—	15	—	343	—	—	—	136	151
KIP-366	—	—	—	—	295	283	328	—	18	—	343	—	—	**84**	136	151
KIP-405	—	—	—	—	257	243	319	—	43	—	346	—	—	**94**	136	151
KIP-414	—	—	—	248	246	236	300	—	46	—	342	—	—	**97**	137	151

AZIMUTH (°), BY PROMINENT TOPOGRAPHIC FEATURE

KIP-567	—	—	—	—	—	—	326	—	31	345	—	—	90	136	151
KIP-1010 (W)	—	—	—	251	242	293	—	37	336	—	—	97	137	152
KIP-1010 (E)	—	—	—	251	242	293	—	37	336	—	—	97	137	152
KIP-1025	—	—	—	297	286	327	—	15	342	—	—	83	136	151
KIP-1151	—	—	—	301	282	344	—	33	352	—	—	—	136	151
KIP-1156	—	—	—	—	293	342	—	35	351	—	—	87	136	151
NAK-27	—	—	—	—	—	290	—	29	333	—	—	—	137	152
NAK-29	—	—	—	—	—	—	—	25	331	—	—	97	137	152
NAK-30	—	—	—	—	—	—	—	24	331	—	—	—	137	152
NAK-34	—	—	—	—	—	—	—	33	—	—	—	—	137	152
MAH-231	—	—	—	258	251	284	—	3	—	—	—	—	138	152
MAW-2	—	—	—	—	—	—	—	—	—	—	—	95	138	153
NUU-79	—	—	—	266	263	277	298	292	—	—	3	—	140	155
NUU-81	—	—	—	265	—	276	297	291	—	47	3	—	—	—
NUU-100	—	—	—	262	259	—	292	—	—	—	359	245	—	141	155
NUU-101	—	—	—	268	264	279	300	—	—	45	3	250	140	155
NUU-175	—	—	—	—	—	—	—	—	—	—	—	—	—	—
KAU-994 (Loʻaloʻa)	—	—	—	—	—	—	284	—	—	8	322	—	143	157
KAU-995 (Kou)	—	—	—	269	—	—	—	—	—	37	354	253	100	141	155
KAU-324 (Pōpōiwi)	—	—	—	—	—	—	—	—	—	358	317	—	—	143	158

Note: Values are obtained from the survey data where possible or else calculated from digital terrain data. The measurements and calculations assume a "bare" topography, lacking in vegetation. Roman type indicates that the feature concerned forms part of the visual horizon; italic type indicates that it is below the horizon. Dark shading indicates that the feature falls within the azimuth ranges (mean azimuth ± semiwidth) identified in table 4.7 and in online supplementary tables S4.1 or S4.2. Light shading indicates that the feature falls within 5° of such a range. Bold type indicates that the range concerned is in the direction the *heiau* faces.

405, KIP-414, and KIP-1025) appear to face this direction, while the remaining three face north. There is no consistency in their elevations: LUA-36 is coastal, KIP-366 and KIP-1025 are somewhat inland at around 100 m, KIP-405 is above 600 m, and KIP-414 above 700 m. The azimuth of Ka Lae o Ka ʻIlio varies from 77° at LUA-36 to 97° at KIP-414; it is well below the sea horizon as seen from the upland sites. A further site not included in the systematic analysis, the unusual KIP-1146, features a standing upright 1.2 m high that, when viewed from a small leveled area likely intended for viewing, is directly below Ka Lae o Ka ʻIlio (see fig. 7.32).

Two sites—ALE-121 and KIP-307—face Kanahau to the north but are also aligned upon Ka Lae o Ka ʻIlio to the east. At ALE-121 the principal axis (mean az. 355°/175°) and the secondary axis (79°/259°) are some 6° out from perpendicular, while at KIP-307 the two axes (345°/165° and 82°/262°) are 7° out. This makes it more possible that both alignments were intentional, rather than one being a fortuitous consequence of the other, because this would not have happened if the walls had been built perpendicularly. The same is true at KIP-405, except that the principal orientation is to the east; here the mean axial azimuths are less well defined but are 12° out from perpendicular.

Another evident trend in table 4.10 is that nine sites (LUA-4, ALE-4, ALE-140, KIP-77, KIP-567, NAK-27, NAK-29, MAW-2, and KAU-324 [Pōpōiwi]) manifest at least an approximate alignment upon the rounded peak of Mauna Loa on Hawaiʻi Island. But several factors urge caution. None of the *heiau* actually faces the direction concerned (south-southeast, between az. 149° and 158°), and in most cases the azimuth range is wide (ALE-4, KIP-567, and NAK-27 are the exceptions). Added to this, Mauna Loa is visible only in clear weather, and no site is oriented even approximately upon the more prominent peak of Mauna Kea 15° to the left. This suggests that this apparent trend is most likely to have arisen as a simple consequence of other factors.

Finally, it is noteworthy that three of the upland Auwahi temples—WF-AUW-359, WF-AUW-403, and WF-AUW-493—are oriented westward almost directly upon Puʻu Pīmoe. Given that there are only three instances, that the temples do not face in this direction, and that the *puʻu* (cinder cone) is relatively inconspicuous from two of the sites (WF-AUW-359 and WF-AUW-493), the intentionality of these alignments remains questionable, although their accuracy (the mean orientation is no farther than ~1° from the peak in each case) is intriguing.

Other than those just noted, the data show no systematic preference for any particular topographic feature. For example, there is no systematic evidence from the structural orientations that three peaks of the Lualaʻilua Hills were a widespread focus of attention, despite being conspicuous on or below the skyline to the east, north, or west from many of the sites. (A possible association between these hills and the Pleiades as viewed from several of the upland Auwahi sites is discussed in the section "Specific Structural Orientations" below.)

The Problem of Ridgeline Cinder Cones

During our surveys, we observed that a number of north-facing *heiau* faced prominent cinder cones (not just Kanahau) on what appears the highest part of the Haleakalā skyline—examples include WF-AUW-176A, KIP-1010 (W and E), and NAK-34—as did northerly wall alignments at some east-facing sites (e.g., LUA-1, KIP-414). Details are given in part II. The systematic analysis, and the data listed in table 4.10, draw attention to the evident danger that, because ridgeline cinder cones are fairly numerous, it is all too easy to spot ones upon which temple walls happen to be aligned while ignoring many other possibilities, thus placing too much credence on the idea that the alignments thus singled out were actually intentional. On the other hand, any systematic analysis of this nature will fail to pinpoint associations that were specific to particular sites, especially given that different ridgeline cinder cones stand out visually as seen from different places.

Topographic Features Associated with Astronomy

LUA-29 faces north-northeast, but in the west-northwesterly direction its secondary orientation (mean az. 300.4°) is more or less midway between two isolated peaks that rise conspicuously behind the smoothly sloping lava ridge in the foreground. One is Pu'u Pīmoe (az. 293.5°), into which both the Pleiades and the summer solstice sun would have set directly around the time of construction (for details, see part II). The other is Pu'u Hōkūkano (az. 313°), a hill called by what is presumably a forgotten star name. There are a number of reasons for supposing the association between the *heiau* and these hills to have been significant (see Kirch, Ruggles, and Sharp 2013, 59–61).

We have also identified, at a number of other sites, various associations between the topography and the astronomy that are of potential significance but are not reflected in the *heiau* orientation. At KIP-1156, for example, the southeastern peak of the Luala'ilua Hills just protrudes above the nearby lava ridge to the west-northwest and coincides with the setting point of the summer solstice sun and the Pleiades, as does the distant peak of Kahua at NUU-100. At NAK-29 the Pleiades rose directly behind a conspicuous bump on the eastern skyline, although the *heiau* walls are aligned 2°–3° to the left. At NUU-79 the December solstice sun rose directly from the foot of the cliffs of Pu'u Māneoneo. At NUU-100, sunset at both solstices coincides with a conspicuous topographic feature, but the *heiau* faces a direction of no obvious significance between them.

We elaborate on cases such as these in the discussion of specific sites in part II, and more is evident in the full horizon profiles included among the supplementary online materials. Systematic analysis is unlikely to have the power to provide any useful evidence as to whether any particular associations of this nature actually had cultural significance: there are simply too many possibilities.

Specific Structural Orientations and Features of Particular Interest

In this section we highlight a number of particular associations between site architecture and the visible landscape and skyscape, some repeated among a small group of sites and some specific to individual sites, that we consider noteworthy. These include certain structural features that align upon topographic and/or astronomical phenomena in directions other than the direction faced or the other principal directions of orientation. They are subjectively selected on the basis of our knowledge and experience of Hawaiian culture; arguments that they were intentional and meaningful must depend on the broader archaeological and historical context. In each case, more details can be found in part II. A still wider range of topographic and astronomical alignment possibilities is given in part II, and yet more can be discerned among the full horizon profiles available online.

The Double *Heiau* KIP-1010

KIP-1010, a large double *heiau*, is important because it was constructed in two phases; the eastern one, which we believe to be later, is a larger "copy" built adjacent to an existing notched *heiau* on the west and has a similar plan with similar orientations, including nonorthogonal walls. This strongly suggests that the layout was not simply a poor attempt at a more regular design but a careful construction from elements that were of cultural significance. Details of the wall orientations and a plan of the site are provided in chapter 7.

The wall orientations in the northern parts of the two *heiau* match to within about 3° (north wall: 87° [W], 85° [E]; east wall 354° [W], 351° [E]), with each *heiau* facing a prominent cluster of cinder cones on the Haleakalā ridgeline between azimuths 350° and 357°. The southeastern corners of both *heiau* are markedly acute rather than orthogonal, and that of the eastern *heiau* is elaborated into a distinct canoe-prow shape. The orientations of the adjacent walls are slightly different between the two *heiau* (east wall: 180° [W], 176° [E]; south wall 100° [W], 99° [E]), but the direction of the diagonal bisecting the corner (not taking into account the corner elaboration in the eastern *heiau*) is within 3° (140° [W], 137.5° [E]) and points directly at Mauna Kea on Hawai'i Island—in the case of the E *heiau*, very close to the summit. The duplication of these alignments reinforces the impression that both the ridgeline cinder cones and Mauna Kea were indeed significant and that their directions were deliberately reflected in the temple design.

Pleiades Sighting Device at ALE-140

The small *heiau* ALE-140, situated in a lava swale, has restricted views. Its structure—with a formal entrance to the east-northeast and a smaller, higher court to the west-southwest—suggests that it faces west-southwest, but the view in this direction is entirely blocked by the side of a swale about 10 m away. On the other hand, from a seated position within the small western chamber, as might have been occupied by an officiating *kahuna*, there is a spectacular view to the east-northeast. A jagged local *'a'ā* ridge just over 25 m distant drops away in front of the smooth slope of Haleakalā beyond. At the junction of these skylines stands a pair of upright boulders, one of which seems to have been artificially set in position (see chapter 6 for details), framing a segment of the distant slope up to about 0.5° wide.

The "window" formed by the boulders spans a declination range of approximately +26.2° to +26.7°, depending on the exact position of the observer. This may be compared with the ranges of +22.6° to +23.1° for the Pleiades at the date of construction, just before AD 1600, and +23.25° to +23.75° for the June solstice sun. In other words, the window formed by the boulders as seen from the chamber is somewhat displaced from the rising position of these astronomical bodies, which would have occurred behind the *'a'ā* ridge to the right.

However, the window did line up with the rising position of the Pleiades around AD 1600 if viewed from a standing position outside the *heiau* on the northern side, lining along the outer face of the north-northwestern wall (fig. 4.9). The details are given in chapter 6. This would seem to imply that this sighting device was used to observe the Pleiades not from inside the *heiau* but from a position just outside it.

Pleiades and/or Hyades Sighting at KIP-275

KIP-275 is a platform, originally rectangular, facing east-northeast, with its west-southwestern side and east-northeastern wall roughly parallel, but with the north-northwestern and southern sides widening toward the east-northeast. The easterly azimuth of its north-northwestern side is 68.3°, while that of the remaining boulder facing on the southern side is 86.5°, yielding a mean orientation of 77.4°. In these directions the distant view of the smooth

FIGURE 4.9. The horizon to the ENE at *heiau* ALE-140, as viewed from a standing position sighting along the NNW wall. A DTM-generated profile is superimposed on a photograph of the local 'a'ā ridge. The solid shaded line indicates the rising path of the sun at the June solstice, and the dotted line the rising path of the Pleiades, in AD 1600. (Composite of photo by Clive Ruggles and graphic by Andrew Smith)

Haleakalā slope running down from the north is blocked by a jagged lava outcrop just 15 m from the platform, which drops away sharply at its left-hand end.

Because of the proximity of the outcrop, its azimuth changes significantly depending on one's position on the platform. However, viewing from the approximate center of the platform, or along past the center from a position at the southern end of the facing on the west-southwestern platform edge, the Pleiades would have been seen to rise up the left side of the outcrop around the date of construction (about AD 1625). (For details, see chapter 7.) The direction of the outcrop from these central positions (left-hand edge ~67°, highest point ~69°) is similar to the orientation of the north-northwestern side, rather than to the mean orientation.

During our survey we also noted a concentration of cobble and coral on the western side of the platform, toward the west-northwestern corner; from a seated position here, a large boulder stacked atop a ground-fast stone and chocked in place appears directly in line with the broad natural outcrop (from here, left-hand edge ~72°, highest point ~74°), at about half the distance. (Again, see chapter 7 for details.) This alignment of features is oriented toward the rising position around AD 1625 of Ka Nuku o Ka Puahi (Hyades-Aldebaran), in which the systematic analysis has already indicated a possible interest at other sites.

It is possible that both Makali'i and Ka Nuku o Ka Puahi were of importance at this temple. While the structural orientation may have been influenced by, or defined by, Pleiades observations made using the nearby natural outcrop, the alignment of tangible features upon the Hyades, including a plausible *kahuna* seating spot, is more suggestive of a sighting device.

Pleiades and/or Hyades Sighting Device at KIP-273B

The temple enclosure KIP-273B, situated just ~80m north-northwest of KIP-275, appears to have had a similar function. It faces a prominent lava outcrop, on a nearby ridge about 100 m away, that has a small but conspicuous notch in its center. From the center of the enclosure this would have marked the rising point of Ka Nuku o Ka Puahi (Hyades-Aldebaran) around the time of construction (early or mid-seventeenth century), but from just outside the enclosure on the south side, paralleling the situation at ALE-140 as described earlier, it could have functioned as a sighting device marking the rising of Makali'i (the Pleiades).

Pleiades Sighting at NAK-30

NAK-30 is a platform *heiau* oriented to the east-northeast upon the point of the Pleiades rise, as already noted in the systematic analysis. An additional feature of interest here is the presence of two upright slabs in the north-northeastern and east-southeastern corners of the platform—0.45 m and 0.65 m high, respectively (fig. 4.10). As viewed from a seated position on the west-southwestern side of the platform, as might well have been adopted by a *kahuna* while performing a ritual, the two uprights demarcate a stretch of horizon running from azimuth ~55° to ~80°, where an *'a'ā* ridge about 10 m away runs along beneath the smooth Haleakalā slope descending from left to right in the distance. Midway along the ridge is a large pointed natural boulder, which acts as a precise marker of a point on the distant slope. The azimuth of this pointed boulder is approximately 67°, consistent with the mean orientation of the platform, yielding a declination of about +23°, which corresponds to the rising position of the Pleiades around AD 1600 (and also that of the summer solstice sun). Unfortunately, there is nothing to indicate a precise observing position, and the proximity of the pointed boulder means that its azimuth (and declination) varies by about 1° for each 20 cm lateral shift of the observer. Thus, while we can be confident of the Pleiades association, the archaeoastronomical evidence cannot improve upon the uncertain date of construction.

Pleiades Sighting at NAK-27

A large craggy *'a'ā* outcrop just ~60 m to the east-northeast of site NAK-27 blocks the more distant Haleakalā slope in the direction of orientation. The south-southeastern wall is aligned upon a small rock with a conspicuous vertical face at the foot of this crag on its southern side. While the direction concerned (az. ~60.5°) is several degrees to the left of the rising position of the Pleiades or June solstice sun, the foot of the outcrop would have marked these risings from locations either within or just outside the *heiau* on its north-northwestern side (for details, see chapter 7).

Although the wall indication and ideal observing positions do not coincide, the sighting devices at other sites strengthen the idea that this *heiau* was also positioned and configured so as to incorporate a natural feature marking the rising position of the Pleiades, thus facilitating their observation for calendrical purposes.

Another possibility is that this *heiau* was placed so that the first light of the rising sun, even at its farthest north at the June solstice, would fall across the entire enclosure rather than just part of it. Further dating evidence and more extensive survey data may help clarify these possibilities.

FIGURE 4.10. View toward the E taken from a seated position on the platform of site NAK-30. Note the two upright stones at the end of the platform, on the left and right, which together frame the pointed, natural outcrop on the near ridge. (Photo by P. V. Kirch)

Possible Sirius Sighting Device at KIP-77

KIP-77 is a classic notched *heiau* facing east-northeast—toward the rising of Nānāhope (Pollux), as already noted in the systematic analysis—but with a distinctive additional feature. An arrangement of small stone blocks and supporting stones atop the east-northeastern wall of the southern court creates the impression of a sighting device. The stones create two side-by-side apertures that offer a clear view out toward the east-southeast when viewed from the opposite direction, along a line that passes diagonally across the main court. A boulder outcrop 1.2 m wide within the main court offers a convenient viewing point, right on this line.

If this was a sighting device, then it would have been suitable for viewing the rising of ʻAʻā/ Hōkūhoʻokelewaʻa (Sirius) over the sea horizon (see chapter 7 for details). The systematic analysis has already indicated a possible interest in Sirius at other sites.

Sighting Navigation Stars at KIP-1151

The platform *heiau* KIP-1151 has the distinctive feature of a small raised platform on the southeastern side, reached by a ramp, where one can well imagine a *kahuna* performing rites while others observed from the main terrace. The site also shares with ALE-140, KIP-275, KIP-273B, and NAK-27 the presence of a nearby craggy lava outcrop to the east. But, whether viewed from the dais or the main platform, the junction is some degrees to the right of the rising point of Makaliʻi. This suggests instead that the craggy top of the outcrop could have been used to mark the rising points of other stars. From the dais, the outcrop spans a declination range from

Landscape and Sky • 119

approximately +18° down to –15°, which includes the majority of Nā Hōkū Hoʻokele, the navigation stars that circle the sky between the Ke Alanui Polohiwa a Kāne ("the black shining road of Kāne") and Ke Alanui Polohiwa a Kanaloa ("the black shining road of Kanaloa")—that is, the celestial Tropics of Cancer and Capricorn, respectively, with declinations +23.5° and –23.5° (Johnson, Mahelona, and Ruggles 2015, 30). See table 4.2 for details of Hawaiian asterisms.

Pleiades and Sun at Sites in Upland Auwahi

The seven upland Auwahi sites are all cardinally oriented to within about 5°. Four of them (WF-AUW-359, 391, 403, and 493) face east, WF-AUW-338 and 574 face north, and the directionality of WF-AUW-343 is undetermined. Five of the seven temples are standard or variant forms of the classic notched design, with the notched corner being either to the northwest or the northeast. The exceptions are WF-AUW-493 and WF-AUW-391, which are simple rectangular (almost square) enclosures, although the exterior wall in the northeastern corner of WF-AUW-391 is shaped in a way suggestive of a notch.

The nearby Lualaʻilua Hills are prominent from all of these sites, conspicuously visible to the northeast and east-northeast. Comparing the DTM-generated visualizations in the relevant direction at each of the seven sites (fig. 4.11) shows that the Pleiades around the time of construction, as well as the June solstice sun, would generally have been seen to rise up from shallow dips and concavities beside and between the hills rather than from the hills themselves. The only exception is the rectangular enclosure WF-AUW-493, where the rising point coincides more or less with the northernmost Lualaʻilua peak.

That a similar phenomenon occurs in several cases provides credible, if not compelling, evidence that the locations were selected deliberately so that the Pleiades (whose acronychal rise was so important for agricultural and calendrical purposes) would rise from a conspicuous and easily locatable point on the horizon. It is strengthened by the fact that all of the sites are situated within an extensive settlement and agricultural zone, which suggests that most if not all of these *heiau* are likely to have been agricultural temples. The trend cannot be a fortuitous consequence of the sites' location in the landscape, for the rising point would not coincide within the dips between the hills from most locations in the general vicinity.

On the other hand, numerous associations could be surmised between topographical features and astronomical events that are unmarked by structural orientations at the temples. This urges caution, not because such associations are unlikely to have been of significance to Hawaiians but because of the large number of possibilities (cf. Ruggles 2014c, 384).

While WF-AUW-403 fits this pattern, it is also one of a number of temples that directly face nearby topographic features with considerable visual impact that span the entire solar rising arc to the east (see the section "Prominent Topographic Features" above). It also incorporates a likely viewing place in the form of a large, flat stone slab laid into the *heiau* floor (see "Sighting the Eastern Horizon at WF-AUW-403" below). These suggest that this *heiau* had a strong association with the rising sun. Other considerations also suggest that WF-AUW-403 was likely a Kāne temple connected particularly with observations of the sunrise, rather than a Lono temple connected with the calendar and the Pleiades.

WF-AUW-493, oriented directly upon one of the Lualaʻilua peaks, also provides a clear view of the entire solar arc and is likely a Kāne, rather than a Lono, temple.

WF-AUW-359 shares with WF-AUW-403 and 493 a westward orientation upon Puʻu Pīmoe,

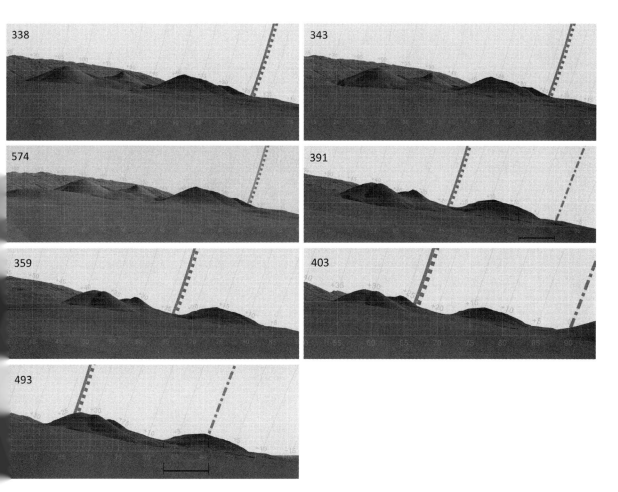

FIGURE 4.11. DTM-generated visualizations of the horizons to the NE and ENE at upland Auwahi *heiau*, showing the rising position of the Pleiades around AD 1600 (*dotted line*) and June solstice sun (*solid shaded line*). The rising path of the sun at the equinoxes (*dash-and-dot* line) also appears in some of the profiles. (Graphic by Andrew Smith)

as noted earlier. Apart from this, there are no notable topographic or astronomical alignments that stand out in the four directions of orientation at any of the five sites WF-AUW-338, 343, 359, 391, and 574. This lack of any obvious topographic or astronomical alternative may in itself add credibility to the idea that they were associated with the Pleiades without this association being reflected in the orientation of the temples.

Sighting the Eastern Horizon at WF-AUW-403

WF-AUW-403 features a large flat *pāhoehoe* slab set into the floor on the western side of the main court, doubtless for use by the *kahuna* while officiating at this temple. The view to the east is framed by nearby Puʻu Hōkūkano (~200 m away) to the right, and the three peaks of the more distant Lualaʻilua Hills (~3 km) to the left. From a seated position on the large floor slab, the eastern wall of the *heiau* mostly obscures the horizon in between and perhaps obscured it completely when built, but from a standing position there is a clear view of the entire rising arc of the sun throughout the year (fig. 4.12).

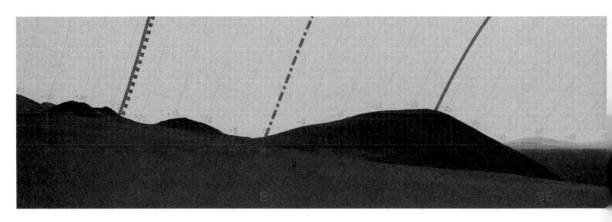

FIGURE 4.12. DTM-generated profile of the eastern portion of the horizon as seen from *heiau* WF-AUW-403. The solid shaded lines indicate the rising path of the sun at the June solstice (*left*) and December solstice (*right*); the dash-and-dot line indicates the rising path of the sun at the equinoxes; and the dotted line indicates the rising path of the Pleiades in AD 1600. The horizontal bar indicates the eastward direction of orientation of the *heiau* walls. (Graphic by Andrew Smith)

For the seated *kahuna*, the combination of distant hills and wall stones could have provided ample reference points to mark sunrise at different dates, raising the possibility that this constituted an intentional "sighting device." (Details are provided in chapter 6.) Whether by design or coincidence, this site also fits the pattern observed at most of the upland Auwahi sites—that the Pleiades rose in a shallow dip between adjacent peaks—but it appears that the main, or sole, astronomical focus of attention at this site was the rising sun.

Striking Topography and Sunrise

WF-AUW-403 is not the only temple facing nearby topographic features with considerable visual impact that span much, if not all, of the solar rising arc. Some instances are shown in figure 4.13. A particularly good example is LUA-1; it directly faces the Luala'ilua Hills, whose closest peak is just ~600 m away. Another is NUU-151, which faces Pu'u Māneoneo, only about 200 m distant. More marginal cases include WF-AUW-493, where two of the Luala'ilua peaks dominate the northern half but not the southern half of the view to the east; KIP-1, where there is a small rounded hill ~200 m from the temple; and NUU-79, where the conspicuous promontory of Pu'u Māneoneo occupies the southern half of the solar arc, looming between the positions of sunrise at the equinoxes and the December solstice.

No similar trend is evident in the opposite (westerly) direction. There is also a clear contrast between east-facing temples, among which there is a complete absence of nearby craggy lava ridges and outcrops in the direction faced, and east-northeast-facing temples, such as KIP-273B, KIP-275, KIP-1151, and NAK-27, together with (opposite to the west-southwestern direction of orientation) ALE-140, where such ridges are not only present but frequently seem to constitute the basis of sighting devices. A possible interpretation is that solar observations generally required only a broad sweep of horizon (whether smooth nearby hills or distant coastal vistas) sufficient for monitoring the movement of sunrise over periods of weeks or months, whereas stellar observations tended to necessitate more precise markers. Only at WF-AUW-403 is there a possible sighting device that seems to be related to the sun rather than to the stars.

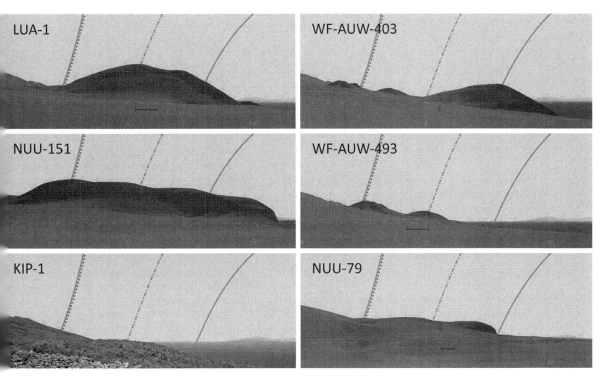

FIGURE 4.13. DTM-generated profile of the eastern horizon at a selection of *heiau* facing nearby topographic features with considerable visual impact. In the case of KIP-1 a photograph of a nearby hillock, too close to show up on the DTM, is superimposed (see also Figure 7.3). The extent of the solar rising arc is indicated by the solstitial rising lines (*solid shaded lines*). The rising path of the sun at the equinoxes (*dash-and-dot line*) and of the Pleiades in AD 1600 (*dotted line*) is also shown. The horizontal bars indicate the direction of orientation of the *heiau* (unmeasurable at NUU-151).

Tracking Sunrise at KIP-1146

An unusual temple site not included in the systematic analysis because of its irregular design, KIP-1146 comprises two stone-faced and stone-paved terraces and other features (see chapter 7), including a leveled area some 2 m in diameter (B in fig. 7.31) and a long wall terminating in a purposely placed upright stone 1.2 m high (A). As seen from the leveled area, the upright stone sits directly below Ka Lae o Ka ʻIlio at azimuth 86° (fig. 4.14), while the straight terrace fronts are oriented with azimuth 96° and 111° (lower and upper terrace, respectively; see fig. 7.33). All of these directions are out to sea but are unambiguously marked by the structural orientations.

This raises the possibility that the site was used to track the position of sunrise for calendrical purposes. The sun would rise in the upper-terrace direction around November 22 and January 19, moving farther right and then returning during the intervening days; in the lower terrace direction around March 6; over Ka Lae o Ka ʻIlio around March 28; and at the point where the slope down from Haleakalā meets the sea horizon, farther to the left, around April 16. This period broadly coincides with the Makahiki season, suggesting that this temple might have been laid out so as to keep track of the timing of the Makahiki, especially its end around late March to April, when the tribute (*hoʻokupu*) would be collected by the chiefs in the name of Lono, during the *akua loa* circuit (see chapter 2).

FIGURE 4.14. Upright stone at KIP-1146 as viewed from leveled area, with Ka Lae o Ka 'Ilio directly behind it. Photograph superimposed on DTM-generated profile. The dash-and-dot line indicates the rising path of the sun at the equinoxes. The solid shaded line indicates the rising path of the sun at the June solstice, and the dotted line indicates the rising path of the Pleiades in AD 1600. (Composite of photo by Clive Ruggles and graphic by Andrew Smith)

Sunrise and Sunset at KIP-366 and KIP-1025

The probable *heiau* KIP-366 is one site where it can be plausibly suggested that observations of sunset were as important as, or even more important than, those of sunrise. Commanding fine views to both the east and the west, it is one of the group of eastward-facing *heiau* identified in the systematic analysis that broadly faces sunrise within a month of the equinox and are also oriented upon Ka Lae o Ka 'Ilio. But its south wall (az. 63°–243°) is significantly skewed from the main orientation (82°–262°). In the west-southwest direction it is aligned upon the place where the ridgeline slope down to Alena Point meets the sea horizon, which is also close to the position of December solstice sunset. Upslope, the Luala'ilua northeast peak, abutting conspicuously above the ridgeline, marks the setting point of the June solstice sun, with the horizon in between spanning the setting arc of the sun through the seasons.

Nearby KIP-1025 is similar in many ways. It also faces Ka Lae o Ka 'Ilio, and the western profile closely resembles that at nearby KIP-366, with the Luala'ilua northeastern peak marking June solstice sunset. But there is no structural indication of the position of the December solstice sunset, which occurs at the very meeting point of the ridgeline slope and the sea horizon. The axial orientation of the temple (85°–265°) is closer to the position of equinox sunset than at KIP-366.

Summary

In table 4.11 we summarize what we consider to be the most likely topographic and/or astronomical features to have influenced the location and orientation of the various Kahikinui and Kaupō temples, taking into account both the systematic analysis and specific features discussed in this chapter.

TABLE 4.11. Summary of the most likely topographic and astronomical influences on the orientation of Kahikinui and Kaupō temples.

SITE	APPROXIMATE ELEVATION (M)	DIRECTION FACED	KOʻA?	NOTABLE TOPOGRAPHIC AND/OR ASTRONOMICAL FEATURE(S)
AUW-9	0	N	—	—
AUW-11	0	NNE	Yes	Approximately faces Kanahau (red cinder cone).
WF-AUW-100	200	NNE	—	Approximately faces Kanahau (red cinder cone). Secondary alignment (ESE) at least approximately upon ʻAʻā/Hōkūhoʻokelewaʻa (Sirius) rise.
WF-AUW-176A	300	N	—	Faces a prominent cinder cone on the highest part of the Haleakalā Ridge.
WF-AUW-338	400	N	—	Makaliʻi (Pleiades) rise in shallow dip between hills.
WF-AUW-343	400	N/E	—	Makaliʻi (Pleiades) rise in shallow dip between hills.
WF-AUW-359	400	E	—	Broadly faces sunrise within a month of the equinox. Opposite alignment (W) upon Puʻu Pīmoe. Makaliʻi (Pleiades) rise in shallow dip between hills.
WF-AUW-391	400	E	—	Broadly faces sunrise within a month of the equinox. Makaliʻi (Pleiades) rise in shallow dip between hills.
WF-AUW-403	400	E	—	Broadly faces sunrise within a month of the equinox. Prominent nearby hills to E; clear view of sunrise throughout year; priest's sighting stone and possible sighting device. Opposite alignment (W) upon Puʻu Pīmoe.
WF-AUW-493	500	E	—	Faces the Lualaʻilua SE peak and broadly faces sunrise within a month of the equinox; clear view of sunrise throughout year. Opposite alignment (W) upon Puʻu Pīmoe.
WF-AUW-574	400	N	—	Makaliʻi (Pleiades) rise in shallow dip between hills.
LUA-1	600	E	—	Broadly faces sunrise within a month of the equinox. Prominent nearby hills to E; clear view of sunrise throughout year. Secondary alignment (N) upon a prominent cinder cone on the highest part of the Haleakalā Ridge.
LUA-3	100	N	—	—
LUA-4	0	NNW	—	Secondary alignment (ENE) at least approximately upon Makaliʻi (Pleiades) rise/June solstice sunrise, raising the possibility that ENE might actually be the direction of orientation.
LUA-29	0	NNE	Yes, converted	Secondary alignment (WNW) [or main alignment upon conversion of earlier koʻa?] between Puʻu Pīmoe and Puʻu Hōkūkano; Makaliʻi (Pleiades) and June solstice sun set into Puʻu Pīmoe.
LUA-36	0	E	Yes	Approximately faces Ka Lae o Ka ʻIlio (Kailio Point). Broadly faces sunrise within a month of the equinox.

(continued)

TABLE 4.11. Summary of the most likely topographic and astronomical influences on the orientation of Kahikinui and Kaupō temples. *(continued)*

SITE	APPROXIMATE ELEVATION (M)	DIRECTION FACED	KO'A?	NOTABLE TOPOGRAPHIC AND/OR ASTRONOMICAL FEATURE(S)
LUA-39	100	N	Yes	Approximately faces Kanahau (red cinder cone).
ALE-1	100	N	—	—
ALE-4	0	NNW	—	View highly restricted; faces center of narrow visible stretch of Haleakalā Ridge.
ALE-121	0	N	Yes	Faces Kanahau (red cinder cone). Secondary alignment (E) upon Ka Lae o Ka 'Ilio (Kailio Point).
ALE-140	0	WSW	—	Faces directly into side of lava swale, blocking view. Opposite alignment (ENE) upon Makali'i (Pleiades) rise, with pair of upright boulders providing artificial sighting device.
ALE-211	600	N	—	*Faces Kanahau (red cinder cone). Secondary alignment (E) upon Ka Lae o Ka 'Ilio (Kailio Point).
KIP-1	600	E	—	Broadly faces sunrise within a month of the equinox.
KIP-75	600	ENE	—	At least approximately faces Nānāhope (Pollux) rise.
KIP-77	600	ENE	—	Faces Nānāhope (Pollux) rise. Possible sighting device for 'A'ā/Hōkūho'okelewa'a (Sirius) rise.
KIP-188	600	E	—	Broadly faces sunrise within a month of the equinox.
KIP-273A	0	N	—	Approximately faces Kanahau (red cinder cone).
KIP-273B	0	ENE	—	Faces nearby outcrop and Ka Nuku o Ka Puahi (Hyades + Aldebaran) rise. Possible sighting device for Makali'i (Pleiades) rise. Opposite alignment (WSW) at least approximately upon 'A'ā/Hōkūho'okelewa'a (Sirius) set.
KIP-275	0	ENE	—	Faces nearby outcrop whose left-hand edge marks Makali'i (Pleiades) rise from center of platform. Secondary alignment (NNW) upon Kanahau (red cinder cone). Possible sighting device for Ka Nuku o Ka Puahi (Hyades + Aldebaran) rise.
KIP-306	0	N	Yes	—
KIP-307	0	NNW		Faces Kanahau (red cinder cone). Secondary alignment (E) upon Ka Lae o Ka 'Ilio (Kailio Point).
KIP-330	0	ENE	Yes	Faces Ka Nuku o Ka Puahi (Hyades + Aldebaran) rise. Opposite alignment (WSW) at least approximately upon 'A'ā/Hōkūho'okelewa'a (Sirius) set. Secondary alignment (NNW) upon Kanahau (red cinder cone).
KIP-366	100	E	—	Faces Ka Lae o Ka 'Ilio (Kailio Point). Broadly faces sunrise within a month of the equinox. Skewed southern wall aligned upon December solstice sunset to WSW; June solstice sun sets into Luala'ilua NE peak; clear view of sunset throughout year.

SITE	APPROXIMATE ELEVATION (M)	DIRECTION FACED	KOʻA?	NOTABLE TOPOGRAPHIC AND/OR ASTRONOMICAL FEATURE(S)
KIP-405	600	E	—	Faces Ka Lae o Ka ʻIlio (Kailio Point). Broadly faces sunrise within a month of the equinox. Secondary alignment (N) at least approximately upon Kanahau (red cinder cone).
KIP-410	600	ENE	—	*Secondary alignment (NNW) at least approximately upon Kanahau (red cinder cone).
KIP-414	700	E	—	Faces Ka Lae o Ka ʻIlio (Kailio Point). Broadly faces sunrise within a month of the equinox. Secondary alignment (N) upon a prominent cluster of cinder cones on the Haleakalā Ridge. Variant canoe-prow corner broadly faces Mauna Kea.
KIP-424	400	N	—	*At least approximately faces Kanahau (red cinder cone).
KIP-567	400	ENE	—	Faces Nānāhope (Pollux) rise.
KIP-728	500	N	—	—
KIP-1010 (W)	700	N	—	Faces a prominent cluster of cinder cones on the highest part of the Haleakalā Ridge. Secondary alignment (E) approximately upon Ka Lae o Ka ʻIlio (Kailio Point). Canoe-prow corner faces Mauna Kea.
KIP-1010 (E)	700	N	—	Faces a prominent cluster of cinder cones on the highest part of the Haleakalā Ridge. Canoe-prow corner faces the summit of Mauna Kea.
KIP-1025	100	E	—	Faces Ka Lae o Ka ʻIlio (Kailio Point). Broadly faces sunrise within a month of the equinox. December solstice sunset where ridgeline meets sea; June solstice sun sets into Lualaʻilua NE peak; clear view of sunset throughout year.
KIP-1146	300	—	—	Possible sighting device for marking sunrise during Makahiki.
KIP-1151	300	ENE	—	Faces craggy outcrop with numerous possibilities for marking navigation stars. Secondary alignment (NNW) approximately upon Kanahau (red cinder cone).
KIP-1156	400	ENE	—	Faces Makaliʻi (Pleiades) rise. Secondary alignment (NNW) of skewed ENE wall upon Kanahau (red cinder cone). Makaliʻi and June solstice sun set into Lualaʻilua SE peak.
NAK-27	600	ENE	—	Faces craggy outcrop, foot of which marks Makaliʻi (Pleiades) rise viewed from NNW side of *heiau*.

(*continued*)

TABLE 4.11. Summary of the most likely topographic and astronomical influences on the orientation of Kahikinui and Kaupō temples. *(continued)*

SITE	APPROXIMATE ELEVATION (M)	DIRECTION FACED	KOʻA?	NOTABLE TOPOGRAPHIC AND/OR ASTRONOMICAL FEATURE(S)
NAK-29	600	ENE	—	Faces ~2° to left of bump on horizon that marked Makaliʻi (Pleiades) rise. Secondary alignment (NNW) upon Kanahau (red cinder cone).
NAK-30	600	ENE	—	Faces Makaliʻi (Pleiades) rise. Natural pointed boulder provides a sighting device.
NAK-34	700	N	—	Faces a prominent cinder cone at the highest part of the Haleakalā Ridge.
MAH-231	500	N	—	Faces Puʻu Pane, prominent below the Haleakalā Ridge.
MAW-2	200	ENE/NNW	—	—
NUU-79	0	E	—	Broadly faces sunrise within a month of the equinox. December solstice sunrise from foot of Puʻu Māneoneo cliffs.
NUU-81	0	WNW	—	Faces Ka Nuku o Ka Puahi (Hyades + Aldebaran) set. Makaliʻi (Pleiades) and June solstice sun set above Puʻu Pane.
NUU-100	200	W	—	Broadly faces sunset within a month of the equinox. Makaliʻi (Pleiades) and June solstice sun set into Kahua; December solstice sunset above Wekea Point.
NUU-101	0	N/E/W	—	Alignment (E) upon Puʻu Māneoneo cliffs and rising of Nā Kao (Orion's Belt).
NUU-175	300	E	—	—
KAU-994 (Loʻaloʻa)	100	ENE	—	Faces Makaliʻi (Pleiades) rise/June solstice sunrise.
KAU-995 (Kou)	0	ESE	—	Faces ʻAʻā/Hōkūhoʻokelewaʻa (Sirius) rise. Opposite alignment (WNW) upon Ka Nuku o Ka Puahi (Hyades + Aldebaran) set.
KAU-324 (Pōpōiwi)	0	NNW	—	Faces highest apparent point on the ridgeline to the NNW. Secondary alignment (ENE) upon Ka Nuku o Ka Puahi (Hyades + Aldebaran) rise. Secondary alignment (WSW) at least approximately upon ʻAʻā/Hōkūhoʻokelewaʻa (Sirius) set. Canoe-prow corner approximately faces Waipiʻo.

Note: "Secondary" alignments refer to one of the directions approximately perpendicular to the direction faced and to the "opposite" direction. Topographic alignments are considered approximate if the target falls within 5° of the azimuth range (mean azimuth ± semiwidth) identified in table 4.7 or online supplementary table S4.1 or S4.2. They are considered "at least approximate" if the target falls within the azimuth range but that range is itself more than 10° wide. Astronomical alignments are considered approximate if the target falls within 2° of the declination range (mean declination ± semiwidth) identified in table 4.9 or online supplementary table S4.3 or S4.4 and are considered "at least approximate" if the target falls within the declination range identified in the same tables but that range is itself more than 4° wide. Information on the elevation and the direction faced is repeated from earlier tables, for convenience. An asterisk (*) indicates that the conclusions are provisional, pending an accurate survey. The table is not exhaustive; further alignment possibilities are listed in part II.

5

Summary and Conclusions

The pioneer of *heiau* archaeology, John F. G. Stokes, after studying more than 200 temple sites on the islands of Hawai'i and Moloka'i in the first decade of the twentieth century, remarked, "A man would be very unwise to attempt to draw a plan of *the* Hawaiian *heiau*. The endless variety in size, shape, and form puzzled me exceedingly" (1991, 21). The fact that it is "almost impossible to find, today, two temple foundations alike" Stokes attributed to the work of the *kāhuna kuhikuhipu'uone*, or temple builders, who were continually experimenting with new designs and architectural innovations. Nonetheless, as Malo (1951, 161) explained, these priest-architects frequently copied the plans of older temples that were known to have been efficacious, so some general trends in *heiau* architecture persisted over time.

Our study of the *heiau* of Kahikinui and Kaupō revealed a similar kind of "endless variety" that puzzled Stokes, yet with enough consistency in overall design that several major categories of *heiau* could be recognized. As described fully in chapter 3, these five groups are notched; square or U-shaped enclosures; platforms or terraces; elongated double-court enclosures; and *ko'a* (fishing shrines) and other small shrines. A sixth group includes a few temples with irregular forms that cannot be accommodated in any of the previous groups. The questions that arise are whether these architectural forms had any chronological progression and whether certain architectural forms had particular functions.

Fornander ([1878–1885] 1996), on the basis of oral traditions concerning *heiau*, advanced the theory that the original form of temple foundation in Hawai'i had been a stone platform or "pyramid." He further proposed that the platform type had been superseded by the walled enclosure form of *heiau* after the arrival of the voyaging priest Pā'ao around AD 1100 (Stokes 1991, 22). Stokes, however, was unable to find support for Fornander's theory in his survey of Hawai'i and Moloka'i temple foundations. Stokes also observed, "[These] different types of foundations... seem to have no connection with the classes of worship to which the *heiau* belonged. Platform and walled *heiau* were used as much for the highest forms of service, where human sacrifices were offered, as for the lesser ceremonies, such as prayers connected with agriculture, etc." (22–23). Michael Kolb (1991, 1992), who excavated at several Maui *heiau*, argued that there had been a chronological progression from early terraced forms to later platform-type structures. Kolb also suggested that temples with walled enclosures were constructed in both early and later periods.

Thanks to extensive ^{230}Th dating of branch corals from *heiau* sites in Kahikinui, we now have high-precision chronological data pertaining to 19 temple sites. Table 5.1 summarizes some key attributes of architectural design (morphology), orientation, and location for these well-dated *heiau*. The earliest date from a notched *heiau* (site WF-AUW-338) is AD 1328, but because the dated sample derives from a surface context, we cannot be certain that the sample relates to the time of temple construction; the coral could have been taken from an older structure and placed on this site later. The earliest "architecturally integral" coral date from a notched *heiau* (site NAK-29) is AD 1603, but four other surface samples date to between AD 1555 and 1592, leaving little doubt that the notched form of *heiau* was in use by the second half of the sixteenth century. The earliest date for an elongated double-court *heiau* (site KIP-1) is AD 1325, but this again comes from a surface context, so we must be cautious. However, KIP-1 also yielded a fairly early radiocarbon date of AD 1297–1495 (see details in chapter 7), so it is conceivable that the elongated double-court is a relatively early *heiau* form. Elongated double-court enclosures continued to be built and used during later centuries, as radiocarbon dates from sites NUU-79, NUU-81, and NUU-100 attest (see chapter 8 for details). Dates for the square or U-shaped enclosure form and for platform *heiau* all indicate that these types were in use from at least the late sixteenth century until the time of European contact. In sum, with respect to chronology, while there is some suggestion that the elongated double-court enclosure may be older than other types of *heiau*, all of the main architectural groups were in use from the late sixteenth through the early nineteenth century. Thus, no clear-cut chronological succession of architectural types is evident in our Kahikinui and Kaupō data.

Heiau orientation likewise exhibits no evident temporal pattern. As seen in table 5.1, the dominant orientation clusters of north-, east-northeast-, and east-facing temples are all represented throughout the entire period in our Kahikinui sample (see the section "*Heiau* Orientations and the Hawaiian Pantheon" below). The only attribute that does strongly correlate with time is inland versus coastal location, with most of the earlier temples (dating prior to AD 1600) being situated inland, while later temples show a strong preference for coastal locations. This matches a general trend of early settlement in the uplands, where rainfall was sufficient to support intensive cultivation of sweet potatoes, with only intermittent exploitation of coastal resources.

Finally, is there any indication that architectural forms correlate with *heiau* function? We believe it is likely that most, if not all, of the notched enclosures were *heiau hoʻoulu ʻai* (agricultural temples) or *heiau hoʻoulu iʻa* (fishing *heiau*, if situated near the coast). This is suggested by their relatively small size, proximity to residential structures, and orientations. However, it is also likely that many of the square or U-shaped enclosures, as well as the smaller platform or terraced temples, also functioned as *heiau hoʻoulu ʻai* or *iʻa*. The largest *heiau*, especially those located in Kaupō and known to have been associated with ruling *aliʻi* (chiefs) such as Kekaulike, functioned as *luakini* (temples dedicated to the god Kū; discussed in chapter 2). These largest temples include an elongated double-court variant (KAU-324), a massive platform (KAU-994), and a U-shaped enclosure (KAU-995), confirming Stokes' view that *luakini* function did not correlate with any specific architectural type.

The Multiple Functions of *Heiau*

The first and foremost function of *heiau* was as places of sacrifice (*hai*) and of prayer (*pule*); *heiau* were sacred (*kapu*) spaces dedicated to such rituals. But did *heiau* have functions other than those of ritual supplication and offerings to the gods? Little consideration has been given by

TABLE 5.1. ^{230}Th-dated temple sites (excluding shrines) with attributes of morphology, orientation, and location (modified after Kirch, Mertz-Kraus, and Sharp [2015, table 2]).

SITE NO.	OLDEST ^{230}TH DATE (AD)	MORPHOLOGY					ORIENTATION				LOCATION	
		NOTCHED	SQUARE/ U-SHAPED	ELONGATED DOUBLE-COURT	PLATFORM/ TERRACE	N-FACING	ENE-FACING	E-FACING	OTHER	COASTAL	INLAND
LUA-4	1696	X	—	—	—	—	—	—	X	X	—
AUW-9	1692	X	—	—	—	X	—	—	—	X	—
KIP-273A	1669	X	—	—	—	X	—	—	—	X	—
KIP-307	1643	X	—	—	—	—	—	—	X	X	—
KIP-275	1625	—	X	—	—	—	X	—	—	X	—
KIP-273B	1618	—	X	—	—	—	X	—	—	X	—
LUA-29	1615	X	—	—	—	—	—	—	X	X	—
NAK-29	1603	X	—	—	—	—	X	—	—	—	X
KIP-405	1601	—	—	—	—	—	—	X	—	—	X
ALE-140	1592	X	—	—	—	—	—	—	X	—	X
LUA-3	1590	—	X	—	—	X	—	—	—	X	—
WF-AUW-359	1588	X	—	—	—	—	—	X	—	—	X
KIP-1010	1580	X	—	—	—	X	—	—	—	—	X
KIP-567	1575	—	—	—	X	—	X	—	—	—	X
KIP-414	1574	—	—	—	—	—	—	X	—	—	X
KIP-728	1572	—	—	X	—	X	—	—	—	—	X
ALE-4	1555	—	—	X	—	—	—	—	X	X	—
WF-AUW-338	1328	X	—	—	—	X	—	—	—	—	X
KIP-1	1325	—	—	X	—	—	—	X	—	—	X

Note: Temples are ordered by the age of the oldest coral date from each site (i.e., the date most closely approximating that of initial construction). Italicized dates are from architecturally integral contexts, which are most reliable as indicators of construction dates; non-italicized dates are from surface contexts.

archaeologists to this question. Our investigations and data lend support to the notion that in addition to serving as ritual spaces, *heiau* had two other important functions—one as gathering places where men assembled not only to witness certain rites but also to partake in craft activities and to share food, and another as places for the systematic observation of the heavens, an essential activity for Hawaiian calendrics.

As related in chapter 2, there were four *kapu* periods in each lunar month during which the men of a neighborhood were required to spend two successive nights within the temple precincts. Although the ethnohistorical sources are somewhat vague, it seems that the *heiau* into which the men retreated during these *kapu* periods were of the *hoʻoulu ʻai* type—that is, agricultural fertility temples. The sources do not tell us what the men did while so sequestered, but our excavations in several *heiau* in both Kahikinui and Kaupō (summarized in chapter 3) provide evidence both for cooking and consumption of a range of foods (pig, dog, bird, fish, turtle, and shellfish; see table 3.3) and for various kinds of craft activities. Lithic knapping, including production and resharpening of adzes, and the manufacture and repair of fishing gear are both evidenced from the artifacts recovered from *heiau* contexts (see table 3.2). Far from simply witnessing rituals conducted by *kāhuna* during these *kapu* periods, the assembled men evidently used some of their time to make and repair objects important to their daily lives. They also cooked and ate together, and no doubt shared information and passed down traditions while sequestered in these sacred spaces.

What has been so striking to us, as we have surveyed *heiau* over several field seasons, is the ways in which temples were so thoughtfully sited and positioned to take advantage of particular viewsheds, often commanding sweeping views of mountain crests, distant headlands, and looming cinder cones. There can be little doubt that the places where *heiau* were built were carefully selected. The phases of the moon were visible to all but, as we have described in detail in chapter 4, *heiau* were typically situated so as to be observing platforms or stations, using the fixed configuration of the landscape (whether in the near, intermediate, or far distance) to index the regular movements of the sun, the Pleiades, or other asterisms. Figures 4.11 and 4.13 offer numerous examples of how the annual progression of the sun from solstice to solstice, or the rising of Makaliʻi, could be readily tracked against the sloping profiles of the Lualaʻilua cinder cones or other prominent topographical features. Our fieldwork has convinced us that *heiau* were positioned where they are precisely in relation to these viewsheds and skyscapes and that these temples functioned as much as observatories as places of ritual activity.

It is not difficult to understand why it would have been essential for the inhabitants of Kahikinui and Kaupō to observe closely the movements of key heavenly bodies that enabled them to maintain their calendar. As described in chapter 2, the Hawaiian lunar calendar regulated not only ritual activity but also the economic and ecological cycles on which life depended. The priestly observation of Makaliʻi, the Pleiades, permitted the annual recalibration of the Hawaiian lunar calendar to the sidereal year. But it is very likely that the priests were attuned to more than just the acronychal rising of Makaliʻi in order to set the timing of Makahiki. In leeward, arid southeastern Maui, rainfall is markedly seasonal (see fig. 2.2). It would have been essential to have prepared the sweet potato fields and planted the seed tubers *prior* to the onset of the Makahiki, so as to take full advantage of the first *kona* rains. Indeed, the available sources indicate that during the first two months of the Makahiki, the newly planted fields were *kapu*, assuring that nothing would disturb the newly sprouting tubers (Maunupau 1998, 152–153).

Careful observation of the sun's progression across the landscape toward the end of the Kau season would have allowed the *kāhuna* to signal when initial field clearing and tuber planting should commence, before the Makahiki began.

Heiau Orientations and the Hawaiian Pantheon

In his careful analysis of the ethnohistorical evidence concerning temples in ancient Hawai'i, Valeri observes that "the hierarchy of sacrifices, the chain of sacrifices that constitutes Hawaiian society, has a correlate in the hierarchy of places in which these cultic acts occur" (1985, 172). Valeri avers that previous attempts by archaeologists to establish typologies of Hawaiian *heiau* (such as that of Bennett [1931]) were flawed "because the types utilized are constituted on the basis of primarily formal [architectural] criteria without a sufficient questioning of the meanings that the Hawaiians attributed to the elements so classified." From a strictly functional point of view, *heiau* clearly fall into two major classes: those associated with war (*heiau kaua*) and those where various rites intended to "produce growth" (*ho'oulu*) were performed (Valeri 1985, 183). Valeri further notes that the second category comprised three subclasses—for producing the growth of food (*heiau ho'oulu 'ai*), for producing the growth of fish (*heiau ho'oulu i'a*), and for inducing rain (*heiau ho'oulu ua*). These functional classes of temple, moreover, correspond with one or more of the major gods in the Hawaiian pantheon. Thus, as Valeri writes, "it seems that on the whole the hierarchy of temple types corresponds to that of the major gods," and he provided a following diagram expressing this hierarchy (Valeri 1985, 184), which we reproduce here as figure 5.1.

We now return to a set of hypotheses, first outlined by Kirch (2004b) and further examined by Ruggles (2007), that links this hierarchy of temples to distinct orientation clusters. A major finding of our investigation of *heiau* in Kahikinui and Kaupō has been that temples are not randomly distributed with respect to their orientations and that—although there are some exceptions—the majority of temples fall into three main orientation clusters around north, east-northeast, and east. The three clusters are clearly evident in figure 4.4 (upland *heiau*), while figure 4.3 shows that the east-northeast and east groups start to merge when coastal temples are included. The empirical evidence is summarized in table 5.2, where it can be seen that of 66 *heiau* for which we are able to determine an axis of orientation, 54 (or 82%) face in one of these directions. (If the north-northeast and north-northwest groups are included as part of a broader north cluster, then fully 62 temples, or 94%, fall into one of the three orientation clusters.) Might these orientation clusters correspond to the hierarchy of gods and their temples as described by Valeri?

In chapter 2, we explored the associations between the principal deities and certain attributes of direction, landscape, and other qualities (see especially table 2.1). Kū, noted primarily

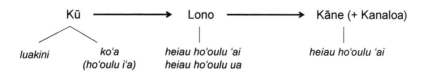

FIGURE 5.1. Hierarchy of temples. (Modified after Valeri 1985, 184)

TABLE 5.2. *Heiau* types and orientations.

HEIAU TYPE	ORIENTATION									
	NNW	N	NNE	N/E	ENE	E	ESE	WSW	W	WNW
Notched	2	7	2	2	5	4	—	1	—	—
Platform/terrace	—	3	—	1	6	—	—	—	—	—
Elongated double-court	1	3	—	1	—	3	—	—	1	1
Square/U-shaped enclosure	—	3	1	—	4	3	1	—	—	—
Irregular forms	1	—	—	1	3	—	—	—	—	—
Koʻa/shrine	—	3	1	—	1	1	—	—	—	—
Total	4	19	4	4	17	14	1	1	1	1

Note: "N/E" means that the orientation is either N or E.

for his association with war, was also a deity of canoe building, and sorcery, as well as the patron deity of fisherman (in his manifestation as Kūʻula). Kū's directional attributes are north, east, and right, and he is associated with high mountains and forest trees. Kāne, representing the male powers of procreation, was also the god of irrigated (taro) agriculture, with directional associations of east and right. Kāne was strongly associated with the rising sun, or as Kepelino put it, "the great sun of Kāne" (*ka la nui o Kane*), and the eastern sky was known as "the flaming path of Kāne" (*ke alaula a Kane*) (Johnson, Mahelona, and Ruggles 2015, 180). The most northerly path of the sun across the sky was known as "the black shining road of Kāne" (30). Lono, deity of dryland agriculture, birth, and medicine—and to whom the sweet potato was sacred—had directional attributes of high and leeward and was also explicitly linked to the acronychal rising of the Pleiades (Makaliʻi) and the onset of the Makahiki, over which Lono presided. Finally, Kanaloa, god of the sea and the underworld, had directional attributes of left, west, and south and was associated with the sunset and death. The western sky was known as "the much traveled red road of Kanaloa" (*ke alanui maʻawe ʻula a Kanaloa*) (179), while the most southerly path of the sun across the sky was known as "the black shining road of Kanaloa" (30).

Given these directional associations of the main Hawaiian deities, well attested in the ethnohistorical sources, it seems reasonable to link the three main orientation clusters indicated by our archaeological studies with the hierarchy of gods. The least problematic association is between the cluster of east-oriented temples and Kāne, who was unambiguously linked with the sun and the east. Note that it would be misleading to describe these temples as "equinoctial"; there is no reason to suppose that the Hawaiians had any interest in the equinoxes as such (see chapter 4). On the other hand, from many such east-oriented *heiau* the priests of Kāne could have observed the progression of the sun throughout the year. Many of the upland temples in Kahikinui (such as those on Heiau Ridge at Nakaohu) have superb viewsheds toward the east, where "the great sun of Kāne" would have risen dramatically over the peninsula of Kaupō. It is also noteworthy that many of the east-oriented temples are situated adjacent to watercourses (even though most of these run only intermittently in this arid terrain). Kāne being the god of flowing waters, such proximity of streams was likely also a conscious choice of the temple builders.

Temples in the east-northeast cluster are generally oriented toward the Pleiades, whose acronychal rising determined the start of the Makahiki season, over which the deity of dryland cultivation, Lono, presided. Although not all temples in this cluster are precisely oriented to the rising position of Makaliʻi, we do have several instances of quite accurate alignments and apparent viewing devices (see the section "Observing Makaliʻi" below). Given the importance of dryland cultivation in Kahikinui and Kaupō, especially of sweet potatoes, Lono must have been an important deity to the people of this region. It seems plausible, therefore, that the east-northeast-oriented *heiau* were *hale o Lono*, temples dedicated to this god. In these temples, the Lono priests could have awaited the acronychal rising of Makaliʻi, ready to announce the onset of the new year. It is equally plausible that they observed the progression of the sun toward the summer solstice.

Finally, for the north-oriented cluster of temples, we need to keep in mind that to be oriented roughly north in Kahikinui or Kaupō is to face the great soaring height of Haleakalā. Thus, these temples were constructed so that their principal axes faced the mountain, either its high summit or, frequently, one of the prominent, reddish cinder cones (*puʻu*) that stand out along the skyline of the mountain's southwestern rift zone. Four of the elongated double-court temples, as well as many of the largest temples, have such a northerly orientation. The most plausible interpretation is that these temples were dedicated to, or associated with, Kū, the deity linked to the high mountains, to the sky, and to forests. Although Kū is often thought of primarily as the god of warfare—indeed, he was certainly the main deity of the ruling chiefs—the worship of Kū in Kahikinui and Kaupō may have focused on other attributes of his, such as canoe building, adze making, and sorcery. The high mountain slopes were the source both of large *koa* trees for canoe hulls and of dense basalt for adzes. Some of the north-oriented Kū temples may also have marked *ahupuaʻa* boundaries.

As indicated in table 5.2, only three *heiau* in our sample have generally westerly orientations. Are these simply anomalous, or are there possible cultural explanations for these orientations? Site ALE-140 appears to be oriented to the west-southwest, in terms of the position of the slightly elevated smaller room (probably an altar area). As the west is associated with Kanaloa, it is conceivable that ALE-140 was dedicated to this deity. However, as we have described in chapter 4, ALE-140 also has a remarkable sighting device that appears to have been designed for the observation of Makaliʻi, making it equally likely that this notched structure was a *heiau hoʻoulu* dedicated to Lono. The two other sites with generally westerly orientations are both located in Nuʻu *ahupuaʻa* of Kaupō (sites NUU-81 and -100). Both are elongated double-court structures with architecturally elaborated altars on their western ends; both have excellent viewsheds looking out over the lands of Nakula to Kahikinui beyond. Nuʻu is the westernmost *ahupuaʻa* in Kaupō, and thus both of these temples may well have served as boundary *heiau*, marking the political division between the Kahikinui and Kaupō Districts. Although not as large as Loʻaloʻa or Pōpōiwi, both NUU-81 and NUU-100 are substantial constructions and could have functioned as war temples (*luakini*) dedicated to Kū. Although in general we believe that Kū temples were oriented to the north, in these cases the orientation facing a potential enemy district would be consistent with a *luakini* function.

A number of southeastern Maui temples appear to incorporate intentional alignments upon distinctive features in the surrounding landscape. This may seem curious given that the more distant of these, particularly on the Haleakalā Ridge to the north and on Hawaiʻi Island to the southeast, are often obscured by clouds during much of the day. As a rule, however, they are clearly visible around dawn, before the sun gets higher in the sky and starts heating the atmosphere, and in the late evening toward sunset. This suggests that the associated rites, at least at Kū temples, most likely took place around dawn.

On the other hand, the eastern and western horizons are generally visible at all times of day; in any case, observations of the rising or setting of the Pleiades or other asterisms, either before dawn or after sunset or in the night, would not have been systematically hampered by weather conditions. Sunrise observations at Kāne temples took place at dawn by definition.

To be sure, we do not have plausible topographic or astronomical explanations for all temple orientations. At AUW-9, LUA-3, ALE-1, KIP-306, KIP-728, MAW-2, and NUU-175 the primary orientation remains a mystery. At some other sites there is no obvious reason for a significant deviation from non-perpendicularity. At KIP-1, for example, the orientation of the eastern wall (2.8°/182.8°) is well defined but deviates from perpendicularity to the east–west walls (themselves remarkably consistent) by 5° to 6°. Yet this cannot be explained by prominent features on the Haleakalā Ridge to the north; for example, the alignment misses the red cinder cone Kanahau and indeed would have been closer to it if the wall *had* been perpendicular.

Taken as a whole, our results suggest that the orientation of the Kahikinui-Kaupō temples were influenced by a combination of visual (topographic) and astronomical factors, differing according to the date of construction, the function of the temple, and the limitations of the chosen location (and even the exact choice of location within broader constraints).

Observing Makali'i

Our systematic analysis identified a cluster of structural alignments upon the rising point of the Pleiades. The most impressive of these is the huge platform at KAU-994 (Lo'alo'a), which was accurately aligned upon the rising position of the Pleiades at the time of its construction or elaboration around AD 1650–1700. The platform at NAK-30 and the enclosure at KIP-1156 also face the rising point of the Pleiades; LUA-4 is oriented upon the Pleiades rise to the east-northeast, raising the possibility that east-northeast might actually be the direction of orientation, although the raised altar platform on the north-northwestern side suggests that this notched *heiau* faced north-northwest. ALE-140 unusually faces west-southwest into the side of a swale, but here we noted an extraordinary device functioning in the opposite direction, making use of a pair of upright boulders, one artificially set in place, to frame the skyline. This would have pinpointed the rising position of the Pleiades as viewed from the northern side of the *heiau*, lining along the outer face of the north-northwestern wall. We also noted that at NAK-30 and other sites (NAK-27, KIP-273B, and KIP-275), nearby boulders or outcrops could have been used as sighting aids for the Pleiades. The coastal sites KIP-273B and KIP-275 feature possible sighting devices that could have been used for either Makali'i or Ka Nuku o Ka Puahi (Hyades-Aldebaran), or both.

There is also a cluster of classic notched enclosures in upland Auwahi as seen from which the Pleiades rose from the bottom of gaps between the Luala'ilua Hills (a consistency that is not expected if the *heiau* had been placed randomly in the local landscape), and there are a number of other sites—including LUA-29, KIP-1156, NAK-29, and NUU-100—where the rising or setting of the Pleiades coincided with a distinctive feature on the horizon, such as a distant peak. At NUU-81 the setting point of the Pleiades was marked by Pu'u Pane, a distinctive cinder cone seen just below a more distant featureless horizon.

None of these associations is convincing in itself because the associations are not "indicated" by structural alignments and there are innumerable possibilities for discovering coincidences between topographic features and astronomical targets from any location taken at random. However, taken together they do suggest that Makaliʻi not only influenced temple orientations and gave rise to practical observing devices but also helped shape the lived-in landscape by affecting the choice of location of temples in relation to the perceived environment.

While there is ample evidence to suggest that the Pleiades had a significant influence on the design and placement of many temple sites in Kahikinui and Kaupō, we cannot rule out the possibility that the solstitial sunrise was the more important event at some of the *heiau*. The fact that the declination of the Pleiades is close to that of the June solstice sun, starting to overlap with it in about AD 1650, means that in many cases we cannot distinguish between structures intended to align upon the Pleiades and those intended to align upon the June solstice sun on the basis of the orientation evidence alone. However, a solstitial explanation cannot account for both of the two sharp peaks between declination +22° and +24° in figures 4.7 and 4.8, whereas they both fit within the declination range of the Pleiades between AD 1500 and AD 1800. It is also likely that the heliacal rise of the Pleiades and the summer solstice were linked; the coincidence of the two rising directions would have been evident, particularly given that the Pleiades started to appear in the predawn sky in the days leading up to the summer solstice and—around the time of the solstice itself—would have been seen to rise ahead of the sun on an almost identical path.

As described in chapter 2, the opening of the Makahiki season was determined by the first appearance of the Pleiades immediately after sunset. In practice, as this season approached, the Pleiades would be seen to rise early in the night, progressively earlier each night. Eventually, the brightness of the sky following sunset would prevent the actual rise being seen, and as the sky darkened, the Pleiades would appear already in the sky. At this point, around the middle of November, Makahiki could begin. We follow convention in referring to this event as the "acronychal rise," although this technical term actually refers to a subtly different phenomenon—the rising of the Pleiades (or any asterism) at sunset. This would not be visible to the naked eye because of the bright sky, and archaeoastronomers generally refer to the observable event as the "apparent acronychal rise" (Ruggles 2014b, 464–465), as explained in chapter 4. It is easy to see how devices such as that at ALE-140 could help a priest identify exactly the right place in the sky to watch for the Pleiades appearing in the east-northeast as the evening sky darkened.

The heliacal rise of the Pleiades, their first appearance in the predawn sky, took place around the end of May. Devices such as that at ALE-140 would also have been ideal for pinpointing this event, although the ethnohistorical record provides little direct evidence that it was of particular importance to Hawaiians. This event does, however, mark the beginning of the year in many parts of southern Polynesia, and phylogenetic analysis suggests that the Proto-Polynesian calendar was divided into two main parts by the heliacal and acronychal rise of the Pleiades (Kirch and Green 2001, 261–267).

The ethnohistorical evidence is less clear as regards observations of the setting of the Pleiades for calendrical purposes. Malo (1951, 30) mentions that Ikiiki, the first month of the summer season (Kau, corresponding to May), was the time when the Pleiades set at sunrise. However, the first setting before sunrise (apparent cosmical setting) occurred around the same time as the last rising after sunset (Ruggles 2014b, 464–465)—that is, in November—and would in fact provide a good alternative indicator of the onset of Makahiki. Malo could possibly mean the time when the Pleiades set at sunset (or, more precisely, their last appearance in the sky after sunset, or heliacal set), which occurred around the end of April, or indeed the time when they rose at sunrise (or, more precisely, their first appearance in the predawn sky, or heliacal rise), which occurred around the end of May. Kirch and Green (2001, 262) list various Polynesian calendars where observations of the heliacal setting of the Pleiades have been recorded as relevant, but it is clear that the observations of the rising (both acronychal and heliacal) are dominant.

As noted above, observations of the (apparent) cosmical setting of the Pleaides could have provided an alternative indicator of the onset of Makahiki equally as good as observations of the acronychal rising. The coincidence between the setting point of the Pleiades and the peak of Puʻu Pīmoe at LUA-29, and the Lualaʻilua southeastern peak at KIP-1156, provides tentative evidence that such observations were indeed made in the Hawaiian Islands. It is also possible that observations were made of the heliacal setting, related to the beginning of the summer season.

Concluding Remarks

One of the most obvious ways in which the material remains of past cultures link to the sky (and certainly the most widely investigated, in a broad range of human societies worldwide) is the orientation of a building or structure such as a tomb or temple so as to face a significant celestial phenomenon, such as the rising point of the sun or a star (Ruggles 2014e). Yet archaeoastronomy has been dogged over the decades by speculative and ethnocentric interpretations of alignment data (Ruggles 2011, 2014c).

Nonetheless, "data-driven" approaches, such as the systematic study of the principal orientations of more than 3,000 later prehistoric tombs and temples in western Mediterranean Europe (Hoskin 2001), have proved crucial in convincing archaeologists that orientation in relation to referents in the visible landscape and the sky was not only of significance but often of prime importance in the design and function of a variety of ceremonial and funerary constructions worldwide. Initially, preliminary studies of this nature in Polynesia seemed unpromising (e.g., Liller's [1989] study of temple orientations on Rapa Nui), but evidence is now gradually emerging from around Polynesia that orientation was not simply a consequence of the lay of the land or the direction of the sea. Thus, the *marae* of the ʻOpunohu Valley in Moʻorea provide an example of temples being oriented upon a particular red mountain peak (Kahn and Kirch 2014, 112–113), while in Mangareva a large platform with cardinal orientations has been documented as an indigenous solstitial observatory, corroborating early missionary accounts (Kirch 2004a).

We believe the work described in this book not only significantly advances such studies in southeastern Maui but that our three-stage approach—systematic analysis followed by contextualized interpretation, together with impartial presentation of the basic archaeoastronomical "facts" (cf. Ruggles 2014a, 413)—provides a useful general strategy for integrating archaeoastronomical considerations into broader investigations of archaeology and landscape in a range of contexts. We believe that continued application of this approach in other regions of the Hawaiian Islands and elsewhere in Polynesia is likely to yield new insights into the relationships between Polynesians and their skyscapes.

II

Catalog of *Heiau* Sites of Kahikinui and Kaupō

Note on Azimuths and Directions

Azimuths are measured clockwise from due north, so that azimuth 0° is due north, azimuth 90° is due east, azimuth 180° is due south, and azimuth 270° is due west. When discussing architectural features, viewsheds, and orientations, we utilize sixteen compass directions:

DIRECTION	MINIMUM AZIMUTH (°)	CENTRAL AZIMUTH (°)	MAXIMUM AZIMUTH (°)
N	348.75	0.0	11.25
NNE	11.25	22.5	33.75
NE	33.75	45.0	56.25
ENE	56.25	67.5	78.75
E	78.75	90.0	101.25
ESE	101.25	112.5	123.75
SE	123.75	135.0	146.25
SSE	146.25	157.5	168.75
S	168.75	180.0	191.25
SSW	191.25	202.5	213.75
SW	213.75	225.0	236.25
WSW	236.25	247.5	258.75
W	258.75	270.0	281.25
WNW	281.25	292.5	303.75
NW	303.75	315.0	326.25
NNW	326.25	337.5	348.75

In the "Architecture," "Viewshed," and "Orientation" sections of our site descriptions and elsewhere where "north," "east," "south," and "west" are abbreviated as "N," "E," "S," and "W," respectively, the term "N" (for example) should be taken to refer to the specific 1/16th part of the compass arc centered on due north—that is, the azimuth range shown above, as opposed to, say, NNW or NNE. When spelled out, "north" and related terms such as "northern," "northerly," and "northward" (and similarly for other directions) are used more loosely, depending on the context.

Note on Radiocarbon and ^{230}Th Dates

All of the radiocarbon dates reported in this catalog are given as "conventional" radiocarbon dates in years "before present" (BP; i.e., before AD 1950), following the protocols of Stuiver and Polach (1977). Radiocarbon dates have been calibrated using OxCal version 4.2.4 with the IntCal13 atmospheric calibration curve (Bronk Ramsey and Lee 2013; Reimer et al. 2013) and are expressed as 2σ ranges (95.4% confidence intervals). Where the confidence interval is spread over more than a single age range, each range is shown with its associated probability in parentheses, adding up to 95.4%. ^{230}Th dates on coral from *heiau* sites are all reported at two standard deviations (2σ), or 95.4% probability. Where precise ^{230}Th dates on corals have been obtained, these are tentatively correlated with the reigns of Maui kings (*aliʻi nui*), based on the chiefly genealogy, with a 20-year interval between rulers (see Kirch 2014, table 2).

Note on Site Access

The *heiau* sites described in this catalog are located on lands controlled by the Department of Hawaiian Home Lands (DHHL) of the State of Hawaiʻi or by private landowners, including ʻUlupalakua Ranch, Kaupō Ranch, and Nuʻu Mauka Ranch. These sites are not open to visitation without prior written permission of the DHHL or of the respective ranches; for this reason we have refrained from providing precise coordinates for the sites. Entry onto DHHL lands or any of the ranches without prior written permission constitutes an act of trespass. Researchers or others with legitimate reasons to visit the *heiau* sites in Kahikinui or Kaupō should contact appropriate authorities or landholders to request a right-of-entry permit.

6

Heiau of Auwahi to Alena, Kahikinui *Moku*

The *Ahupua'a* of Auwahi

Auwahi is the westernmost *ahupua'a* within the *moku* of Kahikinui, lying to the west of the Luala'ilua Hills; the prominent cinder cone of Pu'u Hōkūkano lies in the center of the *ahupua'a* at an elevation of about 400 m above sea level (masl). On the coast, the small bay of Makee provides the best canoe landing along the entire Kahikinui shore; only here would it have been possible for the large double-hulled canoes of the ruling *ali'i* to put ashore on the cobble beach. In the Mahele—the land division of 1848—the *ahupua'a* was claimed by the high chiefess Ruth Ke'elikolani (granddaughter of Kamehameha I), having previously been under the control of her father, Mataio Kekūanāo'a. Immediately inland of the beach at Makee is a substantial village site with several large house enclosures, almost certainly the residences of the *konohiki*, who managed the Auwahi lands for the *ali'i*. Situated within this coastal village cluster are the first three *heiau* described in this section (AUW-6, -9, and -11).

The other nine *heiau* in Auwahi are farther inland, between 200 and 500 masl. These nine sites were all discovered as part of a cultural resources management (CRM) inventory survey conducted by Pacific Legacy Inc. in conjunction with the development of a wind farm energy project (Shapiro et al. 2011). All of these inland sites were subsequently visited by both Kirch and Ruggles in 2010 and 2011, when plane table and GPS maps and total station surveys were made.

The higher-elevation region of Auwahi, *mauka* (inland, toward the mountain) of the Pi'ilani Highway and above 500 masl has not been intensively surveyed for archaeological sites. It is likely that a number of smaller *heiau* remain to be discovered and described in this upland part of Auwahi.

Some of the viewsheds described below have been altered since the time of our field surveys by the installation of the Auwahi wind farm turbines in 2014. This applies not only to Auwahi sites but also, for example, to western views from some of the sites in Luala'ilua and Alena.

Site AUW-6

Location and topographic setting. Site AUW-6 is a small shrine, covered in abundant branch coral, attached to a larger rectangular residential feature at Makee.

Architecture. The shrine consists of a square enclosure of stacked *'a'ā* lava cobbles, 4.5 m across, with walls about 1 m thick. The walls, which are about 60 cm high, have undergone considerable collapse. The interior floor consists of *'a'ā* rubble. The primary indication that this small enclosure had a ritual function is the presence of abundant pieces of branch coral (*Pocillopora meandrina*) as well as waterworn *Porites* coral heads scattered over the walls and the interior floor and in front of the enclosure.

Dating and chronology. No excavations were conducted at AUW-6, but Sharp and Kirch collected a specimen of *Pocillopora* branch coral with well-preserved branch tips wedged in under the northwestern corner of the wall and seemingly deposited there at the time of wall construction. This architecturally integral coral was ^{230}Th dated to AD 1547 ± 2.

Archaeological context. As noted earlier, this small shrine is attached to a larger enclosure that is interpreted as a residential site. Together, these structures form part of the nucleated village at Makee, probably occupied by the *konohiki* and other elites who were likely in control not only of Auwahi but quite possibly of the entire *moku* as well.

Viewshed. The viewshed is essentially the same as that described for site AUW-9 (the next site described).

Orientation. The orientation of this shrine is generally to the north, based on the abundance of branch coral heaped on the north wall. However, no precise observations of orientation were made.

Additional remarks. Based on the single date from branch coral under the wall, this small shrine was constructed in the mid-sixteenth century. It most likely functioned as a household shrine for the occupants of the adjacent house enclosure.

Site AUW-9 (Walker Site 187)

Site AUW-9 was first recorded by Winslow Walker ([1930], 259), who also drew a sketch plan of the site, number 187 in his sequence. Walker called this "a small heiau of the walled enclosure type with a high open platform at the south end." Kirch visited the *heiau* several times, first mapping it with plane table and alidade in January 2003. Kirch and Ruggles conducted a total station survey of AUW-9 on March 24, 2003, while Kirch and Sharp systematically collected coral samples for ^{230}Th dating in January 2008.

Location and topographic setting. This *heiau* site is situated centrally within the Makee village complex, on a substrate of rough *'a'ā* lava. As with the rest of the Makee complex, it lies in a swale or depression in the lee of a high lava ridge to the east.

Architecture. The architecture of AUW-9 is complex, with evidence for two phases of construction (fig. 6.1). On initial inspection, the structure appears to consist of a high platform attached to a well-built U-shaped enclosure with walls up to 2 m high. Broadly speaking, it is cardinally oriented. The well-built, core-filled E wall (built of *'a'ā* cobbles) measures 17 m long and is up to 2 m thick, while the N wall is 10 m long and equally massive. The platform, on the S end of the structure, measures 7 by 10 m and stands between 1.5 and 2 m above the surrounding ground. Most of the platform surface is well paved with *'ili'ili*; two small pits near the center of the platform may be image holes. Basalt flakes and cores were scattered near the pits, while pieces of branch coral were also noted on the platform surface, along with a larger waterworn cobble.

Walker's ([1930]) description refers to a "hole opening into a tunnel which extends under the platform" and adds: "The sides were walled with rock but the hole was too narrow for a man to crawl into with safety, as several large rocks had already fallen in, but nothing could be seen inside." We could find no evidence of this tunnel.

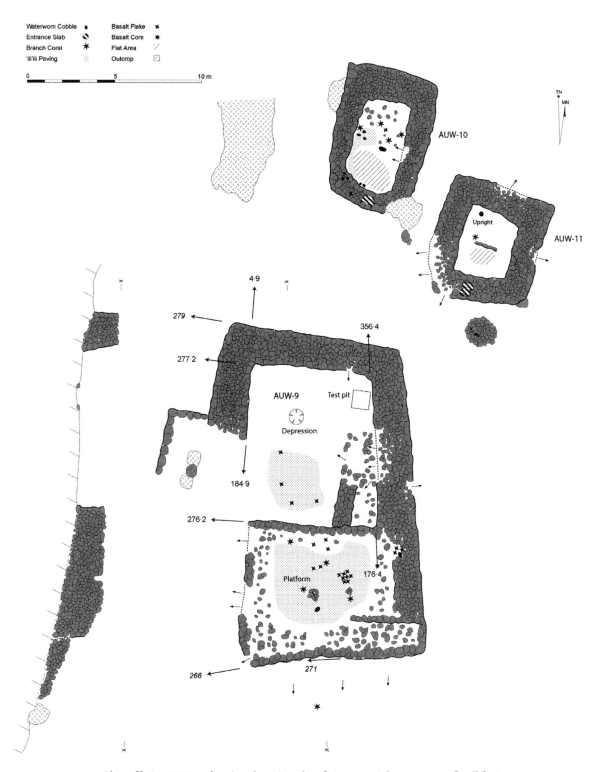

FIGURE 6.1. Plan of *heiau* AUW-9, showing the azimuths of intact straight segments of wall facing as determined by total station (quoted to the nearest 0.1°) or prismatic compass corrected for mean magnetic declination (italicized and quoted to the nearest degree). Structures AUW-10 and AUW-11 are also shown. (Based on plane-table survey by P. V. Kirch)

Careful examination of the walls to the N of the platform revealed that originally the structure was a classic notched *heiau* (with the notch in the NW corner), but that the westernmost wall had been robbed of its stone down to the base course of large *niho* (set or base) stones; the original SW corner is missing entirely. It appears that these W and S walls of the original notched *heiau* were removed in order to construct the platform, which was then abutted to the southern part of the original *heiau*'s E wall.

A small room or chamber is defined by a low, stacked cobble wall in the corner where the platform abuts the main E wall; part of this chamber is now covered by rubble from a partial wall collapse. The relatively level floor of the *heiau* court defined by the main enclosure walls is partly paved with 'ili'ili gravel; several basalt flakes and a basalt hammerstone were noted on this surface, along with a *Pocillopora* coral head. A shallow pit or depression, about 1 m in diameter, is situated near the center of the floor.

Excavation. A 1 m² unit was excavated within the enclosure, in the NE corner adjacent to the inside face of the E wall. Beneath a thin (2–3 cm) layer of overburden, the excavation revealed a single cultural layer of rocky, ashy sediment with considerable charcoal flecks throughout, ranging from 20 to 35 cm in thickness. Aside from charcoal, the cultural deposit yielded 242 basalt flakes, 15 pieces of coral, 183 NISP (number of identified specimens) of fish bone, and 146 NISP (31.07 g) of marine mollusks and sea urchins. In addition, there was one dog incisor, three shaft fragments of a medium mammal (either dog or pig), and one humerus of a chicken. Identifiable fish bones included specimens of convict tangs (*Acanthurus* sp.), goatfish (Mullidae), spiny puffers (Diodontidae), and parrotfish (Scaridae). Identifiable mollusk species included 'opihi (*Cellana exarata*), snake's head cowrie (*Cypraea caputserpentis*), and *pipipi* (*Nerita picea*). There were also significant quantities of sea urchin spines and test fragments (*Colobocentrotus atratus*, *Echinothrix* sp.). Of particular interest is the presence of a bone fishhook and a coral abrader, suggesting that fishhook manufacture or repair was undertaken within the *heiau* enclosure.

Dating and chronology. A sample of charcoal (identified as *Chamaesyce* sp.) from level 4 of test pit 1 (TP-1) was AMS radiocarbon dated (Beta-183146) with a conventional age of 390 ± 40 BP. This yields two possible calibrated age ranges at 2σ: AD 1436–1529 (60.8%) and 1544–1634 (34.6%). Five samples of branch coral from AUW-9 were ^{230}Th dated, four of the corals also having replicate dates. The results of this extensive dating from AUW-9 are detailed in table 6.1. The dates are highly consistent, falling in the decade of AD 1690–1698, with a single exception. The outlier is sample CS-4, which came from the collapsed fill of the platform and yielded an age of AD 1671 ± 2. However, this specimen was slightly water-rounded and therefore was most likely collected from the Makee beach, rather than being taken live from the ocean as the other dated corals were; it therefore is likely to have some inbuilt age. Sample CS-6 was removed from within the main wall fill in a small collapsed area and was certainly placed within the wall during construction.

The single radiocarbon date from TP-1 is slightly older than the suite of ^{230}Th dates obtained from the branch corals. The charcoal sample is of a short-lived species, so there is no expectation of inbuilt age. However, the sample cannot be directly associated with the stone architecture, as it came from the base of a cultural deposit that may underlie the *heiau* wall. It is likely that the radiocarbon date derives from cultural activity on the site prior to wall construction and should therefore be considered as a *terminus post quem* date for the *heiau*. That the main walls of the original notched *heiau* were constructed within the period from AD 1690 to 1698 is solidly confirmed by the suite of ^{230}Th dates. On architectural grounds, the platform was built later, with stones

TABLE 6.1. ^{230}Th dates on coral specimens from site AUW-9.

SAMPLE NO.	SAMPLE TYPE	PROVENIENCE	DATE (AD, AT 2σ)
AUW-9-CS-1	Branch, intact tips	Surface, on floor, near base of old wall	1696 ± 2
AUW-9-CS-1-R	Branch, intact tips	Surface, on floor, near base of old wall	1697 ± 3
AUW-9-CS-2	Branch, intact tips	Architecturally integral, in stones from collapsed old wall	1698 ± 2
AUW-9-CS-2-R	Branch, intact tips	Architecturally integral, in stones from collapsed old wall	1698 ± 3
AUW-9-CS-3	Branch, intact tips	Architecturally integral, under base of old wall	1692 ± 2
AUW-9-CS-3-R	Branch, intact tips	Architecturally integral, under base of old wall	1690 ± 10
AUW-9-CS-4	Branch, slightly rounded	Architecturally integral, in collapsed wall of new platform	1671 ± 2
AUW-9-CS-6	Branch	Architecturally integral, in old wall exposed by collapse	1694 ± 3
AUW-9-CS-6-R	Branch	Architecturally integral, in old wall exposed by collapse	1693 ± 10

Note: In sample numbers, R = replicate.

robbed from the W and S *heiau* walls; how much later this secondary construction phase took place we cannot say on present evidence.

Archaeological context. As noted, AUW-9 stands within the central part of the Makee village complex. Immediately NE of the *heiau* is site AUW-11, a high-walled *koʻa*, or fishing shrine. About 25 m farther NNE of the *heiau* is a very large, rectangular house enclosure (AUW-15), probably an elite residence.

Viewshed. The Haleakalā Ridge is visible from the NNW round to the NE (az. ~335° to ~45°), as is open sea from the ESE round to the SW (az. ~110° to ~230°), with Auwahi Bay in the foreground. However, a lava ridge 0.5–1 km distant restricts the view to the WSW round to the NW, as does the high lava ridge only ~100 m away to the ENE and E.

Orientation. The fact that the raised platform is situated at the S end of this structure suggests that during the later phase the focus was to the S. The substantial N and E walls indicate that the original *heiau* faced one of those directions, and the lack of an open view to the E suggests that it likely faced N.

The E wall and the remaining part of the W wall diverge toward the S: measurements from the intact segments of their inner faces yield azimuths of 176.4° and 184.9°, respectively, with a mean of 180.6°, very close to due S (see fig. 6.1). The N wall and the N side of the platform diverge slightly to the W, their westward azimuths varying from 276° (platform side) to ~279° (outer face of N wall). The platform itself also widens to the W, its S side being oriented with azimuth ~266°, giving it a mean E–W orientation of ~271°, perpendicular to the overall N–S axis. What appears to be a small segment of the original S wall, predating the construction of the platform, shares a similar orientation of ~271°. In short, the structure was cardinally oriented to within about 1°.

During our survey the Haleakalā Ridge to the N was obscured by clouds, so for the northern profile we had to rely on a digitally generated horizon profile. This shows that the wall orientations (in this direction 356.4° and 4.9°, respectively, with a mean of 0.6°) fall somewhat to the left of the highest point of the ridge: the stretch of horizon concerned does not include any particularly noteworthy features. However, at azimuth 8°, the conspicuous summit of Puʻu Hōkūkano, appearing beyond local ground at a distance of 2.4 km, is seen to lie directly below a prominent cinder cone tip at the left end of the highest part of the Haleakalā Ridge.

"Hōkūkano" is not a star name known from the ethnohistorical sources, and in any case there is no reason to suppose that the hill was named for its alignment as seen from this or any other *heiau*. The cinder cone (altitude +12°) corresponds to declination +78°, but the only reasonably bright star rising in this vicinity at around the time of construction was Kochab (β UMi, mag. 2.1), the second-brightest star in Ursa Minor after Hōkūpa'a (Polaris), whose declination was between +75° and +76° around AD 1700. It is not known to have been significant to Hawaiians (see table 4.2).

To the S the *heiau* faces an open stretch of sea, and it is worthy of note that the entire low arc traced across the southern sky by Newe, the Southern Cross (together with the Pointers, α and β Cen), between azimuths ~150° and ~210° would have been clearly visible above the sea horizon from this location.

The various wall orientations to the W span a stretch of lava ridge without any features of particular prominence, while the close lava ridge obscures the more distant horizon to the E.

Additional remarks. AUW-9 is the primary ritual structure centrally located within the important Makee village complex in Auwahi, a cluster of sites thought to be the residences of the *konohiki* and other elites who managed this *ahupua'a* and quite likely the district as a whole. The *heiau* was originally of the classic notched form, built precisely along cardinal directions oriented to the N, with the notch in the NW corner. This first construction phase can be dated to the final decade of the seventeenth century, putting it most likely within the reign of Maui king Ka'ulahea II. Sometime later the W and S walls of the original notched *heiau* were taken down to ground level (leaving only the *niho* stones of the W wall), and the stones were used to construct the high platform in the SE part of the site. The focus appears to have switched at this time from the N, where Hōkūpa'a hung fixed above the Haleakalā Ridge, to the S, where Newe passed across low above the ocean. No other specific topographic or astronomical associations are evident.

SITE AUW-11

This site, being adjacent to AUW-9, was visited on the same occasions.

Location and topographic setting. This *ko'a*, or fishing shrine, is a mere 3.5 m inland (NE) from the NE corner of AUW-9, forming part of a cluster with the latter and also with a small rectangular enclosure (AUW-10) immediately NW of AUW-11 (see fig. 6.1). It is situated on the same rough '*a'ā* substrate as the AUW-9 *heiau* and shares the same viewshed.

Architecture. This is a nearly square enclosure, 6 by 6.5 m, with well-built core-filled walls between 1 and 1.5 m high. Its orientation deviates from cardinality by about 17° in a clockwise sense. There has been some collapse of the outer faces on the WNW, NNE, and ESE walls. There is no doorway to this high-walled enclosure, but a formal entrance over the SSW wall is indicated by the presence of two large, flat lava slabs atop the wall; thus, entrance to the shrine would have been gained by climbing up the front face of the seaward wall and passing over these entrance stones, then descending into the interior.

A small niche is situated at the base of the inner face of the ESE wall. The interior court is divided into two spaces by a low alignment of '*a'ā* cobbles running roughly WNW–ESE, with the inland side slightly elevated. In the NNW corner of the shrine is a large, upright Kū'ula stone of dense, waterworn basalt (see fig. 3.4). Surrounding this upright are other small waterworn stones, many of them broken, with flakes removed. There are also many branch coral pieces and at least one large *Pocillopora* coral head surrounding the upright.

Excavation. A small test excavation, 50 by 50 cm, was dug within the floor of the fishing shrine, in front of the main area of branch coral offerings. A very dark gray (Munsell 7.5 YR 3/1) cultural deposit with considerable charcoal extended from 2 to 20 cm below surface. Beneath this was a more ashy, gray (7.5 YR 5/1) sediment that extended to a depth of 30 cm below surface, below which was culturally sterile subsoil and large ʻaʻā cobbles.

Thirty-two basalt flakes were recovered from the cultural deposit, along with 151 NISP of bone and 14 NISP (10.41 g) of marine shell and sea urchin. The identifiable bones included two dog teeth and one dog bone shaft, two medium mammal bones (probably dog), two marine turtle bones (Cheloniidae), and one bird bone. Identifiable fish bones included convict tangs (*Acanthurus* sp.), triggerfish (Balistidae), jacks (*Caranx* sp.), wrasses (Labridae), eels (Muraenidae), groupers (Serranidae), and parrotfish (*Scarus* sp., *Calotomus* sp.). Invertebrate remains included ʻopihi (*Cellana exarata*), cowrie (*Cypraea caputserpentis*), and sea urchin spines and test fragments.

Dating and chronology. A sample of *Chamaesyce* sp. charcoal from level 4 of the test pit was radiocarbon dated (Beta-183147), yielding a conventional age of 160 ± 40 BP. This yields 2σ calibrated age ranges of AD 1663–1710 (16.8%), 1717–1890 (60.8%), and 1910–1950 (17.8%); the last calibrated range can be rejected on archaeological grounds because there were no postcontact artifacts in the cultural deposit of level 4.

Kirch and Sharp collected two branch coral specimens that were ^{230}Th dated. Sample AUW-11-CS-1 is a branch that we sawed off an entire *Pocillopora* coral head lying on the court floor next to the stone upright; this yielded a date of AD 1762 ± 9. Sample AUW-11-CS-2 is an architecturally integral specimen that we removed from the inner face of the NNE wall, where it had been wedged in a cavity between wall stones. This sample yielded a date of AD 1755 ± 3.

The single radiocarbon date and two branch coral dates from AUW-11 are consistent and indicate that the koʻa was constructed and initial offerings placed in front of the Kūʻula stone around AD 1755–1760. This date would put the construction of the shrine at the end of the reign of Maui king Kamehamehanui or at the beginning of the reign of his successor, Kahekili.

Archaeological context. As already noted, AUW-11 is part of a cluster of structures in the central part of the Makee village complex. It is the only koʻa within this elite residential complex. Quite likely it served the fishermen who were responsible for obtaining fish for the occupants of the high-status house sites immediately inland.

Viewshed. Same as for AUW-9.

Orientation. The entrance on the SSW wall indicates that this structure was oriented to the NNE. The wall orientations in this direction, as determined from the plane-table plan, are ~16° for the ESE wall and ~18° for the WNW wall, giving a mean axial azimuth of ~17°. The NNE wall is perpendicular to this (az. ~107°/287°), but the SSW wall is significantly skewed and runs closer to due E–W (az. 94°/274°).

This *heiau* faces the right-hand half of the highest portion of the Haleakalā Ridge, close to the summit of the prominent red cinder cone Kanahau (az. ~18.5)°. The horizon altitude is close to 12°, and the declination is between approximately +70.5° and +72.5°. This was close to the rising position of Pherkad (γ UMi, dec. +72.9°) in AD 1700. This star forms the opposite end of the modern constellation of Ursa Minor from Polaris, but it is not one of the brightest stars (its magnitude is 3.0) and is not known to have been significant to Hawaiians (see table 4.2). Similarly to AUW-9, this structure is oriented upon open sea to the SSW, an unremarkable part of the lava ridge ~1 km to the WNW, and a very close lava ridge to the ESE.

Additional remarks. This site is a classic fishing shrine, unmistakable with the upright Kūʻula stone and numerous branch coral offerings. The walls are unusually high and well built, perhaps reflecting its location within an elite residential complex. It is NNE-facing, and its orientation deviates by about 17° from the cardinal directions in a clockwise sense, but there is nothing to suggest this was motivated by any specific concerns regarding the visible landscape or the sky.

SITE WF-AUW-100

Site WF-AUW-100 was discovered during a cultural resources inventory survey for the Auwahi wind farm (Shapiro et al. 2011). Kirch plane-table mapped the site in August 2010, assisted by William Shapiro; on November 4, 2011, Kirch and Ruggles made a total station survey along with further observations.

Location and topographic setting. The *heiau* lies in a slight depression within a massive, jagged *ʻaʻā* lava flow, just a few meters *mauka* of an old four-wheel-drive ranch road running east–west across Auwahi. A fairly prominent *ʻaʻā* ridge rises immediately WNW of the structure.

Architecture. This is a notched *heiau* whose orientation deviates from cardinality by about 18° in a clockwise sense. It has six wall segments but is open to the SSW, with the notch in the ENE corner (fig. 6.2). The overall dimensions are roughly 11.5 m (WNW–ESE) by 8 m (NNE–SSW). The walls are of *ʻaʻā* cobbles and boulders with smaller rubble fill, built using the classic core-filled technique. The walls range in height from 0.6 to 1.5 m and are 1–1.5 m thick. There is a small niche in the ENE corner of the inner wall face. A unique feature of this site is a low platform faced with *ʻaʻā* cobbles and paved with smaller clinkers that abuts the main *heiau* enclosure on the SSE. One piece of branch coral was noted on the interior floor of the main court.

Archaeological context. There are no other features in the immediate vicinity of the *heiau*, but a large residential complex lies about 110 m to the southeast of the site.

Viewshed. The nearby lava ridge blocks the view to the W and WNW, while other lava ridges between 200 and 500 m distant restrict the view to the ENE and E. This serves to emphasize the distant Haleakalā Ridge profile from the NW round to the NE. From the ESE round to the WSW there is a clear view down to the sea.

Orientation. The fact that this *heiau* is open to the SSW implies that its orientation is to the NNE. Segments of intact facing on the inner and outer faces of the ESE wall yield northward orientations of 18.7° and 19.1°, respectively, north of the notch and 21.4° and 16.4°, respectively, south of the notch, which gives a mean azimuth of 18.9° in both cases. The WNW wall is convex: while the outer face of its southern part has a northward azimuth of 9.6°, its northern part is oriented similarly to the ESE wall. Thus, we can be confident that the intended direction of orientation was very close to 18.9°.

The long, straight outer face of the NNE wall is oriented with an eastward azimuth of 101.5°, but a straight segment of the inner face yields 105.1°, giving 103.3° as the best estimate of the intended orientation. Intact segments of the two faces of the central wall yield azimuths of 108.6° and 111.9°; the mean figure, 110.3°, matches the orientation of the longer, northern face of the SSW wall.

The Haleakalā Ridge rapidly clouded over after our arrival at the site, preventing points on

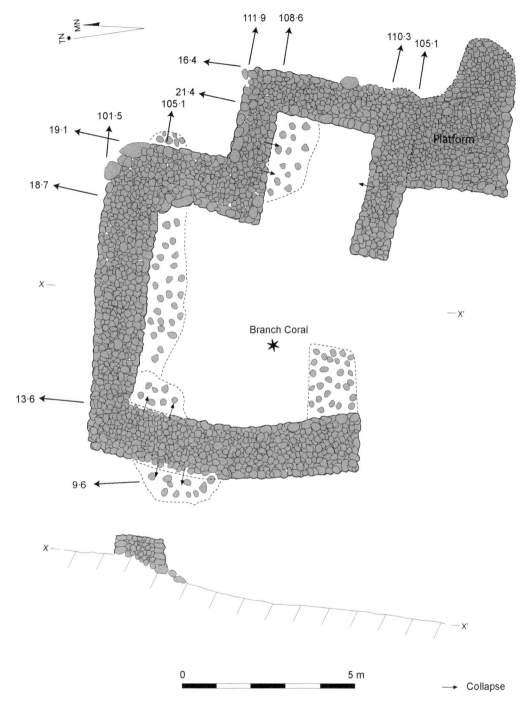

FIGURE 6.2. Plan of *heiau* WF-AUW-100, showing the azimuths of intact straight segments of wall facing as determined by the total station survey. (Based on plane-table survey by P. V. Kirch)

the profile from being surveyed, so we rely on the digitally generated horizon profile, which was "ground-truthed" by matching photographs of foreground features. The direction of orientation (az. 18.9°) aligns with the right-hand slopes of the prominent red cinder cone Kanahau extending from ~15.0° to ~19.5°, which itself forms the right-hand (easterly) extremity of the high central part of the ridge. It is reasonable to suppose that the *heiau* was intended to face this cinder cone.

SITE WF-AUW-176A

This *heiau* was discovered as a part of the cultural resources inventory survey for the Auwahi wind farm (Shapiro et al. 2011). It was plane-table mapped by Kirch in August 2010; Kirch and Ruggles made a total station survey of the structure on November 7, 2011.

Location and topographic setting. The site lies at an elevation of about 290 masl, about 75 m west of an intermittent stream channel. It is situated on the edge of a prominent ʻaʻā ridge that drops down precipitously along the western edge of the *heiau*.

Architecture. This is a small notched *heiau* with six wall segments but is open to the W; the notch is in the NE corner (fig. 6.3). The structure measures 10.5 m N–S by 7 m E–W, and its longer (N–S) axis deviates from cardinality by about 8° in a clockwise sense. The walls are well built of ʻaʻā cobbles and boulders, using the core-filled technique; walls are 1–2 m thick and generally about 60–80 cm high. The walls are distinctly nonorthogonal, so that instead of being oriented E–W, most of the relevant wall faces are in fact oriented WNW–ESE. In addition, the exterior SE corner forms a distinct point, or canoe-prow shape, an architectural feature seen in other southeastern Maui sites. The interior court is level and consists mostly of fine ʻaʻā rubble. There is no wall along the W side of the court, which instead is defined by the steep drop-off down into a natural gully.

The northern (technically, NNE) wall segments define a small room about 1.5 m square, open to the court; the floor of this room is slightly elevated above the main court, possibly demarcating an altar. In the ENE corner of this small room is a waterworn cobble of dense basalt, about 50 cm long; this likely is a representation of an ʻaumakua or other deity. No branch coral was noted on the site.

Archaeological context. WF-AUW-176A is part of a more extensive site complex, including a large residential terrace a few meters to the E (with a slab-lined hearth in its surface). Within a radius of about 300 m there are several other habitation clusters.

Viewshed. The entire Haleakalā ridgeline is visible from the NW round to the NE, with a wide view extending down the western slopes to the Kanaio coastline and Kahoʻolawe Island beyond. Cinder cones Puʻu Hōkūkano, 1.0 km distant, and Puʻu Pīmoe, 5.2 km away, are prominent on this skyline to the NW and W respectively. There is open sea from the ESE round to the WSW. The Lualaʻilua Hills are clearly visible to the NE, with the more distant eastern slopes of Haleakalā falling away behind them, but the view farther down these slopes is blocked by an ʻaʻā ridge little more than 100 m away. A lava outcrop boulder on this ridge, ~40 m from the site, is prominent on this skyline to the E.

Orientation. The small raised room (possible altar) on the north side indicates that this *heiau* was oriented toward the N. The canoe-prow corner aside, the straight outer and inner faces of the E wall are aligned with northward azimuths of 6.4° and 9.8°, respectively, giving a mean orientation of 8.1°; the eastern, outer face of the northward extension of this wall, beyond the notch, has exactly the same orientation. The shorter segments of inner wall on either side of the

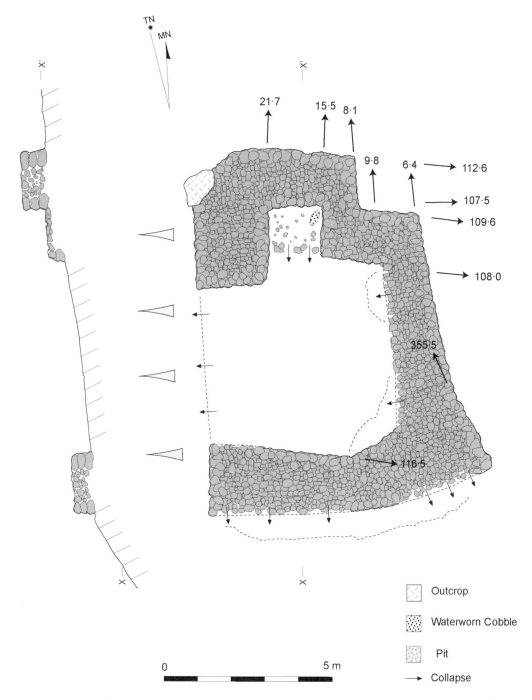

FIGURE 6.3. Plan of *heiau* WF-AUW-176A, showing the azimuths of intact straight segments of wall facing as determined by the total station survey. (Based on plane-table survey by P. V. Kirch)

small northern room have more divergent orientations. We conclude that 8.1° represents the intended, and principal, orientation. The inner and outer faces of the northern wall east of the notch yield eastward orientations of 108.0° and 109.6°, respectively, for a mean azimuth of 108.8°. West of the notch, the intact part of the outer face of the northern wall has an orientation of 112.6°, but the inner face orientations are less reliable; the short part bounding the small raised room yields 107.5°, suggesting that the intended mean azimuth was about 110.0°. The intact straight segment of the inner southern (technically, SSW) wall, west of the canoe prow, has an orientation of 116.5°; the western segment of the outer wall has collapsed but appears to have been straight as well and oriented some 7° to the left, yielding a mean orientation around 113°.

A DTM-generated profile for the half of the horizon from the WSW round to the ENE is shown in figure 6.4. This *heiau* faces a prominent cinder cone at the center of the highest portion of the Haleakalā Ridge (az. 9.8°, alt. 12.9°, dec. ~+77.8°). The orientation to the WNW falls between the two prominent cinder cones, but the estimated mean orientation of the southern (SSW) wall (az. 293.0°, alt. 3.5°, dec. +22.7°) coincides with the setting position of the Pleiades in AD 1600 (dec. +22.6° to +23.1°). To the S, the *heiau* is oriented upon open sea. The lava outcrop boulder 40 m to the E, although conspicuous from the site, is too close to show up on the DTM-generated profile. As determined from the survey, its azimuth as viewed from the NW corner is 95.8°–96.6°, and from the SW corner 85.2°–86.0°; thus, from the approximate center it is due E. The altitude as viewed from ~1.5 m above the current ground level in the center would be approximately +1.5° at the base and +2.5° at the top, but given the lack of any precise marker at the center and the proximity of the outcrop, we cannot be more accurate.

Additional remarks. This smallish *heiau* of classic notched form was part of a *kauhale*, or residential complex; it may have served as the *mua*, or men's house (including *heiau*), for the household group occupying that complex. In the northward direction, it faces directly toward the cinder cone at the center of the highest part of the Haleakalā Ridge, and westward toward the setting point of the Pleiades around the time of construction, assumed to be around AD 1600. A conspicuous natural lava pillar standing ~40 m away marks due E as seen from the center of the *heiau*.

Site WF-AUW-338

This site was discovered as part of the cultural resources inventory survey for the Auwahi wind farm (Shapiro et al. 2011). It was visited by Kirch and Ruggles on March 27, 2008, when a sketch map, a GPS survey, and compass-clinometer observations were made. A more accurate compass-and-tape survey was made by Kirch on February 12, 2016 (fig. 6.5). No detailed plane-table or total station survey has been carried out for this site.

Location and topographic setting. The site lies at an elevation of about 375 masl and is 150 m west of the boundary between Auwahi and Lualaʻilua. The structure was built into the *makai* (toward the ocean) face of a prominent ʻaʻā flow that forms a ridge to the east of the *heiau*. The *heiau* overlooks a broad swale about 100 m wide with good soil, which was doubtless a major garden site.

Architecture. This is a variant of the notched *heiau* form, with a stone platform attached to the main structure on the downslope side; the enclosing walls thus have 10 rather than the usual 6 wall segments (see fig. 6.5). The structure deviates from cardinality by up to 5° in a counterclockwise sense, the notch being on the NW corner. The overall dimensions are 16 m N–S by

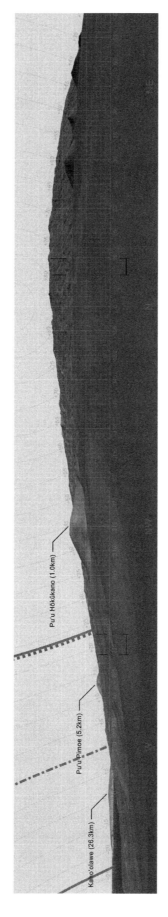

FIGURE 6.4. DTM-generated profile of the western and northern horizon as seen from *heiau* WF-AUW-176A. The solid shaded lines indicate the setting paths of the sun at the December solstice (*left*) and June solstice (*right*); the dash-and-dot line indicates the setting path of the sun at the equinoxes; and the dotted line indicates the setting path of the Pleiades in AD 1600. The horizontal bars indicate the direction of orientation of the *heiau* walls to the W and N. (Graphic by Andrew Smith)

FIGURE 6.5. Plan of *heiau* WF-AUW-338, showing the azimuths of intact straight segments of wall facing as determined by prismatic compass corrected for mean magnetic declination. (Plan based on compass-and-tape survey by P. V. Kirch)

12 m E–W. The walls are constructed of 'a'ā cobbles and boulders using the core-filled technique. The *heiau* appears to be oriented to the N (upslope), with the stone platform (elevated 1 m on the downslope side) forming a lower court; the main (upper) court is enclosed by the notched *heiau* walls. The lower platform is set off from the main court by a step about 30 cm high. Scattered around this division between the two courts are pieces of branch coral and some waterworn 'ili'ili pebbles.

Dating and chronology. Kirch and Ruggles collected two pieces of branch coral, one of which was ^{230}Th dated. This was a branch tip set between some paving stones in the vicinity of the 'ili'ili pebbles at the entrance to the upper court. The coral yielded a date of AD 1328 ± 3. This age is anomalously early in comparison with dates from other inland *heiau* in Kahikinui; we doubt that it corresponds to the actual date of *heiau* construction or use. It is possible that the coral was taken from an older temple elsewhere and placed here as a dedicatory offering.

Archaeological context. WF-AUW-338 is part of a relatively dense settlement pattern of residential and agricultural sites in the Auwahi uplands. As noted, it overlooks a swale that was probably intensively farmed for sweet potatoes or other crops.

Viewshed. There are unimpeded views of the Haleakalā ridgeline stretching away from the WNW (~3.5 km distant) round to the NE (more than 10 km distant) and of the open sea from the ESE round to the WSW. Pu'u Hōkūkano, 1.1 km distant, is prominent on the skyline to the W, while the Luala'ilua Hills, between 1 and 2 km away, are clearly visible to the NE, with the more distant eastern slopes of Haleakalā falling away behind them. To the ENE and E, a local ridgeline 1–2 km distant stretches down to meet the sea horizon.

Orientation. This *heiau* appears to be oriented to the N, as noted. Measurements obtained independently by Kirch and Ruggles, using different prismatic compasses, on three E–W-oriented segments of the enclosing wall (the southernmost being collapsed) yielded true azimuths between 81° and 90°, while measurements on the three wall segments enclosing the main structure (i.e., excluding the sides of the platform) yielded 355° consistently (see fig. 6.5). In the digitally generated profile (see fig. 4.11), it can be seen that around AD 1600 the Pleiades (and the June solstice sun) rose in a shallow dip at the foot of the right-hand slope of the southernmost peak of the Luala'ilua Hills.

On the western horizon, the flat peak of Pu'u Hōkūkano extends roughly over an azimuth range of 278.5° to 280.5°, with an altitude of 4°, corresponding to a declination range of about +9.5° to +11.0°. Several bright stars—Altair (α Aql, dec. +8°), Bellatrix (γ Ori, dec. +6°), Betelgeuse (α Ori, dec. +7°), and Procyon (α CMi, dec. +6°)—set into its left slope in the centuries around AD 1600; but as already noted (see "Site AUW-9"), "Hōkūkano" is not a known star name, and in any case there is no reason to suppose that the hill was named for its alignment as seen from this or any other *heiau*. On the other hand, the progress of the setting sun over the seasons would have been evident against Pu'u Hōkūkano during the summer half of the year and against Kaho'olawe Island to its left during the winter half.

Additional remarks. This relatively small notched *heiau* probably functioned as a *heiau ho'oulu 'ai*, or agricultural temple, given its proximity to a large garden swale. We regard the early ^{230}Th date on branch coral as anomalous; it is most likely that the site was constructed in the sixteenth or seventeenth century, as were most other *heiau* of this type in the Kahikinui uplands. Given both of these considerations, it is noteworthy that around AD 1600 the Pleiades would have been seen from the *heiau* to rise up from the middle of a shallow dip at the foot of the conspicuous Luala'ilua Hills. This is unlikely to have gone unnoticed and may possibly have been a factor in deciding the exact location of the *heiau*, although the eastward orientation of the walls is not in this direction but virtually due E, and the *heiau* itself seems to be facing N.

SITE WF-AUW-343

Site WF-AUW-343 was discovered during the cultural resources inventory survey of the Auwahi uplands (Shapiro et al. 2011). On March 27, 2008, it was briefly visited by Kirch and

Ruggles, who made a sketch map and a GPS survey and took compass-clinometer readings. It has not been plane-table mapped or surveyed by total station.

Location and topographic setting. The site is located at an elevation of about 370 masl, some 115 m W of WF-AUW-338. It lies on a rough, young ʻaʻā lava substrate.

Architecture. This is an aberrant form of the notched *heiau* type, having only five wall segments and being open on both the S and W sides. It is cardinally oriented, with the notch in the NE corner. The main wall, 5.2 m long, runs roughly N–S and is about 1 m thick, with core-filled construction; the open court extends to the W of this wall. Near the S end of the E wall, set into the inner face, is a large upright slab of *pāhoehoe* lava. On the N, three wall segments each about 4 m long define a smaller room or chamber, open to the court. Two pieces of *Porites* coral were noted on the court, in front of the E wall.

Archaeological context. As with site WF-AUW-338, this structure is part of a dense settlement zone, including numerous residential and agricultural features.

Viewshed. As at WF-AUW-338, there are unimpeded views of the Haleakalā ridgeline from the WNW round to the NE and of the open sea from the ESE round to the WSW. Puʻu Hōkūkano, 1.0 km distant, is prominent on the skyline to the W, while the Lualaʻilua Hills are clearly visible to the NE and ENE, with the more distant eastern slopes of Haleakalā falling away behind them. To the ENE and E a local ridgeline, 1 to 2 km distant, stretches down to meet the sea horizon.

Orientation. This *heiau* could have been oriented either to the N or to the E. Measurements obtained independently by Kirch and Ruggles, using different prismatic compasses, on both faces of the long E wall give a mean orientation of 359° (true) at the S end and 0.5° at the N end; the shorter walls on either side of the small chamber have outer faces that converge toward the N, but they also have a mean orientation of 359°. Various measurements on the two E–W-oriented wall segments yielded true azimuths varying from 87° to 93°, indicating that the mean orientation was very close to 90°.

The overall appearance of the horizon is similar to that seen from nearby WF-AUW-338. As at that site, the Pleiades (and the June solstice sun) rose in a shallow dip at the foot of the right-hand slope of the southernmost peak of the Lualaʻilua Hills (see fig. 4.11). Despite a slight shift in the position of the closer Lualaʻilua Hills against the more distant horizon, the rising position of the Pleiades with respect to the horizon features is negligibly different from that at WF-AUW-338.

From this site, the flat peak of Puʻu Hōkūkano extends roughly over an azimuth range of 280° to 283°, with an altitude of ~4.5°, corresponding to a declination range of about +11.0° to 13.5°. As at WF-AUW-338, Altair (α Aql, dec. +8°), Bellatrix (γ Ori, dec. +6°), Betelgeuse (α Ori, dec. +7°), and Procyon (α CMi, dec. +6°) would have set into its left slope around AD 1600. In addition, Regulus (α Leo, dec. +14°) would have set just to the right of the flat summit.

Additional remarks. This small temple with its variant notched form is likely to have functioned as a *heiau hoʻoulu ʻai* (agricultural temple), given its location in the main upland residential and agricultural zone of Auwahi. Despite its simple structure the *heiau* is cardinally oriented with impressive accuracy in both axes. As at nearby WF-AUW-338, in the sixteenth and seventeenth centuries the Pleiades would have been seen to rise up from the bottom of a distinct dip on the horizon.

SITE WF-AUW-359

This site was first noted by Kirch in May 2006 during a reconnaissance trip to Auwahi and was recorded during the wind farm cultural resources inventory survey (Shapiro et al. 2011). It was visited again by Kirch and Ruggles on March 27, 2008, when GPS and compass-clinometer observations were made, and branch coral samples collected.

Location and topographic setting. The *heiau* is at an elevation of about 410 masl, situated immediately northeast of Puʻu Hōkūkano on an *ʻaʻā* lava flow passing directly east of the cinder cone. A prominent intermittent stream channel lies just to the west of the *heiau*.

Architecture. The site is in poor condition, the western portion having been obliterated by a four-wheel-drive roadway that skirts the structure; the interior court and the S wall have also been heavily disturbed by bulldozing, leaving only portions of the N and E walls intact, along with the base *niho* stones of the S wall. Enough remains of the structure, however, to tell that it was of the notched form, broadly cardinally oriented, with original dimensions approximating 16 m E–W by 9.5 m N–S. The extant walls are well constructed of *ʻaʻā* cobbles and boulders, using the core-filled technique. The notch was in the NE corner. Several pieces of branch coral were noted on the court and in the N wall.

Dating and chronology. Two pieces of branch coral, both with intact branch tips, were ^{230}Th dated. The first coral sample (CS-2) was wedged between the pavement stones of the court about 3 m W of the inner E wall face. This yielded a date of AD 1712 ± 3. The second sample (CS-3) was collected from between fill stones of the N wall, where it was presumably placed during wall construction. It yielded a date of AD 1588 ± 4. These dates suggest construction of the temple during the late sixteenth century (during the reign of either Piʻilani or his son Kiha-a-Piʻilani) and continued use of the site into the eighteenth century.

Archaeological context. This *heiau* is part of an extensive complex of residential features that covers the *ʻaʻā* ridge bounding the western side of the large swale on the upland side of Puʻu Hōkūkano. This deep and broad swale, immediately W of the *heiau*, was a major zone of cultivation. The *heiau* likely functioned as an agricultural temple, or *heiau hoʻoulu ʻai*.

Viewshed. There are unimpeded views of the Haleakalā ridgeline and its upper western and eastern slopes, stretching from the W round to the NE. The Lualaʻilua Hills, 2 km away, dominate the horizon between the NE and the E, while a low lava ridge 250 m distant forms the horizon to the ESE. From here round to the SW is open sea. Puʻu Hōkūkano, between 200 and 300 m distant, is prominent on the skyline to the WSW, and the tip of Puʻu Pīmoe, 4.5 km away, can be seen in the distance at an azimuth of around 273°.

Orientation. Given its prominent E wall, this *heiau* was probably oriented to the E. Measurements obtained independently by Kirch and Ruggles, using different prismatic compasses, on surviving wall segments yielded easterly orientation estimates between 87° and 97° and northerly orientation estimates between 354° and 4°. Thus, this *heiau* was cardinally oriented, at least to within 2°–3°. There are no distinctive horizon features in the direction of orientation, but to the W the *heiau* is aligned almost directly upon Puʻu Pīmoe.

As seen from this *heiau*, the Pleiades around AD 1700 would have risen near the base of a shallow dip between the central and southernmost peaks of the Lualaʻilua Hills (see fig. 4.11). The similarity of this phenomenon to that occurring at WF-AUW-338 and WF-AUW-343 provides credible evidence that the locations were selected deliberately so that the Pleiades, whose acronychal rise was important for agricultural and calendrical purposes (see chapter 2), would rise from a conspicuous and easily locatable point on the horizon.

SITE WF-AUW-391

This relatively small and unimpressive structure was discovered during the cultural resources inventory survey for the Auwahi wind farm (Shapiro et al. 2011). Kirch made a plane-table map of the site in August 2010. Kirch and Ruggles made a brief total station survey of the structure on November 4, 2011.

Location and topographic setting. The site lies on the eastern crest of a prominent ʻaʻā ridge, at an elevation of 425 masl, overlooking a large swale that doubtless was used for gardening.

Architecture. The structure consists of a relatively well-built enclosure, nearly square and broadly cardinally oriented, measuring about 8 by 7.5 m on the exterior; the level space defined by the walls measures 4 by 4 m square (fig. 6.6). The E and S walls are the most prominent and are of core-filled construction, roughly 2 m thick. The outer face of the E wall is up to 1.5 m high, buttressed on the downslope side by a rough terrace. There is an angular bend in the exterior face of the E wall that seems to be intentional, giving the appearance of a notch, but it would be an overinterpretation to call this a true notched *heiau*. On the W and N sides the structure is defined mostly by rubble and blends into the rocky ridge. No branch coral was observed on the feature.

Archaeological context. This site lies within the densely settled upland residential and agricultural zone of Auwahi. A number of residential and agricultural features lie within a 200 m radius of the site.

Viewshed. Nestled as it is into the side of the ridge, the only open view is to the E, overlooking the large agricultural swale downslope. The Lualaʻilua Hills, around 2 km away, are clearly visible beyond, their three peaks forming a conspicuous part of the skyline between the NE and the E (fig. 6.7). Higher ground rises immediately to the W of the *heiau*, while, to the N, rising ground partially obscures the view of Haleakalā.

Orientation. The prominent E wall with its exterior buttressing terraces suggests that the *heiau* was oriented to the E. Likewise, the viewshed suggests that any wider focus of interest in the landscape or sky was to the E. Intact segments of the outer and inner faces of the massive E wall are oriented with northward azimuths of 352.1° and 358.1°, respectively, yielding a mean wall orientation of 355.1°. Estimates from the plane-table survey give the orientations of the inner and outer faces of the S wall as 82° and 90°, respectively, yielding a mean of 86°. Thus, the structure was laid out along approximately cardinal directions.

As is the case for each of the previous three sites discussed, the Pleiades would have been seen from this *heiau* to rise in the base of a dip, in this case (as at WF-AUW-359) the dip between the central and southernmost peaks of the Lualaʻilua Hills. The rising path of the Pleiades for AD 1600 is shown in figure 4.11; by AD 1700 it would have shifted slightly to the left (to overlap more with the path of the June solstitial sun).

Additional remarks. The interpretation of this site as a *heiau* rests largely on its topographic location with an open view to the E, its fairly massive E wall of typical *heiau* construction, and its orientation to cardinal directions. No branch coral was found on the structure, but neither was there any shell midden or other evidence of habitation. In our estimation it was

FIGURE 6.6. Plan of *heiau* WF-AUW-391, showing the azimuths of intact straight segments of wall facing as determined by the total station survey (quoted to the nearest 0.1°) and estimates of the orientation of the faces of the S wall obtained from the plane-table survey (italicized and quoted to the nearest degree). (Based on plane-table survey by P. V. Kirch)

FIGURE 6.7. View of *heiau* WF-AUW-391 from the west, showing the agricultural swale to the east and the Luala'ilua Hills beyond. (Photo by Clive Ruggles)

most likely a *heiau ho'oulu 'ai* associated with the large agricultural swale to the E. This likelihood is reinforced by the fact that, as viewed from this site, the Pleiades would have been seen to rise from the bottom of a conspicuous dip in the horizon, something that was also the case at each of three other agricultural *heiau* in the vicinity (WF-AUW-338, -343, and -359).

Site WF-AUW-403

Kirch first visited this site in May 2007 as part of the Auwahi wind farm cultural resources inventory survey. Kirch and Ruggles made a detailed compass-and-tape map and a total station survey on March 27, 2008.

Location and topographic setting. At an elevation of 405 masl this structure sits on the Kealakapu basanite flow, which has both *'a'ā* and *pāhoehoe* surfaces; both kinds of stone are incorporated into the *heiau* structure. The site is about 200 m WNW of the base of the Pu'u Hōkūkano cinder cone.

Architecture. Although it is among the smallest of the Kahikinui *heiau*, this can only be described as an "elegant" structure, evident care having been taken in the stonework and details. It is of classic notched form (fig. 6.8), broadly cardinally oriented, and walled on all sides except the S, which is defined by a terrace face about 60–80 cm high. The notch is in the NW corner. The structure's outer dimensions are 7.5 m N–S by 6.5 m E–W. The walls average about 1 m thick and are of the classic core-filled type with faces of *'a'ā* cobbles and interior rubble fill. However, *pāhoehoe* slabs were also incorporated along some of the inner wall faces, as shown in the plan, giving those faces a sharp, crisp angularity. This is especially the case with the small room or chamber on the northern side of the structure. Moreover, several of these *pāhoehoe* slabs were set upright to create a kind of cist or box within the wall separating the northern room from the main court. This cist may have once held sacred objects or ritual paraphernalia.

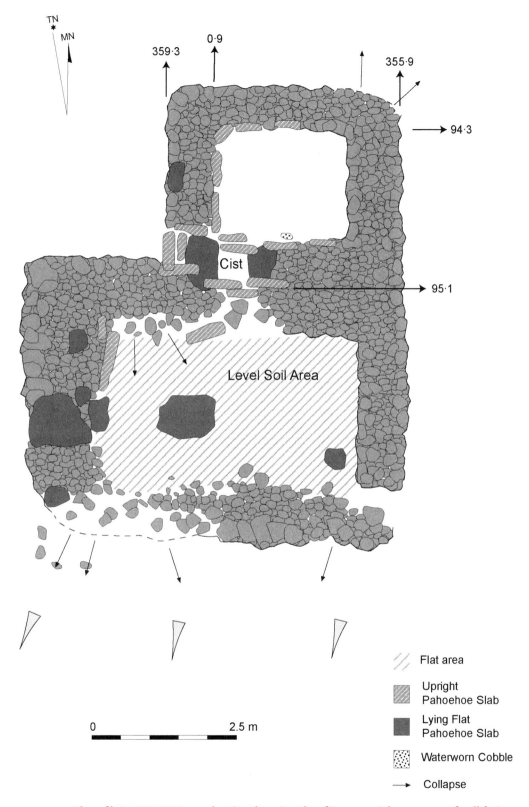

FIGURE 6.8. Plan of *heiau* WF-AUW-403, showing the azimuths of intact straight segments of wall facing as determined by the total station survey. (Plan based on compass-and-tape survey by P. V. Kirch)

The main court consists of the level terrace on the southern part of the site. An obvious entryway onto this terrace is demarcated on the W by a large *pāhoehoe* stepping-stone and several smaller slabs. Near the center of the court is another large *pāhoehoe* slab, nearly 1.5 m across, set firmly in the floor of the court. This can only have been a seat for the priest or officiate at this temple.

Archaeological context. This small *heiau* is part of the extensive upland Auwahi settlement and agricultural zone. It lies a few meters to the W of an intermittent drainage channel that fed water down into a large swale with rich soil, *makai* of the temple; this was doubtless a garden area.

Viewshed. From a standing position within the *heiau* there are unimpeded views in all directions. The peaks of Hawai'i Island, more than 100 km away, would be visible to the SE in good weather conditions, with open sea from the SSE round to the WSW, where most of Kaho'olawe Island is seen 25 km away. The summit of Pu'u Pīmoe, 4 km away to the W, is conspicuous behind the long ridgeline (2 km distant in that direction) that extends up to the Haleakalā rim to the N and NNE and beyond to the NE (20 km distant). The Luala'ilua Hills, around 3 km away, are prominent on the skyline to the ENE, while nearby Pu'u Hōkūkano dominates the horizon to the ESE. Thus, toward the E, the *heiau* offers a spectacular view of Pu'u Hōkūkano to the right, as well as the three peaks of the Luala'ilua Hills to the left. From a seated position on the large *pāhoehoe* floor slab, these hills frame the view with the E wall of the *heiau* rising above the horizon in between.

Orientation. The positioning of the floor slab within the main court is to the W of center but centrally placed in the N–S direction, with the entryway behind it to the W. This strongly suggests that the direction of interest, in which the *heiau* faced, was E. The most reliable indicators of the original orientation are the inner face of the N wall, which has an eastward orientation of 94.3°, and the line of slabs on the southern side of the inner chamber, which yields 95.1°. None of the three longer E–W-oriented wall segments—the southern face of the central wall and the two faces of the S wall—retains any intact segments that would yield a reliable estimate of the intended orientation. The longest straight wall face is the outer face of the E wall, which has a northward azimuth of 355.9°, while intact segments on the outer and inner faces of the wall on the western side of the northern chamber yield 359.3° and 0.9°, respectively, for a mean azimuth of 0.1°. No other segments of N–S oriented walls are measurable. From the available data, then, our best estimate of the mean axial orientation in the N–S direction is 358.0°—that is, 2.0° counterclockwise of cardinal—while the average of the two azimuths estimated from E–W-oriented wall segments, 94.7°, is 4.7° clockwise of cardinal. In other words, the two axes of the structure deviate from orthogonality by some 6°–7°.

The DTM-generated profile for the eastern quarter of the horizon from NE to SE is shown in figure 4.12. The mean easterly orientation is upon a featureless stretch of the lower slope of Pu'u Hōkūkano, although in the opposite direction the mean orientation (274.7°) is precisely upon the summit of Pu'u Pīmoe (see fig. 4.13); this, if deliberate, could possibly explain the displacement from due E–W orientation. The mean northerly orientation (358.0°) is upon the peak of a cinder cone well to the left of the highest part of the Haleakalā Ridge. This is one of several such cones on the ridge and is not necessarily of any significance.

For a priest or officiate sitting on the flat stone, the visible eastern horizon was framed between the northernmost (highest) peak of the Luala'ilua Hills and the broader expanse of Pu'u

FIGURE 6.9. Eastern horizon at site WF-AUW-403 as viewed from a sitting position on the floor slab. The solid shaded lines indicate the rising paths of the sun at the June solstice (*left*) and the December solstice (*right*); the dash-and-dot line indicates the rising path of the sun at the equinoxes; and the dotted line indicates the rising path of the Pleiades in AD 1600. (Composite of photo by Clive Ruggles and graphic by Andrew Smith)

Hōkūkano, providing a clear view of the entire rising arc of the sun throughout the year (fig. 6.9). During the winter half of the year, the sunrise would have progressed up the slope of Puʻu Hōkūkano until reaching just past the summit at the December solstice, before returning. The northern (left) half of the E *heiau* wall, which slopes upward to the left, rises above the dips in the distant horizon, providing a reference against which the motion of sunrise during the summer half of the year could have been tracked. The result is a pleasing visual symmetry about the low point where sunrise occurs at the equinoxes, with the skyline—part natural and part artificial—rising at a similar angle in both directions up to the relevant solstitial rising point. Whether by accident or design, the first gleam of the equinoctial sunrise is seen through a small gap in a wall stone directly E of the seated observer. Two taller uprights placed in the wall, bearing approximately 57° and 84° from the center of the floor slab, do not have any obvious significance in relation to the viewshed and sky.

Additional remarks. Although WF-AUW-403 broadly shares with the previous four temples in this catalog the characteristic that the Pleiades would have been seen to rise from a dip between hills (see fig. 4.11), it also faces an eastern horizon profile whose topographic features have considerable visual impact as well as spanning the solar arc. This suggests that WF-AUW-403 was a Kāne temple, connected particularly with observations of the sunrise, rather than a Lono temple connected with the calendar and the Pleiades. An additional factor in support of the interpretation of this *heiau* as a Kāne temple is its topographic location immediately west of an intermittent drainage channel that would have funneled runoff from *kona* storms into a large agricultural swale downslope from the temple. Kāne was the god of flowing waters (see chapter 2), and many of the eastern-oriented temples in Kahikinui are situated adjacent to such drainage channels.

SITE WF-AUW-493

This site was recorded during the cultural resources inventory survey for the Auwahi wind farm and identified as having a probable "agricultural" function (Shapiro et al. 2011). Kirch visited the site in August 2010 and realized that it almost certainly had a ritual function. The site was studied again on November 4, 2011, by Kirch and Ruggles, when a total station survey and other observations were made.

Location and topographic setting. Located at an elevation of 490 masl and a short distance *makai* of the Pi'ilani Highway, this feature lies on a relatively older basanite 'a'ā lava flow. The site is a short distance to the west of a major intermittent drainage channel that flows down into the large swale behind Pu'u Hōkūkano.

Architecture. The feature is a nearly square enclosure, broadly cardinally oriented, 11.6 m E–W by 11.4 m N–S on the exterior, making it too large for a habitation site. The walls, originally of core-filled construction, are badly collapsed in most places, 40–60 cm high, and 1–2 m wide. The E wall is the most substantial and well constructed, while the opposite W wall is crude, partly incorporating natural 'a'ā outcrops. The S wall might have originally had a secondary terrace face on the exterior, but it is so badly collapsed that this cannot be determined for certain. Also, the SW corner may originally have had the pointed, canoe-prow shape found in some Kahikinui *heiau*. The interior court consists of the gently sloping natural surface. One piece of water-rolled *Porites* coral was noted on the E wall.

Archaeological context. The site lies within the dense upland settlement of Auwahi; numerous residential and some agricultural features are located within a 200 m radius of the *heiau*.

Viewshed. The Haleakalā ridgeline is visible from the NW round to the NE. To the W and WNW, the view extends to a ridgeline about 2 km away, devoid of any outstanding features. The Luala'ilua Hills, at a distance of just over 2 km, are prominent to the ENE and E. The peak of Pu'u Hōkūkano, only 0.8 km distant, is conspicuous downhill, directly to the S, but well below the sea horizon. Pu'u Pīmoe, 4.4 km away, is visible at an azimuth of 266°, though not very conspicuous.

Orientation. The more substantial nature of the E wall suggests that E was the direction this structure faced. A short segment of intact outer facing at the eastern end of the S wall yielded an azimuth of 83.0°, while a compass estimate of the orientation of the western part of the same wall (corrected for magnetic declination) gave 91°, for a mean of 87°. The inner face of the E wall yielded a best-fit orientation of 358.6°, while the outer face of the W wall gave 2.6°, for a mean of 0.6°. Thus, the structure was laid out close to true N–S and within a few degrees of true E–W.

The structure is aligned upon the left slopes or peak of the southern summit of the Luala'ilua Hills; this peak is almost exactly due E of the site, and the equinoctial sun would have risen slightly to its right, while the June solstice sun (and also the Pleiades around AD 1600) rose just to the left of the northern summit (see fig. 4.11). To the S, the walls are aligned upon Pu'u Hōkūkano, but this is below the sea horizon, as already stated, and unlikely to have been of any significance. To the W, the temple is aligned almost directly upon Pu'u Pīmoe (see fig. 4.13). There are no notable horizon features in the direction of orientation of the structure walls to the N.

Site WF-AUW-574

This site was recorded as part of the cultural resources inventory for the Auwahi wind farm (Shapiro et al. 2011), then briefly visited by Kirch and Ruggles on March 27, 2008. We made a sketch map and took GPS and compass-clinometer observations but did not have time to conduct either plane-table or total station surveys.

Location and topographic setting. An important aspect of this structure's setting is that it lies no more than 30 m W of the eastern boundary of Auwahi with Lualaʻilua and may therefore have functioned as a boundary temple. It is at an elevation of about 380 masl, on an older ʻaʻā lava substrate.

Architecture. The structure consists of a U-shaped enclosure defined on three sides by stacked walls of ʻaʻā cobbles and rubble laid out on cardinal directions but open to the S (*makai*); it measures roughly 12 by 11.5 m. The E and W walls are smaller, each about 1 m thick, whereas the N wall is 4 m thick, thus forming a kind of platform. In the western end of this N wall/platform is a depression or pit, and next to it a low cairn, or *ahu*, of stacked rocks. On the eastern end of the N wall/platform is a waterworn basalt upright. One basalt core was noted on the platform. A search for branch coral was made, but none was observed. The court defined by the three walls has a level earthen floor; at the southern end of this court is a low stone mound.

Archaeological context. At a short distance to the north of the site is a large complex of features identified as being of agricultural function; other residential or agricultural features are also found within a 200 m radius of the site.

Viewshed. The Haleakalā Ridge is visible from the NNW round to the NE, as is open sea from the ESE round to the SW. The Lualaʻilua Hills are conspicuous to the NNE and NE: the two more northerly peaks are just below the Haleakalā skyline, while the southernmost peak forms a prominent skyline feature to the NE. To the ENE and E, the horizon is formed by a featureless ridge about 1 km distant, while from the WSW round to the NW a closer ridge, 50–100 m away, blocks the more distant view.

Orientation. This structure, open to the S, clearly has a northerly orientation. Measurements obtained independently by Kirch and Ruggles, using different prismatic compasses, yielded a mean azimuth of 356° (true) for the W wall and 2° for the E wall, giving a mean axial orientation of 359°. Only the western part of the outer face of the N wall was sufficiently well preserved to yield a rough estimate of the easterly orientation: the azimuth obtained was 85°.

The *heiau* faces the tip of a cinder cone forming a small but prominent feature at the left end of the highest part of the Haleakalā skyline. From this location, Hōkūpaʻa (Polaris) would have hung in the sky almost directly above this feature, close to the celestial pole (the center of the declination circles representing stellar rising and setting arcs in the DTM-generated north profile [see supplementary online materials]).

To the E the June solstice sun would have risen just to the right of the southeasternmost Lualaʻilua Hills summit (see fig. 4.11). The horizons to the W and S, in the directions upon which the structure walls are aligned, are all featureless.

Puʻu Hōkūkano

Although the prominent cinder cone Puʻu Hōkūkano is a natural geographic feature and not a constructed site, we include it here because scattered over the summit of the *puʻu* are pieces of branch coral. The presence of these offerings suggests that ritual activity of some kind took place on top of the cinder cone. One coral branch, with slightly rounded branch tips, was collected and ^{230}Th dated, yielding a date of AD 1557 ± 3 years. While the name of the hill is suggestive of a stellar association, "Hōkūkano" (literally, proud star) (Pukui and Elbert 1986), is not a star name known from the ethnohistorical sources (Johnson, Mahelona, and Ruggles 2015).

The *Ahupua'a* of Luala'ilua

Luala'ilua is the first land section to the east of Auwahi, sharing its eastern boundary with Alena. It is dominated by the cluster of prominent *pu'u* (cinder cones) collectively known as the Luala'ilua Hills. Whether Luala'ilua had the status of an *ahupua'a* is not completely certain; however, the name appears on both the 1838 Lahainaluna map (the "Kalama map"; see Kirch 2014, 269) and the 1885 Hawaiian Government Survey map of Maui, suggesting that it was a prominent land section. Aside from the immediate coastal strip, most of Luala'ilua has not been intensively or systematically surveyed for archaeological sites.

SITE LUA-1, KOHŌLUAPAPA (WALKER SITE 186; STATE SITE 50-50-15-1386)

This important *heiau* was reported by Walker, who drew a schematic map (Walker [1930], 256, fig. 59). He refers to it as "Heiau at Koholuapapa," implying that Kohōluapapa ("The Hōlua Slide Flat") is the name of the locality rather than of the *heiau* proper; it is likely that there was once a *hōlua* sledding course down the nearby face of the Luala'ilua cinder cone. Walker was impressed with the "massive" walls; he notes the "central court roughly paved in which the principal ceremonies were probably carried on." He refers to an adjoining "series of walls forming irregular enclosures extending to a large dwelling site on the point," which he speculates may have been the "house of the *kahu*."

In 1996 Michael Kolb and his field school students excavated 13.5 m² at site LUA-1 (Kolb and Radewagen 1997, 66–67, fig. 5.4). No report of this excavation is available, although three radiocarbon dates have been published (Kolb 2006). Ruggles visited LUA-1 in April 2002, undertaking compass-clinometer observations. Kirch and Ruggles investigated LUA-1 together on March 24, 2003, making a plane-table and alidade map as well as a total station survey and a GPS survey.

Location and topographic setting. The site occupies a relatively flat area of *pāhoehoe* and *'a'ā* lava, at 610 masl and about 150 m W of the base of the large, northern cinder cone of the Luala'ilua Hills. It sits about 350 m E of the current boundary and fence line between Auwahi and the Hawaiian Home Lands holdings in Kahikinui. When one is standing at the site, the cinder cone looms up directly E; the *heiau* clearly seems to have been oriented to face the *pu'u* (discussed under "Viewshed"). Between the *heiau* and the base of the *pu'u* is a deeply incised intermittent stream channel that fed water from *kona* storms into numerous swales doubtless used for farming, farther *makai*.

Architecture. This is among the more complex of the Kahikinui *heiau* from an architectural viewpoint, not the least because it has had some secondary modifications, probably in the postcontact era. The ground plan (fig. 6.10) does not readily follow any of the common Kahikinui *heiau* configurations, although the central part of the structure resembles the elongated double-court type. This central sector deviates from cardinality by about 7° in a clockwise sense, measuring roughly 24 m E–W by about 12 m N–S. The N wall defining this central section of the *heiau* is in places more than 2 m thick and over 1 m high, built using the core-filled technique. This N wall has a notch, or right-angle jog, about midway along its length. The W and S walls are less well defined. At the eastern end, this central section terminates in a stone face that abuts an elevated natural platform of *'a'ā* boulders and outcrop (what Kolb and Radewagen refer to as a "faced bedrock altar"). Sitting on this natural *'a'ā* platform or altar, about 6 m E of the eastern edge of the court, is a large stone upright, presumably the representation of a deity.

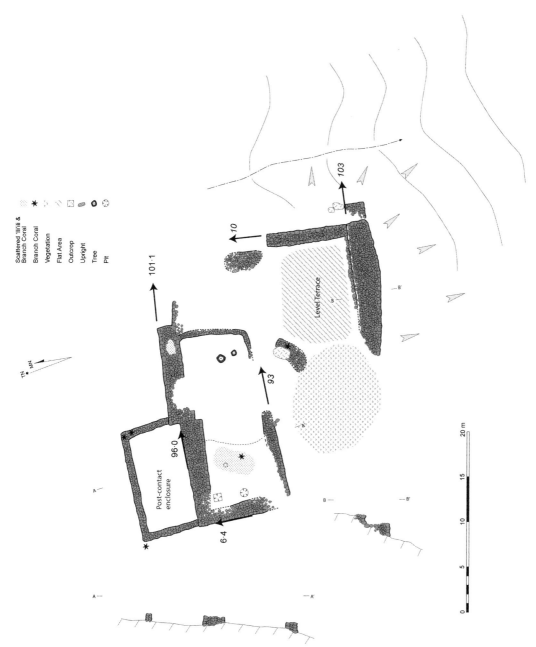

FIGURE 6.10. Plan of *heiau* LUA-1, showing the azimuths of intact straight segments of wall facing as determined by total station (quoted to the nearest 0.1°) or prismatic compass corrected for mean magnetic declination (italicized and quoted to the nearest degree). (Based on plane-table survey by P. V. Kirch)

The court defined by these walls is partly obscured by a thick mat of grass and two large mango trees, although the western part of the court is open. There the court surface was covered in considerable quantities of *'ili'ili* pebbles; several pieces of branch coral were also noted. In the NW corner of the court there is a square pit about 1 m wide and 40 cm deep; a second, circular pit lies in front of the middle of the western wall. These may both be pits for the disposal of sacrificial offerings, situated as they are at the western end of the temple.

Abutting this massive central section of the *heiau* on its N side is a rectangular enclosure, 13 by 7 m, defined by stacked cobble walls about 1 m high. This is without doubt a later construction, for the walls abut to the massive core-filled wall of the main *heiau*. Given the stacked technique, it is likely that this enclosure was a postcontact addition; it may be contemporaneous with the set of irregular enclosure walls S of the *heiau* site (visible in figure 5.3 of Kolb and Radewagen [1997]).

To the S of the central sector is a broad, roughly level terrace extending about 12 m and terminating in a double terrace facing, with the outer face up to 1 m high. On its eastern side the level court defined by these terraces is demarcated by a freestanding, core-filled wall 1 m wide and 9 m long.

Excavation. Michael Kolb excavated several test units in this *heiau* in 1996, although no detailed report is available. Kolb and Radewagen (1997, 66–67) give the following summary of the main features exposed in their excavations:

> Three large features were located: two burn episodes (19, 29) and one faced bedrock altar. Material culture recovered includes fish bone, volcanic glass, pig bone, bird bone, land snails, basalt flakes, shellfish, and many waterworn (*'ili'ili*) stones. An area of intensive burning (19) is located behind the faced bedrock altar. This may have been the location of the oven house of the *heiau*. The second burn episode (29) is a small area that may signify an earlier use of the *heiau* area. The large quantity of pig bone recovered from the site, and the architectural similarities to two other Maui *luakini heiau* previously excavated by Kolb (1991), suggest that Koholuapapa was eventually converted to a *luakini* war temple.

As we will argue, this interpretation of LUA-1 as a *luakini heiau* is questionable (see "Additional Remarks").

Dating and chronology. Kolb (2006, table 1) reports three radiocarbon dates from charcoal samples obtained during his excavations at LUA-1. Unfortunately, he does not indicate whether any of these were identified as being of short-lived taxa; he states that only 31% of his samples were botanically identified prior to radiocarbon dating. The oldest sample (Beta-95909), from a context noted as "basal," has a date of 650 ± 70 BP, yielding a calibrated 2σ age range of AD 1252–1424. A second sample (Beta-95908), also noted to be "basal," has a date of 300 ± 80 BP, yielding 2σ calibrated age ranges of AD 1436–1690 (81.7%), 1729–1810 (10.3%), and 1925–1950 (3.4%). Finally, a third sample (Beta-95910) from a "firepit" had an age of 50 ± 70 BP, yielding 2σ calibrated age ranges of AD 1675–1778 (32.4%) and 1799–1942 (63.0%).

Of the two samples noted to be in "basal" contexts, we reject the oldest as being a valid date for *heiau* construction; most likely it represents either an early phase of pre-*heiau* land use and burning, or indeed it may be from a piece of older dryland hardwood with an inbuilt age. The second basal sample, with the highest-probability calibrated age of AD 1436–1690, is quite consistent with most other Kahikinui *heiau* dates, both radiocarbon and ^{230}Th, and is therefore accepted as a valid age for the initial construction of the Kohōluapapa *heiau*. The third sample, from the fire pit, indicates continued use of the *heiau* into the later precontact or even early postcontact era.

Archaeological context. No systematic archaeological survey has been conducted in the vicinity of LUA-1, so the larger settlement pattern context remains unknown. A cluster of features, including two irregular stacked stone enclosures, lies a short distance to the south of LUA-1; these are likely to be postcontact residential sites. In a reconnaissance survey, Kirch observed several extensive agricultural swales and clusters of residential features to the west of LUA-1 in adjacent Auwahi. The only Mahele-era land claim (*kuleana*) in Kahikinui, that of Makaole (Land Commission Award 5404), is situated about 0.5 km south of the *heiau*.

Viewshed. The nearby Luala'ilua cinder cone dominates the horizon to the ENE, E, and ESE (see fig. 4.13). There are open views in all other directions. Where the southern slope of the cinder cone drops to the sea to the SE, around azimuth 130°, Mauna Kea would have been visible in the distance on clear days, its peak being at around 136°. Likewise, the shallow summit of Kaho'olawe Island would have been visible behind the foot of the slopes down from Haleakalā to the WSW, around azimuth 258°.

Orientation. The elongated design of the central structure with its "bedrock altar" at the eastern end clearly indicates that this *heiau* was oriented to the E. Total station readings show the western and eastern segments of the outer (N) face of the N wall of the central structure to be oriented with azimuths of 96.0° and 101.1°, respectively, while compass measurements of the intact outer face of the surviving western segment of the S wall yield 93° (see fig. 6.10). The mean of these three values is 96.7°, but given the lower accuracy of the last measurement, a best estimate of the intended orientation might be between about 96.4° and 97.0°. This accords with the azimuth of the intact segment of outer face of the W wall, 6.4°—that is, at a right angle to 96.4°.

The *heiau* is clearly oriented toward the highest point of the Luala'ilua Hills (fig. 6.11). The sun rises close to this rounded summit at the equinoxes, its rising position moving to the foot of the northern slope at the June solstice and halfway down the southern slope at the December solstice (see fig. 4.13). In other words, the sun's rising position moves to and fro along the hills over the seasons.

To the W, the direction of orientation is toward the lower slopes extending down from Haleakalā, missing the position of equinoctial sunset by a few degrees to the right; at the December solstice the sun sets just above the southern point of Kaho'olawe. Pu'u Hōkūkano is prominent farther to the left, at azimuth 223°, but is well below the skyline. The W wall is oriented to the N upon the prominent cinder cone at the center of the highest portion of the Haleakalā Ridge, which from here has an azimuth of 7°, an altitude of ~14.5°, and a declination +81°; in the opposite direction, to the S, is open sea.

Additional remarks. LUA-1 is clearly an important temple; with a basal area of 512 m^2 (excluding the later attached enclosure), it is in the same size range as several other *heiau* with elongated double-court architecture, ALE-1, ALE-4, and KIP-1 among them. However, we see no evidence to support Kolb and Radewagen's claim that LUA-1 was a *luakini* temple. *Luakini* were the primary war temples used exclusively by the *ali'i nui* (king). Walker's informants gave no indication that Kohōluapapa was a *luakini*, nor does it figure anywhere in the traditional *mo'olelo* (oral histories) such as that of Kamakau (1961). Indeed, there is not a single reference in any of the

FIGURE 6.11. View eastward along the N wall of LUA-1, showing the Luala'ilua Hills beyond. (Photo by Clive Ruggles)

ethnohistorical accounts of a Maui *ali'i nui* ever residing or conducting ritual activities in Kahikinui. Kolb and Radewagen's interpretation appears to rest primarily on the presence of pig bones in their excavations, but pigs were regularly offered as sacrifices at many different kinds of *heiau*, and we have recovered pig bones from other *heiau* sites in the district.

We believe it more likely that LUA-1 served as the principal upland agricultural temple (*heiau ho'oulu 'ai*) for the western part of Kahikinui (i.e., the area to the west of the Luala'ilua Hills). The temple's easterly orientation, with the towering Luala'ilua Hills framing the annual sunrise arc, as well as its proximity to a major intermittent watercourse, both suggest that it was dedicated to Kāne. A significant zone of swales with rich agricultural soils, as well as a dense settlement pattern of residential sites, lies just to the west of LUA-1 in Auwahi. The occupants of this fertile upland zone may well have constituted the community of officiants at this temple. Based on the available radiocarbon dates, LUA-1 was constructed between the end of the fifteenth century and the mid-seventeenth century.

SITE LUA-3, KALUAKAKALIOA (WALKER SITE 183)

Walker ([1930], 253, fig. 56) gave the proper name "Kaluakakalioa" to this "good sized walled heiau" and drew one of his better site plans of it, with two cross sections. Kolb visited the site in 1996 and conducted limited excavations over a four-day period (Kolb and Radewagen 1997, 65, fig. 5.4). Kirch made a detailed plane-table map of the *heiau* in March 1997 and a GPS survey in January 2005. Ruggles undertook a compass-clinometer survey of LUA-3 on April 13, 2002, and Kirch and Ruggles made a total station survey on March 23, 2003. In January 2008, Kirch and Sharp collected branch coral specimens from the site for ^{230}Th dating.

Location and topographic setting. LUA-3 sits on the 100 m contour line, a few meters east of a jeep road descending to Hanamauloa and Niniali'i. Geologically, the *heiau* lies at the intersection of an older basanite lava flow on the west, a younger ankaramite flow with *pāhoehoe* surface on the east, and immediately adjacent to a small pocket of alluvial sand and gravel.

Architecture. This *heiau* exemplifies the square enclosure form, deviating from cardinality by about 10° in a counterclockwise sense. It is partly open to the S (figs. 6.12 and 6.13), measuring 15 m E–W by 14.5 m N–S. The massive, well-constructed walls use the core-filled technique but are faced with large angular blocks of *pāhoehoe* lava, giving them an especially sharp and crisp appearance. The main enclosure walls range from 1.5 to 2.5 m thick and are up to 2 m high in places. With its entryway on the S, the *heiau* is oriented to the N.

As one approaches the structure from the S, there is a low stone-faced terrace, 4 m wide and 9 m long, abutting the main enclosure. This terrace is filled with *'a'ā* rubble and covered with abundant *'ili'ili* gravel and scattered branch coral pieces. Several flat *pāhoehoe* slabs on the eastern end of the terrace define a formal entryway into the *heiau*.

Entering the main enclosure, one stands on the lowest court of level soil. To the E of this is a raised stone platform, with a lower sunken room in the SE corner. Along the northern edge of the main court is a separate, raised stone terrace elevated 20–30 cm above the court; this appears to be an altar or place for offerings.

Walker refers to "image holes" in the tops of the walls, and his plan shows seven of these, in the E, N, and W walls. Kirch observed one clear pit 50 cm deep in the SE corner of the main wall and another, less well-defined pit in the NE corner. The other holes noted by Walker were not seen, but because Walker's map was not available at the time we made our survey, we did not explicitly search for them.

Excavation. Kolb excavated 4 m^2, finding a "burn episode" and material remains including shellfish, coral, bone (type not specified), an adze fragment, and a drilled coral piece (Kolb and Radewagen 1997, 65).

Dating and chronology. A charcoal sample from one of Kolb's test pits (Beta-95904) yielded a "modern" age. Kirch and Sharp collected three branch coral specimens from the *heiau* and dated one of them (CS-2), a branch found on the floor of the small room or chamber in the SE corner, against the interior wall base. This gave an age of AD 1590 ± 2, which falls at the beginning of the reign of Maui *ali'i nui* Kiha-a-Pi'ilani, according to the genealogical chronology.

Archaeological context. The area surrounding site LUA-3 has not been intensively surveyed, so nothing is known of its larger settlement context. *Heiau* site ALE-1 lies about 245 m NE of LUA-3.

Viewshed. There are clear views of the Haleakalā Ridge from the NW round to the NE and of open sea from the E round to the SW. To the ENE the view is restricted by the Alena lava flow, some 200 m distant and lacking any distinctive features. The view to the WSW and W is blocked by a lava ridge about 100 m distant, and to the WNW by another ridge between 500 m and 1 km away.

Orientation. As already stated, this *heiau* faces N. The inner faces of the walls are generally better preserved and less irregular than the outer faces and thus give the most reliable estimates of the intended orientation. In the northward direction, the inner face of the W wall, which is entirely straight, has an azimuth of 347.9°. The inner face of the E wall is convex, but the inner corners at each end of it are well preserved, and the corner-to-corner azimuth is 352.2°. Thus, the walls diverge toward the N, the mean azimuth being 350.0°. The western edge of the stone platform and the step into the sunken room form an interior division with an azimuth of

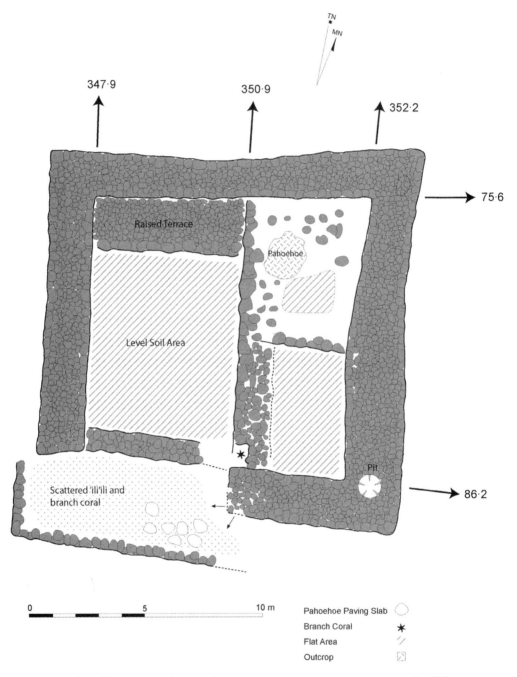

FIGURE 6.12. Plan of *heiau* LUA-3, showing the azimuths of intact straight segments of wall facing as determined by the total station survey. (Based on plane-table survey by P. V. Kirch)

FIGURE 6.13. View of *heiau* LUA-3 from the south, with the Lualaʻilua Hills and the Haleakalā rim in the distance. (Photo by P. V. Kirch)

350.9°, close to the mean axis as estimated from the walls. In the eastward direction, the inner face of the N wall has an azimuth of 75.6° and that of the S wall, E of the entrance, 86.2°. This is a greater divergence than that of the W and E walls to the N, but the mean direction, 80.9°, is precisely perpendicular to the interior division. Overall, the *heiau* deviates from cardinality by some 9°–10° in the counterclockwise sense.

The *heiau* faces a stretch of the Haleakalā Ridge to the left of the highest point and above the Lualaʻilua Hills (see fig. 6.13). There is no particular landmark on the ridge itself at this point, and no evident astronomical association. To the E, the structure is oriented upon a featureless stretch of the Alena lava flow, which forms a horizon about 500 m distant, just above where it falls below the sea horizon, around due E. To the S is open sea, and to the W the view is blocked by steeply rising terrain.

Additional remarks. The coral date from LUA-3 places this structure fairly early in the chronology of Kahikinui *heiau*, quite likely constructed during the reign of Maui king Kiha-a-Piʻilani. Its location suggests that the *heiau* might have functioned as a boundary temple between the *ahupuaʻa* of Alena and Lualaʻilua.

Site LUA-4 (Walker Site 184; State Site 50-50-15-1164)

Walker ([1930]) first recorded this small coastal *heiau* and made a plan of it, but his description is minimal. Kolb investigated the site in 1996, excavating one test pit. Kirch plane-table mapped the *heiau* in May 1996. On March 22, 2003, Kirch and Ruggles made both total station and GPS surveys, along with compass-clinometer observations. Kirch and Sharp visited LUA-4 in January 2008 to collect branch coral samples for ^{230}Th dating.

Location and topographic setting. LUA-4 is situated just 20 m inland of the shoreline bluffs a short distance to the E of Kiakeana Point. It lies on a relatively young and completely barren ankaramite lava flow with a *pāhoehoe* surface.

Architecture. The LUA-4 structure is a classic notched *heiau* whose orientation deviates from cardinality by about 25° in a counterclockwise sense, with six wall segments and with the notch located in the WNW corner (fig. 6.14). The exterior dimensions are 7.8 m ENE–WSW by 9 m NNW–SSE. The walls average about 1 m wide and are about 80 cm high. An unusual aspect of wall construction at LUA-4 is the use of large, vertically set *pāhoehoe* slabs to form the exterior faces of the SSE and ENE walls. Along the exterior of the SSE wall, for about half its distance, is a low stone-faced terrace, the surface of which is paved with small coral *'ili'ili*; this has the appearance of a burial that was added to the *heiau*, possibly in the early postcontact era.

The interior of the *heiau* displays a number of interesting architectural features. A low, stone-faced platform runs the length of the interior WSW wall, forming a kind of bench. The main court of the *heiau*, with its *pāhoehoe* lava floor, is divided into two parts almost equally by a NNW–SSE trending single-course wall, demarcating two distinct space cells. The smaller room, which forms the northern part of the *heiau*, is also set off from the main court by a two-course high terrace face, with the floor of the room roughly paved with stones. A coral file was found on the surface in this northern room. The elevation of this northern room above the level of the main court suggests that it was an altar or sanctum sanctorum of the *heiau*. Pieces of branch coral and small coral heads were scattered both on the *heiau* walls and within the court and altar areas.

Excavation. Kolb and Radewagen (1997, 65–66) note that a 1 m² unit was excavated, but because the *heiau* had been constructed directly on a *pāhoehoe* lava flow, there was virtually no cultural deposit. They did recover "bits of coral, bone, shell, and charcoal."

Dating and chronology. Kolb (2006, table 1) reports a single radiocarbon date of 130 ± 60 BP (Beta-134659) from a "pavement" context, presumably from his test excavation on the exposed floor of the *heiau*. This sample has two 2σ calibrated age ranges of AD 1666–1784 (40.1%) and 1796–1950 (55.3%). Kirch and Sharp collected two branch coral samples and dated CS-1, a branch tip that was sawed off of a larger coral head situated on the NNW wall above the altar area; this yielded a corrected age of AD 1696 ± 2. This ^{230}Th date corresponds well with the earlier intercept range of the calibrated radiocarbon date; together the two samples indicate that the *heiau* was probably built and in use at the close of the seventeenth century, during the reign of Maui *ali'i nui* Ka'ulahea II.

Archaeological context. LUA-4 is isolated from any other archaeological site by a buffer zone of between 100 and 120 m. To the W lies the small cluster of residential sites and a small *ko'a* at Wekea Point. A few small structures that have yet to be formally recorded also lie about 100 m inland and also to the ENE.

Viewshed. There are open views of the Haleakalā Ridge and its upper slopes from the NW round to the ENE. At an azimuth of 70° the eastern slopes disappear behind a closer ridge about 300 m away, which in turn falls below the sea horizon around an azimuth of 80°. There are open sea views from the E round to the WSW. A lava ridge ~400 m away forms the horizon to the W, and another ridge 2–3 km distant forms that to the WNW. The conspicuous Luala'ilua Hills lie below the Haleakalā Ridge between azimuths 337° and 353°.

Orientation. The altar on the NNW side suggests that this *heiau* was NNW-facing. It may be that the outer face of the ENE wall (with its vertically set *pāhoehoe* slabs) is the best indication of the intended orientation: its northward azimuth is 334.9°. Two of the shorter NNW–SSE-oriented wall faces—the outer face of the WSW wall and the inner face of the WSW wall of the smaller northern room—have similar orientations, with azimuths of 334.4° and 334.1°, respec-

FIGURE 6.14. Plan of *heiau* LUA-4, showing the azimuths of intact straight segments of wall facing as determined by the total station survey. (Based on plane-table survey by P. V. Kirch)

tively. However, the remaining three NNW–SSE-oriented wall faces are less consistent (see fig. 6.14). The mean orientations of the three NNW–SSE-oriented wall segments, from WSW to ENE, determined by averaging their inner and outer face orientations, as judged mainly from corner-to-corner alignments, are 336.7°, 338.3°, and 333.6°, respectively. The last of these, being the longest, is presumably the most reliable indication of the intended orientation; a weighted average, assigning double weight to the ENE wall, yields 335.5°.

As can be seen in figure 6.14, the azimuths of the six WSW–ENE-oriented wall segment faces range from 58.2° to 71.9°, the latter figure being that of the *pāhoehoe*-slabbed outer face of the SSE wall. Among the WSW–ENE-oriented wall segments there are three, rather than two, longer faces; assigning double weighting to the three, the weighted average of the orientations comes out at 65.5°, which is perpendicular to that of the NNW–SSE-oriented segments. Thus, 335.5° and 65.5° are our best estimates of the intended orientation to the NNW and ENE, respectively.

The *heiau* faces a stretch of the Haleakalā Ridge to the left of the highest point and just to the left of the Lualaʻilua Hills. The azimuth range 333° to 338° corresponds to the declination range +61.5° to +66.5° (to the nearest half degree). Dubhe (α UMa), or Hiku Kahi—the first of the seven stars in Nā Hiku, Ursa Major (Johnson, Mahelona, and Ruggles 2015, 150)—would have set in the center of this stretch around AD 1700 (dec. +63.5°). There are several small cinder cones on this part of the ridge, but the closest lies on the limit of the stretch concerned, at azimuth 338° and declination about +66.5°.

To the ENE, the *heiau* is oriented upon the lower part of the distant ridge sloping down from Haleakalā, just above where it disappears behind the much closer ridge less than 1 km away. The center of the horizon range in question (az. 65.5°, alt. +1.7°, dec. +23.4°) corresponds both to the rising position of the sun at the June solstice and the Pleiades, although the individual walls are oriented to either side of this. The declination range corresponding to the sun at the June solstice is +23.2° (lower limb) to +23.7° (upper limb), while that of the Pleiades in AD 1700 is +23.0° to +23.5°.

To the WSW is open sea, with the center of the range (az. 245.5°) being close to December solstice sunset (az. 245.5°, alt. −0.1°, dec. −23.1°). To the SSE, the walls align upon the southern end of Hawaiʻi Island, but it is likely to have been visible only during exceptional conditions.

Additional remarks. Although the *heiau* apparently faces NNW, the orientation to the ENE—upon the position of June solstice sunrise and the rising of the Pleiades—might be significant.

SITE LUA-29

This *heiau* forms part of a coastal site complex at Hanamauloa, including a unique *pānānā*, or sighting wall, described in detail elsewhere (Kirch, Ruggles, and Sharp 2013). Here we consider only the *heiau* proper. This was first plane-table mapped by Kirch in March 1997 as part of an investigation of the larger complex. Kirch and Ruggles made a first survey of the site on March 23, 2003; on November 5, 2011, we returned to the complex, making both a total station survey and more detailed plane-table and GPS maps of the *heiau*. Kirch and Sharp also visited the complex in January 2008 to collect a series of coral samples for ^{230}Th dating.

Location and topographic setting. The *heiau* sits about 30 m inland of the low cliffs defining the shoreline at Hanamauloa Point, which happens to be the most southerly place on Maui Island. It lies on an *ʻaʻā* flow and is slightly protected from the incessant easterly winds by a low rise immediately to the east.

Architecture. A first view of the site plan (fig. 6.15) suggests that LUA-29 is a typical notched *heiau*, skewed from cardinality by about 30°–35° in a clockwise sense, with the notch in the NNW corner and an opening on the SSW side. The maximum external dimensions are 14 m WNW–ESE by 9.5 m NNE–SSW. A more detailed inspection, however, reveals crucial differences between this and most notched *heiau*, stemming from the fact that LUA-29 was constructed in

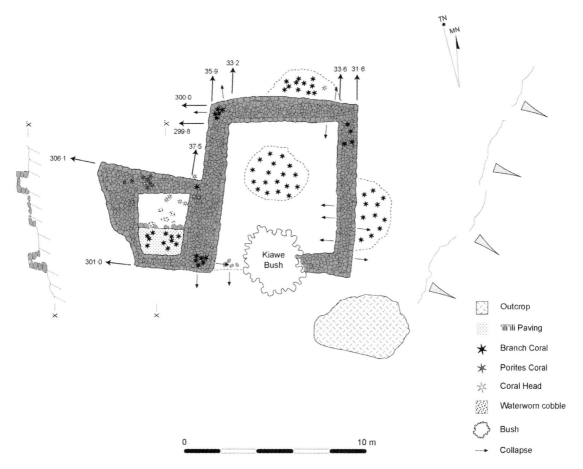

FIGURE 6.15. Plan of *heiau* LUA-29, showing the azimuths of intact straight segments of wall facing as determined by the total station survey. (Based on plane-table survey by P. V. Kirch)

two distinct phases. Note that the smaller room or enclosure on the western side does not open onto the main court in the eastern enclosure, as would be the case in most notched *heiau*. Instead, the western enclosure is actually a *ko'a* (fishing shrine) that apparently existed as a free-standing structure prior to the construction of the larger eastern enclosure.

Taken by itself, the smaller western enclosure is a classic *ko'a* in size and architectural details. Like other *ko'a* along the Kahikinui coast, its interior is subdivided into two small courts, with the inland, or northern, court elevated, the two being separated by a single-course stone facing. The seaward court is paved with small *'ili'ili* gravel and littered with branch coral and waterworn coral pieces. Several larger waterworn basalt cobbles sit on the upper court, along with several entire coral heads; a cluster of *Porites* corals sits on the NNE wall. A curious architectural detail is the extension of the NNW corner of the *ko'a* enclosure into a distinct canoe-prow point.

The larger eastern enclosure is defined by core-filled walls 1–1.5 m thick and about 1–1.25 m high. Close examination of the contact of the enclosure's WNW wall with that of the *ko'a* indicates that the eastern enclosure was added to and abutted against the *ko'a*. The SSW wall of the eastern enclosure has an opening that seems to be a formal entryway. The interior courtyard defined by the eastern enclosure is covered with scattered branch coral and *'ili'ili* pebbles. On the

Auwahi to Alena • 179

exterior of the NNE wall, there is also a concentration of branch coral pieces. There are also branch corals at various places on top of the enclosure walls.

Dating and chronology. Kirch and Sharp collected seven samples of branch coral from LUA-29 and dated three of these (Kirch, Mertz-Kraus, and Sharp 2015). Sample LUA-29-CS-1, embedded within the wall stones at the WSW corner of the ko‘a and presumably placed there during construction, yielded a date of AD 1615 ± 3. LUA-29-CS-4 is a small branch tip that was tightly embedded in the wall at the junction of the ENE corner of the ko‘a with the abutted eastern enclosure; it presumably was placed there when the eastern enclosure was added. This sample yielded a date of AD 1658 ± 2. The third dated sample, LUA-29-CS-6, is a branch tip embedded in the ESE wall of the eastern enclosure; it had to be "excavated" by removing some wall stones and clearly was a part of the original wall construction. This sample yielded a date of AD 1660 ± 2, nearly identical with sample CS-4.

The ^{230}Th branch coral dates from LUA-29 are all from samples that were architecturally integral and should date to the time of wall construction. The age of AD 1612–1618 from the ko‘a indicates that this smaller western enclosure was built about 45 years before the eastern enclosure was added, in about AD 1656–1662. In terms of the genealogical chronology of Maui ali‘i nui, the ko‘a would have been constructed during the reign of Kamalālāwalu, while the eastern enclosure would have been added during the reign of Kalanikaumakaōwakea.

Archaeological context. This heiau is part of a site complex that includes a unique notched wall—a pānānā, or sighting wall—that may have served as a navigational indicator for voyages to, or a memorial of voyages from, islands to the south, the "Kahiki" of Hawaiian mo‘olelo (fig. 6.16). The pānānā complex and its significance are fully described in Kirch, Ruggles, and Sharp (2013). The age of the pānānā is uncertain, although one coral date associated with this complex suggests that it may predate the heiau by about two centuries.

Viewshed. There are open views of the Haleakalā Ridge to the N and its western slopes, never closer than ~5 km, extending down to Auwahi Bay in the W, with Kaho‘olawe (28 km distant) beyond. Pu‘u Pīmoe, 7.7 km away, protrudes conspicuously above this slope at an azimuth of 293°; Pu‘u Hōkūkano, although closer (3.8 km), is rather less prominent upslope at 313° (fig. 6.17). The Luala‘ilua Hills lie well beneath the Haleakalā skyline to the NNW. Toward the ENE and E, roughly between azimuths 68° and 95°, the horizon is formed by local ground within 300 m of the site and currently by vegetation on the low rise immediately adjacent to the site. On clear days the peak of Mauna Kea is visible across the ‘Alenuihāhā Channel at azimuth 134°.

Orientation. The northern elevated court of the ko‘a indicates that it faced NNE; the southern formal entryway indicates the same for the enclosure. However, the NNW-facing canoe-prow corner of the ko‘a also suggests an interest in that direction; as noted in chapter 4, canoe-prow corners are in almost all cases found on the right with respect to the kapu direction, which would suggest an interest in the WNW.

The outer and inner faces of the ESE wall are aligned with northward azimuths of 31.6° and 33.6°, respectively, giving a mean orientation of 32.6°, while those of the WNW wall have azimuths 35.9° and 33.2°, respectively, for a mean wall orientation of 34.5°. Overall, the enclosure is slightly narrower at its northern end, with a mean axial orientation of 33.6°. To judge from the inner face of its ESE wall, the northward orientation of the ko‘a is slightly farther round to the NE, at azimuth 37.5°. The structures are not aligned upon the pānānā, which sits somewhat to the left, between azimuths 23° and 28°. Instead, they face a relatively featureless stretch of the Haleakalā Ridge,

FIGURE 6.16. Aerial photo showing *heiau* LUA-29 in the middle foreground and the notched *pānānā* wall farther inland. (Photo by P. V. Kirch)

FIGURE 6.17. Horizon to the WNW viewed along the southern wall of the older, *koʻa* part of *heiau* LUA-29. Puʻu Pīmoe is conspicuous on the horizon in the center of the photograph, while Puʻu Hōkūkano is just visible up the slope on the right of the photo. (Photo by Clive Ruggles)

well to the left of the only conspicuous feature, a sharp dip, at azimuth 40°. The declination corresponding to azimuth 33.6° is +55.8°, indicating that the enclosure faced the rising point in AD 1660 of the three stars Schedar, Caph, and Tsih (α, β, and γ Cas) (decs. +54.7°, +57.2°, and +58.9°, respectively), probably the three stars of Mūlehu (Johnson, Mahelona, and Ruggles 2015, 195), as well as the "panhandle" end of Ursa Major, Nā Hiku. To the SSW, the walls face open sea.

The outer and inner faces of the NNE enclosure wall are almost parallel; the wall has a mean westward orientation of 299.9°. Although the SSW enclosure wall east of the entrance was too short to yield reliable measurements, the outer face of the SSW wall of the *ko'a* is consistently oriented at 301.0°. On the other hand, the outer face of the NNE wall of the *ko'a* is clearly skewed with regard to the remainder of the structure, as part of the canoe-prow construction; its orientation is 306.1°. Pu'u Pīmoe is a few degrees downslope from the point indicated by the main structural orientation (300°–301°); viewed from the *heiau*, the Pleiades (dec. +22.8° to +23.3° in AD 1650) would have set directly into the summit of this *pu'u* (az. 293.5°, alt. +3.9°, dec. +23.3°), as would the June solstice sun (dec. +23.2° to +23.7°) (fig. 6.18; see also Kirch, Ruggles, and Sharp 2013, 59–61). To the ESE, the *heiau* walls align upon open sea a few degrees beyond (to the left of) the northern limit of Hawai'i Island.

Additional remarks. This structure is of special interest in that it shows the transformation of an original small fishing shrine (*ko'a*) into a notched *heiau*. This *heiau* has several possible astronomical associations, including with the June solstice sun, the Pleiades (unusually, at sunset rather than sunrise), and other Hawaiian stars. The Pleiades association is interesting in relation to the dating evidence, as the Pleiades began to descend directly into the summit of Pu'u Pīmoe only from about AD 1630 onward (by which we mean that the cluster, whose setting path is about 0.5° wide, would have straddled the summit as it set)—that is, it did so at the time when the *ko'a* was converted into a *heiau* but not at the time of construction of the *ko'a* itself. The first setting before sunrise (apparent cosmical setting) occurred around the same time of year as the

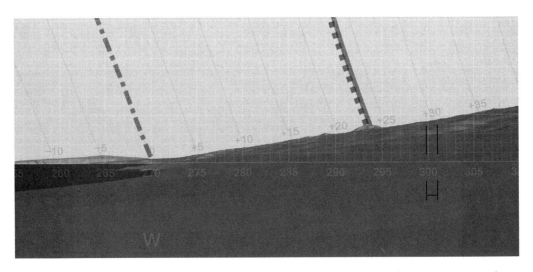

FIGURE 6.18. DTM-generated profile of the horizon to the W and WNW as seen from *heiau* LUA-29. The horizontal bar indicates the direction of orientation of the *heiau*. The solid shaded line indicates the setting path of the sun at the June solstice; the dash-and-dot line, the setting path of the sun at the equinoxes; and the dotted line, the setting path of the Pleiades in AD 1650. (Graphic by Andrew Smith)

last rising after sunset (apparent acronychal rising), in November (see chapter 4) and would therefore make a possible alternative for determining the onset of Makahiki.

Site LUA-36

Site LUA-36 is a *ko'a* (fishing shrine) at Wekea Point, first recorded by Kirch as part of a general survey and plane-table mapping of the Wekea Point cluster of sites in 1996. Kirch and Sharp collected branch coral specimens from the site in January 2008; Kirch and Ruggles made a sketch map and compass-clinometer observations of the site on November 5, 2011.

Location and topographic setting. The structure sits at the edge of the low cliffs, just 20 m from the shoreline at Wekea Point, on an *'a'ā* lava substrate. A bronze plaque marking the Hawaiian Government Survey's Shore D triangulation station is 17 m NW of the *ko'a*.

Architecture. The structure has been somewhat disturbed in modern times by use as a fishermen's campsite, with campfires lit on the court near the E wall. However, the main architectural features appear to be intact. The site consists of a U-shaped enclosure, skewed from cardinality by about 10° in a counterclockwise sense, open to the W, with exterior dimensions of 10 m N–S by 7 m E–W. The walls, which are about 1 m wide and stand between 70–90 cm high, are of stacked *'a'ā* and *pāhoehoe* cobbles and boulders. An interesting detail of the E wall is the incorporation of a large tabular boulder (~80 cm tall) of conglomerate coral sand and basalt pebbles into the middle of the exterior facade, like an upright (see fig. 3.2); a second conglomerate slab is incorporated into the facade near the SE corner. A low platform of *'a'ā* rubble, about 2 by 3 m, abuts the interior face of the S wall; this may be a burial platform. To the W of the main structure, below the drop-off edge of the lava flow and hence at a lower level than the *ko'a*, is a small semicircular enclosure of stacked rocks.

Several large *Pocillopora meandrina* coral heads are situated in or on the E and S walls. The floor of the court, though somewhat disturbed, is paved with *'ili'ili* gravel and littered with pieces of branch coral.

Dating and chronology. Two specimens of branch coral were collected, and one (LUA-36-CS-2) was ^{230}Th dated. This specimen came from a cavity in the S wall and yielded a date of AD 1656 ± 3. The coral appears to have been placed within the wall during construction; hence, the *ko'a* was probably built during the reign of Maui *ali'i nui* Kalanikaumakaōwakea.

Archaeological context. LUA-36 is part of a cluster of sites at Wekea Point, including three substantial walled-house enclosures within a radius of about 50 m. These house sites, however, probably date to the postcontact era (based on their architecture and the presence of European artifacts on their surfaces). Thus, the *ko'a* presumably occupied Wekea Point well before the residential cluster was established.

Viewshed. There are open views all round.

Orientation. Given that the structure is open on the W side and that the E wall is substantial and contains a large conglomerate boulder, this *ko'a* appears to be facing E. Several measurements obtained independently by Ruggles and Kirch, using different prismatic compasses, on the N and S walls give a mean (true) eastward orientation of 81° in both cases. A number of measurements on the eastern wall yield a mean northward orientation of 346°. The *ko'a* faces open sea just beyond the tip of the eastern slopes coming down from Haleakalā. It is aligned upon open sea also to the S, upon Kaho'olawe Island (but not its peak) to the W, and to the N to the Luala'ilua Hills with the Haleakalā Ridge behind, to the left of the highest part.

Site LUA-39

LUA-39, a small *koʻa*, was recorded by Kirch and Ruggles on November 5, 2011. A compass-and-tape map and compass-clinometer observations were made at that time.

Location and topographic setting. The site lies a few meters east of the jeep track descending to Hanamauloa, slightly more than 0.5 km inland and at an elevation of 70 masl. The substrate is an old *pāhoehoe* lava flow.

Architecture. The site consists of a nearly square enclosure, broadly cardinally oriented, measuring 3.25 E–W by 3.5 N–S (exterior dimensions), built up against the seaward face of a fractured *pāhoehoe* lava outcrop (figs. 6.19 and 6.20). The walls are of stacked *pāhoehoe* cobbles, about 1 m wide and 50–60 cm high. There is a formal entryway on the S. The interior is divided into two levels by a terrace face, 20 cm high, of *pāhoehoe* slabs. The upper (inland) court is covered in branch coral pieces. A waterworn basalt cobble (~25 cm long) on this upper terrace may have been a Kūʻula stone. The lower court is rough paved. Two whole *Pocillopora* coral heads sit just inside the entryway; other coral heads are situated on the upper court and the N wall, along with a large waterworn *Porites* coral cobble.

Archaeological context. No systematic archaeological survey has been conducted in the vicinity of this site, so the larger settlement pattern context is unknown.

Viewshed. There are open views of the Haleakalā Ridge from the NW round to the NE and out to sea from the E round to the WSW, with local ridges ~80m distant and 150–300 m distant restricting the views to the ENE and to the W and WNW, respectively.

Orientation. Given the formal entryway on the S side and the higher-level court covered in branch coral pieces on the N side, this *koʻa* is clearly oriented to the N. Various measurements obtained independently by Ruggles and Kirch, using different prismatic compasses, on the W and S walls show the deviation from cardinality on both the E–W and N–S axes to be less than 2°. The *koʻa* faces the highest part of the Haleakalā Ridge to the N. It is aligned upon open sea also to the S, the tip of the slope meeting the sea to the E, and the nearby ridge to the W.

The *Ahupuaʻa* of Alena

That Alena held the status of an *ahupuaʻa* within the *moku* of Kahikinui is strongly suggested by its appearance on both the 1838 Lahainaluna map (the "Kalama map") and on the 1885 Hawaiian Government Survey map. However, the exact boundaries of Alena with Lualaʻilua to the west and Kīpapa to the east were never established (see Kirch [2014, 73–74] regarding the lack of *ahupuaʻa* boundaries in Kahikinui). It seems that Alena largely corresponds with the massive basanite *ʻaʻā* lava flow named Alena-2 by Sherrod et al. (2007). This is the youngest lava substrate within the Kahikinui District, dating to between 5,000 and 10,000 years ago; because of the flow's young age, its surface has been only minimally weathered and especially in the coastal region consists of rough, jagged lava. For the purposes of archaeological survey, Kirch placed the boundary between Alena and Kīpapa at a lava rock wall that runs from the coast all the way inland to above Manukani. This wall was built in the late nineteenth century by Portuguese cattle ranchers and may or may not have anything to do with the original *ahupuaʻa* boundary.

Almost all of Alena remains archaeological terra incognita. A State of Hawaiʻi Historic Preservation Division team surveyed part of the coastal strip (Van Gilder, Nagahara, and

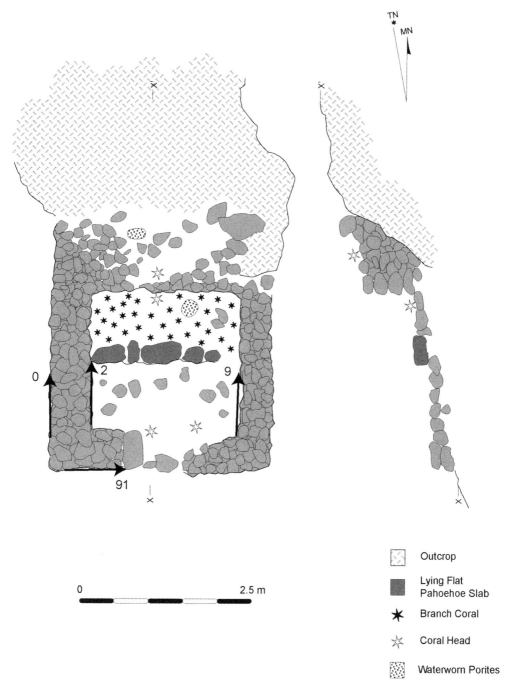

FIGURE 6.19. Plan of *ko'a* LUA-39, showing the azimuths of wall segments as determined by prismatic compass corrected for mean magnetic declination. (Plan based on compass-and-tape survey by P. V. Kirch)

FIGURE 6.20. View of *ko'a* LUA-39 from the south. (Photo by P. V. Kirch)

Hodgins 1999), while Kirch conducted a limited one-day reconnaissance in the Alena uplands, discovering one previously unknown *heiau* (ALE-211). Consequently, the record of *heiau* sites within Alena is almost certainly incomplete and awaits further fieldwork.

SITE ALE-1 (STATE SITE 50-50-15-1157)

This site was first identified during the 1973 "statewide inventory" survey conducted by the State of Hawai'i Historic Preservation Division. Kirch made a detailed plane-table map of the site in August 1999; Ruggles carried out a compass-clinometer survey on April 13, 2002; and Kirch and Ruggles made a total station survey of the *heiau* on March 22, 2003.

Location and topographic setting. The site lies about 120 masl, some 125 m east of the jeep track descending to Hanamauloa. ALE-1 sits on a gently sloping, older *pāhoehoe* lava flow (the Kīpapa-2 ankaramite of Sherrod et al. 2007), but its eastern edge abuts the younger massive Alena-2 basanite *'a'ā* flow. Extensive use was made of slablike *pāhoehoe* blocks in the *heiau* walls and other features.

Architecture. As can be seen in the plan map (fig. 6.21), ALE-1 is a large *heiau* (basal area of 700 m²) of the elongated double-court form (fig. 6.22). Its orientation deviates from cardinality by about 5° in a counterclockwise sense; its exterior dimensions are 40 m N–S and 11–21 m E–W, the S wall being much longer than the N wall. The W wall is divided into two distinct segments, with a notch separating them. The SW wall along with the S wall are buttressed with an exterior, secondary facing. The S wall is also breached midway with a formal entryway. The N, W, and S walls are all about 1–1.5 m wide and 1 m high, well constructed with the core-filled method but utilizing the brick-like *pāhoehoe* slabs, which give them a particularly angular look with clean vertical faces. The eastern side of the *heiau*, in contrast, incorporates the steeping rising slope of the *'a'ā* flow with a facade of stacked *'a'ā* cobbles; this facade has collapsed in many places, leaving the eastern edges of the courts indistinct.

ALE-1 is among the more interesting Kahikinui *heiau* because of the number of architectural features visible in the interior, partly due to the near absence of vegetation. Beginning at the southern end, immediately after entering through the 1 m wide formal entryway, one

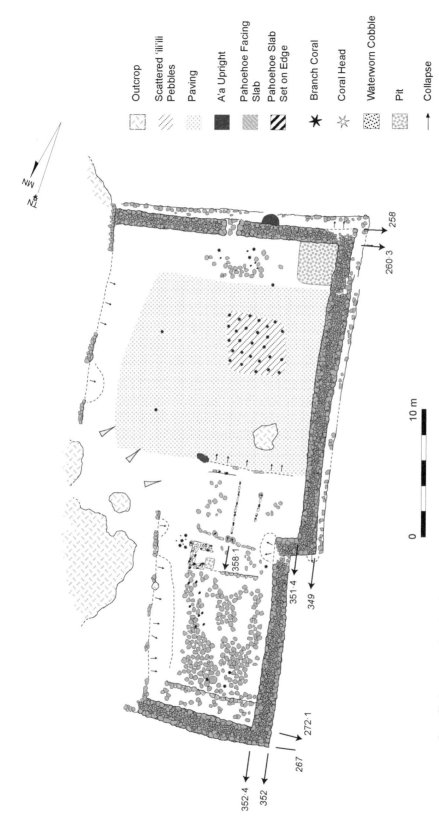

FIGURE 6.21. Plan of *heiau* ALE-1, showing the azimuths of intact straight segments of wall facing as determined by total station (quoted to the nearest 0.1°) or prismatic compass corrected for mean magnetic declination (italicized and quoted to the nearest degree). (Based on plane-table survey by P. V. Kirch)

FIGURE 6.22. View of *heiau* ALE-1 from the southeast. (Photo by P. V. Kirch)

encounters an area with *pāhoehoe* slab paving, on which are scattered a few pieces of branch coral. To the left, in the SW corner of the court, is a depression about 3 m on a side; it may be a filled-in pit, possibly for the disposal of offerings. Proceeding N across the lower court (which is paved with *'a'ā* clinkers), one encounters a roughly square area, about 5 by 5 m, with a concentration of *'ili'ili* pebbles and branch coral. This is apparently the interior paving of a former perishable structure of pole and thatch that once stood here. Toward the N end of the lower court is a low, single-course alignment of *pāhoehoe* slabs running E–W, terminating on the E at a large *'a'ā* boulder standing about 80 cm high. This alignment is the facing for a low terrace, elevated about 25 cm above the lower court.

Stepping up onto the low terrace, one proceeds through a formal pathway, defined by parallel single-course alignments of carefully set *pāhoehoe* slabs, about 1.5 m apart, running N–S. This pathway, about 5 m long, terminates at another E–W terrace facing, again constructed with *pāhoehoe* slabs set upright. This face retains a low terrace, partly paved with flat *pāhoehoe* slabs, about 2 m wide. On the eastern side of this small terrace is a well-constructed, small U-shaped enclosure open to the W, the walls formed of two parallel courses of *pāhoehoe* slabs. This enclosure, about 2 by 2 m, was without doubt the foundation for a small thatched structure, possibly the *hale waiea*, which was said to be small and was the locus for the *'aha* rite (see chapter 2). (A disturbed area, 1 m square, immediately in front of this U-shaped enclosure seemed to be a test pit dug by Kolb in 1997.) A small concentration of branch coral pieces was noted just outside the SE corner of the enclosure.

Continuing northward, yet a third E–W trending alignment of *pāhoehoe* slabs set on edge defines the lower end of the uppermost court. This court, about 9 m long, is remarkably well paved with carefully set *pāhoehoe* slabs. At the N end, about 1.75 m in front of the N wall, is a terrace or bench running the width of the court, again faced with vertically set *pāhoehoe* slabs, about 40 cm high, evidently a kind of altar. On the court, just in front of this terrace, are several pieces of branch coral.

Excavation. Kolb excavated at least one test pit in ALE-1, as he obtained a radiocarbon date (see below), but no details are available.

Dating and chronology. Kolb (2006, table 1) reports a single radiocarbon date of 230 ± 60 BP (Beta-134658), from a "pavement" context. This corresponds to five possible calibrated age ranges: AD 1490–1603 (18.3%), 1611–1706 (28.4%), 1720–1819 (31.9%), 1832–1882 (5.0%), and 1915–1950 (11.7%); the latter two age ranges can be rejected because there is no evidence for postcontact use of the site.

Archaeological context. No systematic archaeological survey has been carried out in the immediate vicinity of ALE-1.

Viewshed. From the *heiau* interior, the view from the NNE round to the SE (between az. ~20° and ~130°) is blocked by the younger lava flow on which the site abuts, while the view to the W, WNW, and NW is restricted to a ridge under 1 km away. Part of the Haleakalā Ridge can be seen to the NNW and N, while open sea is visible from the SSE round to the WSW, with the southern tip of Kahoʻolawe Island visible at around 255°. The leftmost two of Lualaʻilua peaks (one almost behind the other) just break the skyline at azimuth 338°; in good weather conditions at the peak of Mauna Kea, on Hawaiʻi Island, is visible at 134°.

Orientation. This *heiau* is clearly oriented to the N. The total station survey shows the inner face of the W wall to be oriented with a northward azimuth of 352.4° to the N of the notch and of 351.4° to the S of it, with the compass survey confirming a similar orientation for the outer faces. The better preserved side of the formal pathway in the interior yields a slightly different orientation, 358.1°. To judge by their straighter inner faces, the S wall has a westward orientation of 260.3°, while that of N wall is 272.1°.

To judge from the alignment of the W wall, the *heiau* faced a point on the Haleakalā Ridge just to the left of one of the larger cinder cones; the formal pathway is aligned to the right of that cone. There is no obvious astronomical target. The directions of the westward alignment of the two walls bracket an unremarkable stretch of the lava ridge slope, again without obvious astronomical possibilities since the mean alignment of the two walls, 266.2°, is several degrees to the left of the direction of equinox sunset, which is more or less due W.

Additional remarks. The focus for the orientation of this *heiau* appears to be a prominent topographic feature, without astronomical significance. In addition to the temple's orientation toward the rim of Haleakalā, the red color of the large cinder cone in question reinforces the idea that this might be a temple dedicated to Kū, the deity associated with high mountains, forests, and the color red. This *heiau* may also have marked the boundary between the *ahupuaʻa* of Lualaʻilua and Alena (as with site LUA-3).

Site ALE-4, Wailapa (Walker Site 178; State Site 50-50-15-178, Feature 541)

Walker ([1930], 248) first recorded this site, with a cursory description. Kolb and Radewagen (1997, 67–68) mention the site briefly, underestimating its internal area (they give an area of 288 m², whereas our GPS-based calculation of basal area is 455 m²). Kolb (2006, table 1) reports a single radiocarbon date, presumably from a sample obtained during excavation in 1997 (no report is available). Van Gilder, Nagahara, and Hodgins (1999) report the site as feature 541 in their coastal survey of Alena. Kirch mapped ALE-4 with plane table and alidade in August 1999, also making a GPS survey. Kirch and Sharp visited the site to collect branch coral samples for ^{230}Th dating in January 2008. On November 7, 2011, Kirch and Ruggles conducted a total station survey and further GPS mapping and made compass-clinometer observations.

Location and topographic setting. ALE-4 sits in a large natural swale or valley within the massive Alena-2 'a'ā lava flow, about 180 m inland of the coast. A steep lobe of the lava flow rises abruptly on the eastern side of the *heiau*.

Architecture. Although it has a distinct, angled notch along the WSW side, ALE-4 more closely resembles an elongated double-court *heiau*. This configuration, however, is the result of a second building phase, so that in its original configuration the temple was a square enclosure open to the SSE. Its orientation deviates from cardinality by some 30°–32° in a counterclockwise sense.

As can be seen in the plan map (fig. 6.23), ALE-4 consists of two distinct courts, with overall exterior dimensions of 28 m NNW–SSE and 19 m ENE–WSW (fig. 6.24). The northern, upslope court is defined on the WSW, NNW, and ENE by well-constructed, core-filled walls 2 m wide and up to 1.6 m high. The ENE wall, built up against the steep face of the adjacent 'a'ā lava flow, is really more of a retaining wall than a freestanding one. Along the inner face of the NNW wall there is a low shelf or altar, as is found in a number of other *heiau*. Perpendicular to the NNW wall and extending down the middle of the court for 3 m is a low, single-course alignment that may once have defined the edge of a formal pathway (as at site ALE-1). A basalt whetstone fragment was found on the surface near this alignment. The SSE edge of the northern court is well defined by a single-course alignment of 'a'ā cobbles that forms a low terrace face, setting the northern court off from the southern court.

The wall bounding the southern court on the WSW, SSE, and ENE contrasts markedly with that surrounding the northern court. It not only is narrower and lower (about 1 m wide and 60–80 cm high) but was crudely constructed using a stacking method, rather than being core-filled. The walls are also irregular and crooked, having the appearance of being hastily thrown up. The WNW corner of this wall clearly abuts to the SSW corner of the northern court, which was therefore preexisting. The lower court is roughly paved with 'a'ā clinkers, and there are some scattered 'ili'ili pebbles. A coral abrader was found in the ESE corner of the court, on the surface.

There can be little doubt that the northern court and its well-built core-filled walls formed the original temple, which was of the square-enclosure form, open to the SSE. At some later date the southern court was added, apparently rapidly and without much labor investment given the crude construction of the walls. We had the distinct impression that this lower court might have been built for just a single ritual event.

Excavation. Kolb apparently conducted some test excavations in ALE-4, but no report is available.

Dating and chronology. Kolb (2006, table 1) reports a single radiocarbon date of 100 ± 50 BP (Beta-122893) from a "firepit" context; this has 2σ calibrated ranges of AD 1676–1777 (33.9%) and 1799–1941 (61.5%). Although the latter range has a higher probability, it falls into the postcontact period after *heiau* use was abandoned.

Kirch and Sharp obtained two branch coral specimens, both from surface contexts, and dated one (ALE-4-CS-1) with a ^{230}Th age of AD 1555 ± 4. This specimen had a slightly rounded tip, so it might have some inbuilt age.

These two dates suggest that ALE-4 may have first been constructed as early as the late sixteenth century, possibly as part of the major phase of *heiau* construction that we associate with

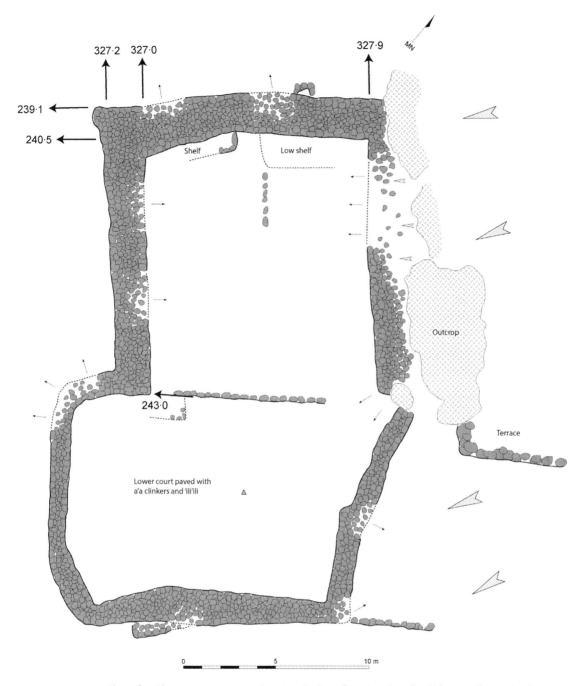

FIGURE 6.23. Plan of Wailapa Heiau (ALE-4), showing the best-fit azimuths of wall faces as determined by the total station survey. (Based on plane-table survey by P. V. Kirch)

the reigns of Piʻilani and Kiha-a-Piʻilani. At a later date the *heiau* was expanded with the addition of the southern court; the radiocarbon date suggests that the *heiau* was in use in the early eighteenth century, roughly around the time of Kekaulike's rule.

Archaeological context.   No other sites are immediately adjacent to ALE-4, but a number of smaller structures lie around the site at a radius of about 100 m. These include various small

FIGURE 6.24. View of Wailapa Heiau (ALE-4) from the west. (Photo by P. V. Kirch)

shelters and enclosures that may be temporary residential features. The Alena-2 ʻaʻā lava flow is far too arid and barren at this low elevation to grow crops, presumably ruling out an interpretation of ALE-4 as an agricultural temple.

Viewshed. For the most part the view is limited to the barren ʻaʻā on both sides of the swale—up to ~600 m distant from the SW round to the NW, but only up to ~50 m distant from the N round to the E. A small section of more distant horizon is visible to the NNW, between azimuths of ~315° and ~345°, depending on one's position within the *heiau*, as is a narrow stretch of open sea from the ESE round to the SSW, between azimuths of ~120° and ~200°. The NNW horizon comprises part of the Haleakalā Ridge to the right of azimuth 327°, and the tips of the Lualaʻilua Hills to the left of it. The southerly "window" includes, in good weather conditions, a view of the summit of Mauna Kea at 135°.

Orientation. This *heiau* is clearly oriented to the NNW. Sets of measurements on the outer and inner faces of the WSW wall of the northern court yield best-fit orientations of 327.2° and 327.0°, respectively in the northward direction, for a mean of 327.1°. Similarly, the inner face of the ENE retaining wall yields 327.9°. Thus, we obtain 327.5° as our best estimate of the intended axial orientation of the original square enclosure. For the NNW wall, best-fit estimates from intact facing segments yield westward azimuths of 239.1° and 240.5° for the outer and inner faces, respectively, for a mean of 239.8°. The westward orientation of the cobble alignment separating the two courts is 243.0°.

The *heiau* faces the point where the Lualaʻilua Hills meet the more distant Haleakalā profile in a small notch whose declination is +56.7°. The star Mizar (ζ UMa), or Hiku Ono—the sixth of the seven stars in Nā Hiku, Ursa Major (Johnson, Mahelona, and Ruggles 2015, 150)—would have set into this notch in AD 1660, as would Caph (β Cas), possibly one of the three stars in the asterism Mūlehu (195) in AD 1570.

The NNW wall alignment to the WSW yields a declination of around −27°, somewhat

192 • CHAPTER 6

outside the solar setting arc, while the ENE and WSW walls align to the SSE upon the northern slope of Mauna Loa, some 160 km away, which would only occasionally have been visible low above the sea horizon.

Additional remarks. This *heiau* was constructed in two phases, with the first phase consisting of the well-built northern court with its massive walls. The later addition of the southern court, much more crudely constructed, probably resulted from a rededication of the *heiau* at a later time, possibly associated with one of the kingly successions on Maui.

As regards orientation, this *heiau* deviates significantly (by more than 30°) from cardinality, perhaps because of the topographic constraints of placing it within the swale. Another possibility is that this was in order to align upon a visible part of the distant Haleakalā Ridge, since the view to the highest part, toward due N, is blocked. Since Kū is associated with high mountains as well as the north direction (see chapter 2), this is consistent with the idea that ALE-4 might have been dedicated to Kū, possibly functioning as a boundary temple marking a division between land units.

Site ALE-121 (State Site 50-50-15-4892, Feature 488)

This small *ko'a* (fishing shrine) was first recorded during a survey of the Kahikinui coastal zone by staff of the State of Hawai'i Historic Preservation Division (Van Gilder, Nagahara, and Hodgins 1999). Kirch and Sharp visited the site in January 2008, collecting two branch coral samples for ^{230}Th dating. Kirch and Ruggles made a GPS survey, a sketch map, and compass-clinometer observations on November 7, 2011.

Location and topographic setting. ALE-121 sits atop a prominent ridgeline on the Alena-2 *'a'ā* lava flow, about 100 m from the coast. On the eastern edge of the site the terrain drops away precipitously into a deep gully.

Architecture. The site consists of a simple C-shaped shelter, cardinally oriented to within 10°, open to the S, and built up against a lava outcrop on the NE side (fig. 6.25). The walls are stacked, about 1 m wide and 60–70 cm high. The space enclosed by walls is divided into two levels, with a slightly elevated terrace or shelf on the northern side, upon which are several waterworn basalt cobbles along with several entire *Pocillopora* coral heads. Just to the S of this shelf the floor is paved with *'ili'ili* pebbles, and branch coral is scattered over this surface as well.

Dating and chronology. One sample of branch coral (with intact tips), recovered from within the E wall of the *ko'a* and thus presumably placed there during construction, was ^{230}Th dated to AD 1699 ± 2.

Archaeological context. The *ko'a* is part of a cluster of about one dozen structures situated on the same *'a'ā* ridge. These include several small enclosures that may have been temporary habitations.

Viewshed. The ridgeline setting provides excellent open views from the NW round to the ENE (Haleakalā Ridge), to the E along the Alena coastline, and seaward from the ESE round to the SSW. From the SW round to the NW the view is restricted to the more local *'a'ā*, up to ~300 m distant.

Orientation. The *ko'a* clearly faces N. The inner faces of the three walls are rounded, so the straighter outer faces give a better indication of the intended orientation. As determined approximately by compass measurements and from the GPS survey, the N wall has an eastward orientation of ~79°, while the shorter E wall has a northward orientation of ~350°, so they are perpendicular to within the accuracy of our measurements. The W wall, on the other hand, is ori-

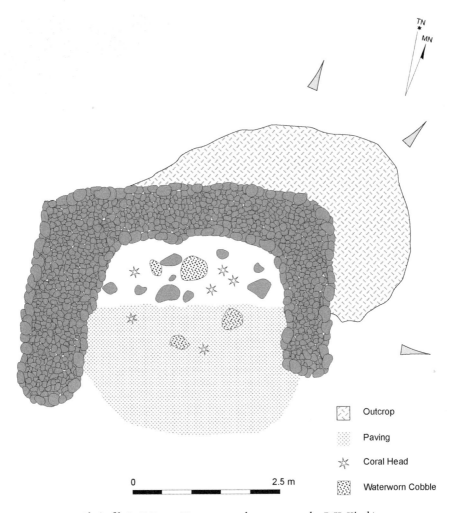

FIGURE 6.25. Plan of *ko'a* ALE-121. (Compass-and-tape survey by P. V. Kirch)

ented due N (az. ~0°), so the mean of the E and W walls is ~355°. The red cinder cone Kanahau marks the highest point on the northern horizon as seen from here and is at azimuth 355°. The tip of the eastern slope extending down from Haleakalā to the sea at Kailio Point is at ~78.5°.

There are no obvious astronomical connections, but it is possible that both the topographic alignments were significant.

Site ALE-140 (State Site 50-50-15-4902, Feature 482)

This classic notched *heiau* was first recorded during a survey of the Kahikinui coastal zone by staff of the State of Hawai'i Historic Preservation Division (Van Gilder, Nagahara, and Hodgins 1999). Kirch mapped the site with plane table and alidade in August 1999. In January 2008, Kirch and Sharp visited the *heiau* to collect branch coral for ^{230}Th dating. Kirch and Ruggles made a total station survey, a GPS survey, and compass-clinometer observations on November 7, 2011.

Location and topographic setting. The *heiau* lies within a narrow trough-like swale or depression in the massive Alena-2 *'a'ā* lava flow, about 150 m from the coast. The structure is built against the western side of this swale, which rises steeply above the *heiau* (fig. 6.26).

FIGURE 6.26. *Heiau* ALE-140 viewed from the NE, with the total station survey in progress. (Photo by P. V. Kirch)

Architecture. The structure is a notched *heiau*, skewed from cardinality by 25°–30° in a counterclockwise sense, with exterior dimensions of 10 m NNW–SSE by 10 m ENE–WSW; the notch is in the SSW corner (fig. 6.27). The walls, about 1 m thick and up to 1.2 m high, are well constructed with the core-filled technique. The walls are freestanding on the SSE, WSW, and NNW; on the ENE the retaining wall drops off into the floor of the lava depression. There is a formal entryway through the ENE retaining wall, leading to a small area paved with fine *'ili'ili* gravel. Adjacent to this paved area were two waterworn basalt cobbles. The main court is itself paved with *'ili'ili* and has waterworn coral scattered over it.

On the WSW side is the smaller room or chamber formed by the notched WSW wall. The smaller court is slightly elevated above the lower court; two waterworn basalt cobbles were situated here.

While surveying the *heiau* in 2011, we noted that if one is seated within the smaller, western chamber—as a *kahuna* might have been when officiating at the temple—the view to the ENE focuses on the intersection of the long, broad slope of Haleakalā in the distance, with the angular outline of an *'a'ā* ridge in the foreground, about 27 m away from the viewer. Just where these two skylines meet, two prominent *'a'ā* boulders on the near ridge form a kind of "window" that frames the Haleakalā slope (see fig. 4.9). To the right of these two boulders, the ridge is modified by the facade of a stacked stone terrace.

Examining at close quarters the ridgeline with its two prominent boulders, we were fascinated to see that while one of the boulders (the northernmost) is in a natural position in the lava flow, the other had been artificially set in position, on a base of three other smaller boulders. Thus, the window formed by the two boulders when viewed from the *heiau* had been in part artificially constructed, presumably for this very purpose. The stone terrace just to the south of the set boulder is about five courses high, well constructed against the steep slope of the natural *'a'ā* flow. It is not clear whether this terrace conceals a burial or whether it was constructed for some other reason.

Dating and chronology. A branch coral tip exposed by a partial collapse of the NNW wall was ^{230}Th dated, yielding an age of AD 1592 ± 3. Because the coral had been placed within the wall fill, it provides a good age for the construction of this *heiau*.

Archaeological context. Immediately north of ALE-140 is a nearly square enclosure,

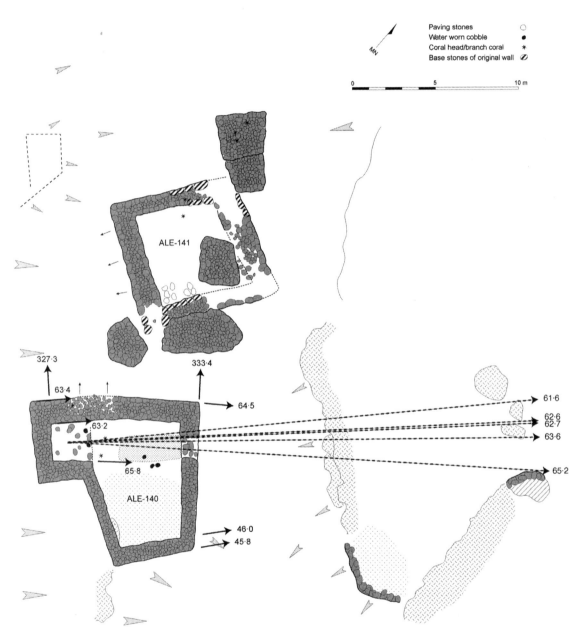

FIGURE 6.27. Plan of *heiau* ALE-140, showing the azimuths of intact straight segments of wall facing or corner-to-corner alignments as determined by the total station survey. Alignments from the western chamber to the boulders on the ridge to the ENE are also shown. (Based on plane-table survey by P. V. Kirch)

measuring 7 by 8 m, with core-filled walls. The construction style of this feature (ALE-141, state site 50-50-15-4902, feature 486) suggests that it was built at the same time as the notched *heiau*, although its function is not certain (a house for the priest of ALE-140 is a possibility). The enclosure walls have been partly robbed of stone to construct five separate, subrectangular stone platforms, almost certainly burials, on top of and adjacent to this feature.

In addition, about 8 m directly east of the *heiau*, on the eastern side of the natural *'a'ā* gully, the surface of the lava ridge has been artificially leveled and paved with *'a'ā* clinkers; part of the edge of this rough terrace has been faced with cobbles as a retaining wall. There is *'ili'ili* paving in one area. This leveled space seems to have been a seating area, possibly for participants or observers of the rituals taking place on the ALE-140 *heiau*.

Viewshed. There are restricted views from this *heiau*, with the sides of the swale limiting visibility to within some 30 m to the E and even closer from the S round to the W. The distant profile of Haleakalā is visible from the NNW round to the NE, between azimuths ~325° and ~65°, depending on the observer's exact position. From a seated position within the western chamber (as described under "Architecture"), the eastern slope of Haleakalā descends to meet the local *'a'ā* ridge at 64.0°, but from a standing position the Haleakalā slope extends down to about 64.5°. To the SE and SSE, the sea is visible between ~115° and ~160°, again depending on one's exact position. This includes the whole of Hawai'i Island, which lies between 126° and 160°, with the peak of Mauna Kea at 135°, although this would be visible only in clear weather conditions.

Orientation. The placement of the formal entrance to the ENE, and of the smaller, higher court to the WSW—probably the altar area—strongly suggests that this *heiau* is WSW-facing. This orientation is anomalous in comparison with all the other *heiau* we have documented in this and the adjacent *ahupua'a* and is all the more surprising given that the distant view is entirely blocked in this direction, while there is a spectacular view in the opposite direction to the ENE.

The straight eastern section of the NNW wall yields an eastward azimuth of 64.5°, while the shorter straight segments of wall on the northern and southern sides of the western chamber yield 63.3° (the mean of the outer- and inner-wall orientations shown in the figure) and 65.8°, respectively, for a mean axial orientation that is also 64.5°. The ENE wall, oriented with a northward azimuth of 333.4°, is almost perpendicular to this. The SSE wall of the *heiau*, however, is significantly skewed, at 45.9° (NE).

As mentioned earlier, from a seated position in the western chamber, the two prominent *'a'ā* boulders on the nearby ridge frame part of the distant Haleakalā slope. This occurs at an azimuth between 62.6° and 62.7° according to the ground plan survey, or 62.7°–62.9° when comparing the DTM-generated Haleakalā profile and the position of the local ridge obtained from photographs. As viewed from this position, however, this window has no obvious astronomical significance: the Pleiades did not rise here around AD 1600; rather, they rose somewhat farther to the right, above the steep slope to the south of the boulders. (The same is true for the June solstice sun.) The situation is not improved by standing rather than sitting in the western chamber, whereupon the distant profile appears well above the two boulders, with the Pleiades (and solstitial sunrise) still occurring well to the right. However, the window did line up with the rising position of the Pleiades around AD 1600 if viewed from a standing position outside the *heiau* on the northern side, lining along the outer face of the NNW wall, which is aligned to within 1.5° of the relevant direction.

Additional remarks. This *heiau* is unusual both in facing WSW and in facing a direction where the view is blocked. Yet in the opposite direction it incorporates an elegant device—part natural and part artificially constructed—for identifying the precise spot on the distant horizon where the Pleiades rose at the time of construction, from a standing position just outside the *heiau* on the NNW side, sighting parallel to the wall.

SITE ALE-211

ALE-211 was discovered by Kirch during a reconnaissance transect in the Alena uplands, *mauka* of the Pi'ilani Highway, on January 16, 2005. Only a rough sketch map was made, and GPS points taken, along with a few notes on the viewshed. Ruggles has not visited this site. The description given below should be regarded as very preliminary.

Location and topographic setting. The *heiau* sits atop a long ridge of the young Alena-2 'a'ā lava flow, at an elevation of about 640 masl and about 0.5 km east of the base of the Luala'ilua cinder cones. It overlooks a large swale to the west, which probably was a major planting area for sweet potatoes.

Architecture. This site is a variant of the elongated double-court type, broadly cardinally oriented, with a notch or offset along the E wall. The structure is about 25 m long (N–S) and 9 m wide (E–W). The northern part of the site consists of a rectangular enclosure with core-filled walls. Joining this on the S is a square paved court, with walls only on the N and E. A second paved area lies slightly farther down the ridgeline to the S. The western edge of the structure lies astride the ridge edge; the terrain drops off precipitously into the deep swale to the west.

Archaeological context. This upland part of Alena has never been systematically surveyed, so the broader settlement pattern context of ALE-211 remains unknown.

Viewshed. There are good views to the E, N, and W (Luala'ilua Hills). To the E there is a clear view of the point of Ka Lae o Ka 'Ilio in Kaupō, bearing 92° (corrected for magnetic declination). To the W, the highest summit of Luala'ilua bears 256°; the summit of the middle *pu'u* bears 223°, and that of the lowest *pu'u* bears 210°.

Orientation. The E walls were observed to have an orientation of 3°, and the *heiau* is oriented to the N. The ridgeline of Haleakalā was shrouded in clouds during Kirch's brief visit, but the computer-generated profile shows that the *heiau* is oriented toward a prominent cinder cone that forms the highest point on the profile as it appears from this location, at an azimuth of 2°. This is Kanahau, the red cinder cone noted from a number of Kīpapa sites.

7

Heiau of Kīpapa to Manawainui, Kahikinui *Moku*

The *Ahupua'a* of Kīpapa and Nakaohu

The land section of Kīpapa, lying to the east of the Luala'ilua Hills, occupies the central core of Kahikinui District. The name appears on the 1838 "Kalama map," and thus Kīpapa was a prominent territorial unit, almost certainly an *ahupua'a*. (The name also is listed as an *ahupua'a* in Royal Patent Grant 2901, dated to 1855 [Kirch 2014, table 6]). Nakaohu, adjoining Kīpapa to the east, is first shown on the 1882 Hawaiian Government Survey map by W. R. Lawrence, where its name was handwritten in fine pencil sometime after the original map was drafted. It is doubtful that Nakaohu had the status of an *ahupua'a* (it also is listed in the Royal Patent Grant 2901 to Kanakaole in 1855; see Kirch [2014], table 6); it is possible that Nakaohu was an *'ili* segment (a land unit smaller than an *ahupua'a*) within Kīpapa. Because the boundaries of Kīpapa and Nakaohu are unknown, for the purposes of archaeological survey Kirch combined all the sites within both Kīpapa and Nakaohu under the three-letter code "KIP." The western boundary of Kīpapa with Alena is taken to be defined by a long stone wall, running *mauka–makai*, or roughly north–south, from the coast up to the base of the Manukani cinder cone. The wall is not an actual boundary marker but, rather, a cattle control wall built by Portuguese ranchers after 1872, but it provides a convenient demarcation.

The geological substrate in Kīpapa and Nakaohu consists of several Hāna volcanic series lava flows ranging between 10,000 to 50,000 years old (Sherrod et al. 2007). The relatively high nutrient status of these younger flows made them ideal for intensive dryland cultivation of sweet potato and other crops, especially in the inland zone between 400 and 600 masl (Kirch 2014, 98–114). Consequently, the archaeological settlement pattern of upland Kīpapa and Nakaohu is one of densely concentrated sites, including numerous residential complexes and *heiau* (61–74). A total of 29 sites within Kīpapa-Nakaohu have been identified as potential *heiau*, although we will argue that 3 of these had other functions (as an oven house associated with a temple group, a *kahua* platform for *hula* performance, and a chiefly residence).

The Kīpapa-Nakaohu region has been more extensively surveyed for archaeological sites than any other part of Kahikinui *moku*. Initial survey work began in 1966 with Bishop Museum's project under the direction of Peter Chapman (Kirch 2014, 18–26), when 544 structures or features were recorded. Kirch recommenced the Kīpapa-Nakaohu survey in 1995, while Boyd Dixon (Dixon et al. 2000) surveyed the uppermost reaches of this area for the Department of Hawaiian

Home Lands. In total, 1,418 distinct archaeological features have been recorded for the Kīpapa-Nakaohu area (Kirch 2014, table 1).

KIP-1 (Walker Site 181; State Site 50-50-15-181; Bishop Museum Site MA-A35-1)

Site KIP-1 was first recorded by Walker ([1930], 251), who gave it site number 181 and described it as "a double terraced platform with a wall around the upper terrace" (see also Walker [1929], 20). Walker noted the presence of a pit that he dug out within the western "terrace," finding "ashes and charcoal to a depth of 4 feet." The Bishop Museum 1966 survey team relocated the site and sketch-mapped it with compass and tape. Kirch revisited the site in 1995, mapping it with plane table and alidade at a scale of 1:200. Kolb and Radewagen (1997, 68–69) briefly describe the site; Kolb later excavated at KIP-1, obtaining two radiocarbon dates (Kolb 2006, table 1). Ruggles visited the site on April 11 and 16, 2002, conducting compass-clinometer surveys; Kirch and Ruggles conducted GPS and total station surveys of KIP-1 on March 25, 2003.

Location and topographic setting. Site KIP-1 is situated in the Kīpapa uplands at an elevation of 640 masl. The temple was constructed on the relatively level tongue of an Alena basanite lava flow, immediately west of a small hillock (*puʻu*) formed on an adjacent flow of Kīpapa ankaramite.

Architecture. KIP-1 is a substantial structure of the elongated double-court type (fig. 7.1). It consists of two enclosures, each roughly square in plan view and broadly cardinally oriented,

FIGURE 7.1. Aerial photo of *heiau* KIP-1, viewed from the SW. (Photo by P. V. Kirch)

adjoining each other (and sharing a common joining wall) but offset by about 6 m (fig. 7.2). Both enclosures have a combined footprint of about 790 m², making this one of the larger temples in Kahikinui. The E enclosure (19 m E–W by 17 m N–S) is substantially constructed, the outer face having two steps for a buttress-like effect. The easternmost wall is the strongest, with a width of nearly 3 m and an exterior height of 2.2 m above the adjacent ground. An opening in the middle of the low W wall served as an entryway into the E enclosure. Within the E enclosure, the courtyard is divided into two sections, the western part covered with scattered ‘ili‘ili pebbles and coral, while the eastern half is roughly paved with ‘a‘ā cobbles. The SE corner exhibits a rectangular pit or depressed area defined by a low cobble facing and abutting the interior faces of the enclosure walls; two other more irregular depressions or pits were also noted in this eastern half of the court.

The W enclosure is most clearly defined on the E and S (where the stacked cobble wall is about 2 m wide and 1 m high) and is open on the W, but the enclosure is much less impressive than the E enclosure. A low, poorly constructed wall defines the NW corner. Three prominent ‘a‘ā uprights from 60 cm to 80 cm high stand in a line near the SW corner; these may represent deities. To their north is a low, paved terrace. The interior of the W enclosure has scattered ‘ili‘ili pebbles and branch coral. Near the center of the W enclosure is the circular depression or pit (1–1.5 m diameter) mentioned by Walker as having contained charcoal and ash; it was presumably either an earth oven or a pit for the disposal of offerings.

Dating and chronology. Kolb (2006, table 1) reports two radiocarbon dates from KIP-1,

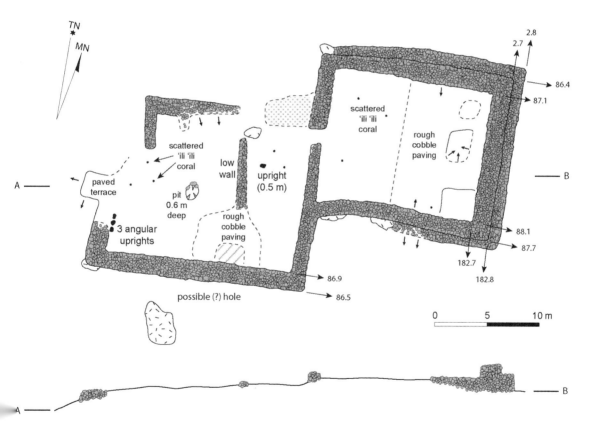

FIGURE 7.2. Plan of *heiau* KIP-1, showing the azimuths of intact straight segments of wall facing as determined by the total station survey. (Based on plane-table survey by P. V. Kirch)

both from "basal" contexts, although no further contextual details are provided. Sample B-122891 (500 ± 60 BP) has 2σ calibrated age ranges of AD 1297–1495 (94.1%) and 1601–1615 (1.3%). Sample B-122892 has a date of 350 ± 60 BP, yielding a calibrated date range of AD 1444–1649. Kirch and Sharp dated one branch coral offering (with a slightly rounded branch tip, KIP-1-CS-1; see Kirch, Mertz-Kraus, and Sharp [2015], table 1), from the surface of the E enclosure. This coral sample yielded a ^{230}Th age of AD 1325 ± 3, which overlaps with the older calibrated age range of radiocarbon date B-122891. Given the agreement between the radiocarbon and coral dates from KIP-1, we cautiously suggest that this temple was initially constructed in the fourteenth century, prior to the main phase of temple expansion in Kahikinui. This would make it one of the earliest known *heiau* in the district.

Archaeological context. There are no structures immediately adjacent to site KIP-1, which may be an indication of the *kapu* nature of the structure. The closest structures are part of a residential complex (*kauhale*) situated on a knoll about 100 m upslope from the *heiau*. Other residential clusters lie about 200 m to the west and 250 m to the east.

Viewshed. There is an unimpeded view of the sweep of the northern ridgeline from the WNW (~4 km distant) upward to the Haleakalā rim and beyond down to the coast 20 km to the E (az. ~80°). Around due E, just to the right of the foot of this slope, is a low rounded hillock, or *puʻu*, at a distance of about 20 m. There is open sea from the ESE round to the WSW, where the tip of the most northeasterly of the Lualaʻilua Hills, 2.0 km distant, appears prominently on the skyline behind a more local ridgeline between ~500m and ~800m away.

Orientation. The main axis of orientation of KIP-1 is clearly to the E, given the elaborate development of the eastern facade and the more prominent construction involved in the E enclosure. The mean orientations of the N wall of the E enclosure and the S wall of the W enclosure are exactly the same, 86.7°, while that of the S wall of the E enclosure, which is rather more sinuous, is 87.9°. From the perspective of a priest or other observer situated within the court of the E enclosure, the most prominent landscape feature is the low *puʻu* to the E (fig. 7.3). It spans about 15° in azimuth from the dip on its left to its rounded summit, the right-hand slope dropping off well below the level of the sea horizon beyond. From the southern side of the court, the azimuth of the summit is around 90° (i.e., due E), while from the northern side it increases to about 102°: it appears either just below or just obscuring the sea horizon, depending on the exact observing position and whether the observer is seated or standing. It follows that the sun would have been seen to rise over the *puʻu* from the court for a period of a few weeks leading up to and around the spring equinox, and around and after the autumn equinox.

The only wall yielding a reliable indication to the N is the E wall, whose mean azimuth is 2.8°—that is, deviating by some 5°–6° from perpendicularity with the E–W-oriented walls. It is not aligned upon any particularly prominent topographic feature, thus reinforcing the impression that the northern orientation was of no particular significance. The range of wall alignments to the W (az. 266.5° to –268.0°, alt. 4.0°, dec. –2.0° to –0.5°, all quoted to the nearest 0.5°) corresponds closely to the setting position of Orion's Belt—which either constituted (or was within) the Hawaiian asterism Nā Kao (Johnson, Mahelona, and Ruggles 2015, 198)—at any time from the fourteenth century onward (see table 4.2). It is also quite close to, although not exactly upon, the setting position of the equinoctial sun (az. 268.2°–268.7°), while the winter solstice sun would have been seen to set just to the left of the northwestern Lualaʻilua peak and above the limit of the sea horizon.

Additional remarks. While the consistency of orientation of the E–W-oriented walls at

FIGURE 7.3. View eastward along the S wall of the eastern enclosure of *heiau* KIP-1, showing the low *puʻu* about 20 m to the E. (Photo by Clive Ruggles)

this temple is striking, there is no evidence of any precise topographic or astronomical alignment to the E. Instead, in the direction faced by the temple, the sun would have been seen to rise for a few weeks at around the time of the spring and autumn equinoxes over the low but prominent *puʻu* immediately to the E.

The "Heiau Ridge" Complex (KIP-75, -76, -77, -115, -405, and -410)

Six sites form an extensive complex of stone temple foundations and associated features situated on a prominent ridge of *ʻaʻā* lava in the land section of Nakaohu, roughly 3 km inland of the coast, at an elevation of about 400 masl. We colloquially named this complex "Heiau Ridge" because it occupies that prominent landscape feature (Kirch 2014, 204–207). This temple complex lies within the main upland agricultural and habitation zone of Kahikinui *moku*, as described in Kirch et al. (2004). The *ʻaʻā* ridge slopes gently from northwest to southeast, bounded on the north by an intermittent stream gully that follows the boundary between two lava flows. The temple complex was constructed on the younger Nawini ankaramite *ʻaʻā* lava flow (between 10,000 and 30,000 years old) and overlooks an older Kahikinui basanite lava flow to the north and east. As Kirch et al. (2004, 2005) and Hartshorn et al. (2006) have demonstrated, this 53,000-year-old Kahikinui substrate was ideal for intensive agriculture, especially sweet potato and dryland taro farming. The individual *heiau* within the Nakaohu temple complex arguably include structures dedicated to the Hawaiian gods Kāne and Lono, based on their architectural styles and orientations (Kirch 2004b).

An overview plan map of the main part of the Heiau Ridge temple complex is given in figure 7.4 (with the exception of KIP-410, which lies beyond the map boundary to the north), while figure 7.5 is an aerial view of the central part of the complex. At the top of the complex is

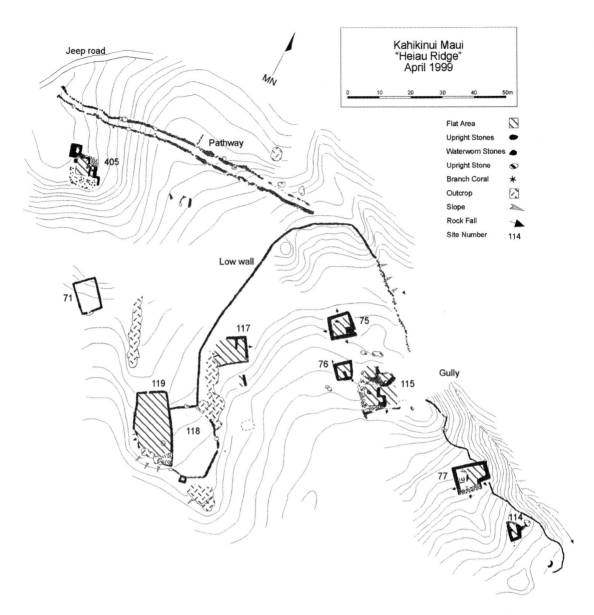

FIGURE 7.4. Map of the Heiau Ridge complex in Nakaohu, upland Kahikinui. (Based on plane-table survey by P. V. Kirch)

KIP-410, a small notched *heiau*. Descending from KIP-410, one next encounters KIP-405, a *heiau* that occupies a prominent knoll with commanding views over the landscape and overlooks the lower temples. A stone-lined pathway, oriented east–west, runs from this knoll down to an intermittent watercourse. The main cluster of structures is bounded on the north and west by a low wall (~40–60 cm high) of stacked ʻaʻā boulders. Site KIP-75 is a square *heiau* with a raised platform in the southeast. KIP-76 is a smaller enclosure that excavation showed to be an oven house, probably for cooking sacrificial offerings to be presented in KIP-75 and possibly other temples. To the west of KIP-75 and KIP-76 is a well-constructed habitation site (KIP-117) that was excavated and interpreted by Kirch et al. (2010) as a residence of a priest (*kahuna*) who was likely

FIGURE 7.5. Aerial photo of a portion of Heiau Ridge, from the NE. In the middle ground, the three structures from nearest to farthest are KIP-75, KIP-76, and KIP-115. In the distance, site KIP-77 is visible. (Photo by P. V. Kirch)

responsible for the entire Heiau Ridge complex. West of KIP-117 are two large, level spaces defined by enclosing walls, which may have served as assembly areas for persons observing rites at Heiau Ridge. Site KIP-115 is a low *heiau* foundation that may not have been completed. Next, site KIP-77 overlooks the nearby intermittent watercourse prominently and is a classic notched *heiau* oriented to the ENE. Below this is KIP-114, a small enclosure of uncertain function. Site KIP-71 (also shown on fig. 7.4) is a rectangular enclosure, possibly a postcontact house, that is not part of the Heiau Ridge complex.

It is likely that this close association of *heiau*, unique within the entire Kahikinui District, formed a sort of acropolis with temples dedicated to a number of different gods and cults. The following sections describe the six main structures in detail, beginning with the highest on the ridge (KIP-410) and descending downslope in order.

KIP-410 (STATE SITE 50-50-15-4365; BISHOP MUSEUM SITE MA-A35-410)

This notched *heiau* with associated terrace was first recorded during the 1966 Bishop Museum survey, at which time Kirch and Kikuchi mapped it with plane table and alidade. Kirch visited the site on March 26, 2003, making both GPS and compass-clinometer surveys. Kolb and Radewagen (1997, 70) briefly mention the site, while Kolb (2006) reports a single radiocarbon date from excavations that remain unpublished.

Location and topographic setting. KIP-410 sits astride a ridge of quite young ʻaʻā lava (the Kanahau basanite, 5,000–10,000 years old), at an elevation of about 650 masl.

Architecture. The site consists of a small notched *heiau* enclosure, its orientation deviating from cardinality by about 15° in a counterclockwise sense, situated on the ʻaʻā ridge, with two paved terraces immediately to the WSW at a lower elevation down the sloping face of the lava ridge. A crude ramp connected the main *heiau* enclosure with the terraces, although this has largely collapsed. The entire complex has a combined basal area of 220 m², and the notched enclosure an area of 145 m² (Kolb and Radewagen [1997, 70] erroneously give the "internal area" of the *heiau* as 400 m²).

The notched *heiau* enclosure has external dimensions of 14.3 m ENE–WSW by 11.2 m NNW–SSE, with core-faced, well-constructed walls 1–2 m thick and about 1 m high. Numerous ʻiliʻili pebbles are scattered over the *heiau* floor in its western sector. The notch is situated in the WNW corner; there is no wall on the WSW side, where two or more facings (now largely collapsed) descend to the adjacent terraces. The upper terrace is about 3 m wide, separated from the broader lower terrace (6 m wide) by a low, single-course facing of ʻaʻā cobbles. These terraces likely formed a seating area for persons witnessing the rituals conducted within the *heiau* enclosure proper.

Dating and chronology. Kolb (2006, table 1) reports a single radiocarbon date (Beta-122898) of 500 ± 60 BP, with 2σ calibrated age ranges of AD 1297–1495 (94.1%) and 1601–1615 (1.3%). The sample is reported as coming from a "basal" context, but no other stratigraphic details are available. This date is considerably earlier than most radiocarbon or ^{230}Th dates from Kahikinui *heiau*; the "basal" context suggests that it may represent burning or land use prior to construction of the temple itself (the date, from unidentified charcoal, may also have an inbuilt age factor). We therefore regard this date as a *terminus post quem* and reject it as a date for the actual construction or use of the temple.

Archaeological context. Site KIP-410 is fairly isolated from other structures. A possible burial site lies about 70 m to the E, and a small C-shaped structure sits on the same ridge about 75 m upslope.

Viewshed. The Nawini flow ridge dominates the view, but there is still a full sweep of distant horizon from the Haleakalā rim extending eastward down to the Kīpahulu Valley and the sea. All the other Heiau Ridge sites are visible downslope. The Lualaʻilua Hills, 2.8–3.4 km away, rise above the ridgeline to the WSW.

Orientation. The lower terraces and the lack of a wall on the WSW side suggest that this *heiau* was oriented toward the ENE. Compass measurements on the NNW wall yield a best estimate of the eastward orientation of 73° (corrected for mean magnetic declination), while readings on the ENE wall yield 348° as the best estimate of the northward orientation.

The *heiau* is broadly oriented along the local contours. It faces the eastern slopes of Haleakalā at an altitude of ~2° and a declination of ~+16.5°. In the opposite (WSW) direction the walls are aligned at least approximately toward the northern Lualaʻilua peaks (around az. 253°, alt. 2°, dec. −15°). The orientation to the NNW is upon the Haleakalā rim, quite close to the red cinder cone Kanahau (az. ~345.5°); to the SSE it is upon open sea just beyond the southernmost point of Hawaiʻi Island. Without a more accurate survey, it is impossible to be more specific about the topographic or astronomical possibilities.

KIP-405 (STATE SITE 50-50-15-4364; BISHOP MUSEUM SITE MA-A35-405)

This *heiau* of rather unusual morphology was first recorded during the 1966 Bishop Museum survey and was plane-table mapped by W. K. Kikuchi. Kirch prepared a new plane-table map of the structure in July 1995. Ruggles visited the site on April 16 and June 16, 2002, conducting compass-clinometer surveys; and on March 26, 2003, Ruggles and Kirch carried out both GPS and total station surveys. Kolb and Radewagen (1997, 69) described the site as a "notched enclosure." Kolb (2006) reports a single radiocarbon date from otherwise unpublished excavations.

Location and topographic setting. KIP-405 was constructed on the lip of a knoll on the prominent Nawini ʻaʻā lava flow, at an elevation of 625 masl. The knoll commands a superb view eastward all the way to Kaupō, as well as westward to the Lualaʻilua Hills; it also looks down on the other structures making up the Heiau Ridge complex.

Architecture. As can be seen in figure 7.6, KIP-405 has a unique and quite irregular plan (it is neither an enclosure nor notched, as claimed by Kolb and Radewagen [1997, 69]). The site consists of a leveled terrace, roughly paved with ʻaʻā rubble, bounded on the N and E by a network of walls and platforms. Broadly speaking, it is cardinally oriented. The main terrace is approximately 15 m N–S by 13 m E–W, while the prominent platform on the N measures 7 by 7.5 m. The entire structure has a footprint of about 320 m^2 (as opposed to the 154 m^2 estimate of Kolb and Radewagen [1997, 69]). The square platform on the N has an internal, sunken compartment 4 by 2 m in area with an earthen floor, which is likely to have been covered over with a thatched structure when the *heiau* was in use; this was test excavated (see "Excavation").

A noteworthy feature, situated in the NE corner of the main terrace, is what appears to be an altar—a small level terrace paved with fine clinkers upon which several pieces of branch coral were set (fig. 7.7). This terrace is protected from the strong prevailing winds by an adjacent windbreak wall about 1.5 m high. Also notable are two rectangular spaces or "rooms" built into the wide eastern wall; one of these has an upright stone in front of it. These spaces may have been roofed over with thatched structures.

Excavation. A single test pit (1 m^2) was excavated in August 1999 within the sunken compartment situated in the stone platform on the N side of the structure. Beneath a 10 cm thick layer of overburden, the excavation revealed a single cultural deposit consisting of gravelly loam flecked with charcoal throughout; the layer varied in thickness but had a maximum depth of 35 cm. Five basalt flakes and two flakes of volcanic glass were recovered from the cultural deposit, along with small quantities of marine shell (15 pieces), unidentified bone (38 pieces), and weathered coral (2 pieces).

Dating and chronology. Kolb (2006, table 1) reports a radiocarbon date (Beta-122896) of 320 ± 70 BP from a "basal" context; the exact location and stratigraphic position of the sample remain unpublished. This sample has 2σ calibrated age ranges of AD 1438–1675 (91.5%), 1777–1800 (2.8%), and 1941–1950 (1.0%). Kirch's excavation within the northern platform produced a sample (AA-38652) of identified wood charcoal (*Sida* sp., a short-lived taxon) with an age of 159 ± 34 BP; this yields calibrated age ranges of AD 1664–1708 (16.6%), 1718–1888 (60.6%), and 1911–1950 (18.2%).

Kirch and Sharp (2005) obtained a ^{230}Th date on one of the pieces of branch coral deposited on the altar feature of KIP-405; the corrected age is AD 1601 ± 7. This age falls within the range of Kolb's "basal" radiocarbon date, and the coral most likely was placed on the altar as a dedicatory offering. It seems likely that the temple was constructed at the beginning of the seventeenth

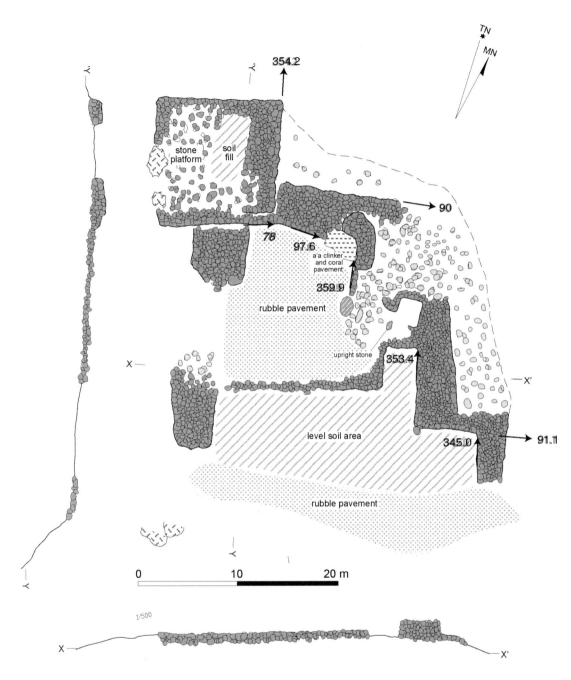

FIGURE 7.6. Plan of *heiau* KIP-405, showing the azimuths of intact straight segments of wall facing and platform edges as determined by total station survey (quoted to the nearest 0.1°) or prismatic compass corrected for mean magnetic declination (italicized and quoted to the nearest degree). (Based on plane-table survey by P. V. Kirch)

FIGURE 7.7. Branch coral offerings on the altar at KIP-405. Note also the waterworn cobble. (Photo by P. V. Kirch)

century (coincident with the reigns of Maui kings Kiha-a-Piʻilani and Kamalālāwalu); it probably continued in use up through the early postcontact period.

Archaeological context. There are no structures in the immediate vicinity of KIP-405. As noted earlier, the site is part of the Heiau Ridge complex, overlooking sites KIP-75, -76, and -115 a short distance downslope.

Viewshed. Northward from KIP-405 there is an unimpeded view of the Haleakalā ridgeline, while to the E are superb views of the lower slopes of Haleakalā running down to Nuʻu Bay and Ka Lae o Ka ʻIlio in Kaupō. To the WSW, the Lualaʻilua Hills feature prominently on the skyline beyond where the sloping ridgeline, about 1.5 km distant, meets the sea horizon. There is open sea horizon from the E round to this point.

Orientation. While the altar in the NE corner indicates that this temple faced E or N, the irregularity of this structure makes it unclear which of these was in fact the case. East seems more likely because of the superb viewshed and slight architectural elaborations on that side.

An intact segment of the inner face of the N wall, W of the altar, perhaps gives the most reliable estimate of the E–W orientation as 97.6°, but other less reliable wall segments suggest orientations around 90° to 91°, and the stone platform has an anomalous azimuth of around 78°. The E side of this platform (az. 354.2°) is on an orientation similar to that of the inner face of the central one of the three wall segments stretching away to the SE (353.4°), but intact segments of the other two north–south-oriented walls yield azimuths of 345.0° and 359.9°, respectively, so there is wide variation. The various orientations are shown in figure 7.6.

The range of possible orientations to the E spans a stretch of sea horizon immediately above Ka Lae o Ka ʻIlio, in line with sunrise for a few days before and up to the spring equinox, as

Kīpapa to Manawainui • 209

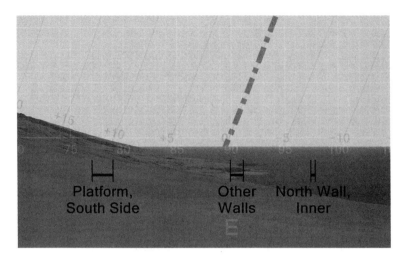

FIGURE 7.8. DTM-generated profile of the eastern horizon as seen from *heiau* KIP-405. The horizontal bars indicate the various directions discussed in the text. The dash-and-dot line indicates the rising path of the equinoctial sun. (Graphic by Andrew Smith)

well as at and after the autumn equinox (fig. 7.8). It also includes the rising position of Orion's Belt (Nā Kao) or part of it (Johnson, Mahelona, and Ruggles 2015, 198). The range to the N spans a segment of the Haleakalā rim, including the prominent red cinder cone Kanahau peaking at azimuth 346°, upon which the southernmost north–south-oriented wall segment (in this case, to be precise, NNW–SSE) is approximately aligned (but not the other walls).

To the W, the winter solstice sun would have set into a small stretch of sea horizon visible just to the right of the southernmost of the Luala'ilua peaks, at azimuth 245°, but the wall orientations are well to the right of this.

Additional remarks. If, as seems likely, this *heiau* was oriented to the E, then it faced a prominent landmark—Ka Lae o Ka 'Ilio, visible below the sea horizon—which was broadly in the direction of equinoctial sunrise. To the N, at least one of the walls is aligned upon a prominent topographic feature: the conspicuous red cinder cone Kanahau on the Haleakalā ridge.

KIP-75 (STATE SITE 50-50-15-4362; BISHOP MUSEUM SITE MA-A35-75)

This square enclosure was first recorded during the 1966 Bishop Museum survey, although it was not recognized as a *heiau* at that time. Kirch mapped the site with plane table and alidade in June 1996, carrying out a test excavation at the same time. Kolb and Radewagen (1997, 69) briefly report the site. Ruggles and Kirch conducted both GPS and total station surveys on March 26, 2003.

Location and topographic setting. KIP-75 is situated on a gently sloping section of Heiau Ridge, in the lee of a prominent line of '*a*'*ā* outcrops a few meters to the east. The elevation is approximately 590 masl.

Architecture. KIP-75 is a square enclosure 15 m on each side, with a basal area of 225 m² (fig. 7.9). Its orientation deviates from cardinality by 30° in a counterclockwise sense. The core-filled walls are 1–2 m thick and were originally up to 1 m high, although many of the original

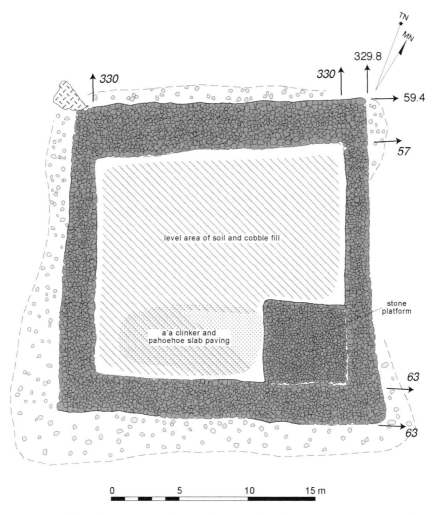

FIGURE 7.9. Plan of *heiau* KIP-75, showing the azimuths of intact straight segments of wall facing as determined by the total station survey (quoted to the nearest 0.1°) and estimates of the orientation of other wall faces obtained from the plane-table survey (quoted to the nearest degree). (Based on plane-table survey by P. V. Kirch)

faces have suffered collapse. The interior consists of an earthen floor except for the southern part, which exhibits evidence of having been paved with *'a'ā* and *pāhoehoe* slabs. A low stone platform, 4 by 4 m, occupies the ESE corner of the enclosure; this may have been the foundation for either a small thatched structure or an oracle tower (*'anu'u*). On the ENE wall near its center point we observed an unusual, square-sectioned basalt stone 35–40 cm long that is likely to be a fallen upright.

Excavation. Two grid units (M9 and M11), each 1 m², were excavated in the interior of KIP-75. In both units a shallow cultural deposit, with much charcoal included, extended to a depth of 28–32 cm below surface. In unit M11 a combustion feature was uncovered near the base of the cultural deposit, with a basin-shaped pit extending into the subsoil to a depth of 69 cm below surface. The feature, which was most likely an earth oven (*imu*), had a maximum diameter of 80 cm and a depth of 30 cm. The feature also contained some fire-altered stones.

The excavations yielded basalt and volcanic glass flakes, along with pieces of branch coral and both marine shell and vertebrate faunal remains (table 7.1). The vertebrate fauna included small numbers of bones of pig, dog, and chicken, as well as marine turtle (*Chelonia mydas*) and several species of fish (*Scarus* sp., *Calotomus* sp., Serranidae). These faunal remains are all likely to have derived from offerings or from foods consumed by individuals performing rituals or feasting within the enclosure. There were also significant numbers of bones of the Pacific rat (*Rattus exulans*), which may have been scavenging on the remains of offerings left within the temple enclosure.

Dating and chronology. Two radiocarbon dates were obtained from the unit M11 test excavation in KIP-75, one from level 2 and one from feature 1. The sample from feature 1 (AA-38637) was on identified wood charcoal from short-lived *Sida* species, with an age of 147 ± 33 BP; this yields 2σ calibrated age ranges of AD 1667–1783 (45.5%), 1796–1891 (32.7%), and 1909–1950 (17.2%). The second sample (AA-38638) was on identified wood charcoal from short-lived *Chenopodium* sp., with an age of 136 ± 33 BP and calibrated age ranges of AD 1670–1780 (41.6%), 1798–1893 (38.6%), and 1907–1944 (15.2%). Both radiocarbon dates overlap and together suggest that KIP-75 was in use from the late seventeenth to the eighteenth century. This period corresponds to the reigns of the later Maui *ali'i nui* Ka'ulahea II, Kekaulike, and Kamehamehanui (Kirch 2014, table 2).

Archaeological context. KIP-75 is part of the Heiau Ridge complex. Site KIP-76, an oven house, is situated just 13 m to the south and likely provided the cooked food offerings used in KIP-75. The large priest's house, KIP-117, is situated about 50 m southwest of KIP-75.

Viewshed. There are clear views of the Haleakalā rim to the N and its upper slopes on each side, from NW round to ENE. From ESE round to SW is the open sea, with Hawai'i Island visible to the SE in good weather conditions. Both westward, between azimuths of ~230° and

TABLE 7.1. Cultural content of KIP-75 excavations.

MATERIAL	UNIT M9	UNIT M11	FEATURE 1 (UNIT M11)
Basalt flakes	4	51	11
Volcanic glass	1	3	1
Coral	9	6	2
Charcoal (g)	17.2	69.6	37.4
Pig bone (NISP)	1	—	1
Dog bone (NISP)	—	1	—
Medium mammal bone (NISP)	1	1	7
Chicken bone (NISP)	3	11	12
Rattus exulans bone (NISP)	9	76	6
Turtle bone (NISP)	—	2	—
Fish bone (NISP)	31	69	22
Unidentified bone (NISP)	20	88	25
Marine shell (NISP)	40	25	1

~305°, and around due E, depending on one's exact position, the ridge on which the *heiau* sits restricts the view to within about 100 m.

Orientation.   The low platform in the ESE corner and the likely fallen upright on the ENE wall suggest that this enclosure was ENE-facing. An intact segment of the outer face of the NNW wall is oriented with an eastward azimuth of 59.4°, but corner-to-corner estimates of the other ENE–WSW-oriented wall faces yield estimates of 58° for the NNW wall (by averaging those for the inner and outer walls, 57° and 59°, respectively) and 63° for the SSE wall, for a mean axial orientation of 60.5°. An intact segment of the outer face of the ENE wall is oriented with a northward azimuth of 329.8°; consistent values (330°) are obtained for the other NNW–SSE-oriented wall faces.

The *heiau* faces an evenly sloping stretch of the Haleakalā eastern slope—with the closer ridge just below it—between a conspicuous dip at azimuths 45°–46° and the spot, between azimuths 66° and 67°, where the Pleiades rose around AD 1700. To the NNW, the walls are oriented toward a hill junction to the left of a wide cinder cone, but although this is a distinctive horizon feature (az. 329.8°, alt. 14.1°, dec. +60.5°), no bright stars set here around AD 1700 (see table 4.2). To the WSW, the wall orientations are toward local ground, while to the SSE they are toward the broad northern slopes of Mauna Loa on Hawai'i Island, visible only in clear weather conditions.

KIP-76

KIP-76 is a small enclosure (measuring 10 by 11.5 m) situated just 14 m downslope from KIP-75. Initially we were uncertain whether KIP-76 was another square *heiau* enclosure, but excavations in the site in 1996 demonstrated that it contains a series of earth ovens (*imu*). We therefore interpret the structure as an oven house for preparing and cooking offerings to be presented at adjacent KIP-75 and possibly at other nearby *heiau* such as KIP-115 and KIP-77. We mention the site here because it is an integral part of the Heiau Ridge complex.

We obtained two radiocarbon dates from the excavations in KIP-76, both on wood charcoal from relatively short-lived taxa. Sample AA-38639 yielded an age of 328 ± 36 BP, with a 2σ calibrated age range of AD 1470–1645, while sample AA-38640 yielded an age of 348 ± 43 BP, with a calibrated age range of AD 1456–1640. These similar dates confirm that site KIP-76 was utilized during the same time frame as other sites in the Heiau Ridge complex.

KIP-115 (BISHOP MUSEUM SITE MA-A35-115)

Site KIP-115 was first recorded during the 1966 Bishop Museum survey. Kirch mapped the structure with plane table and alidade in April 1999; three test pits were excavated at the same time. A partial GPS survey was made in March 2003, when the site was briefly visited by Kirch and Ruggles.

Location and topographic setting.   Part of the Heiau Ridge complex, site KIP-115 lies just 10 m SE of KIP-76 on the same spur of relatively young Nawini ankaramite 'a'ā lava. Its elevation is approximately 580 masl.

Architecture.   The architecture of KIP-115 is unimpressive, with the standing walls low and somewhat irregular in configuration. The site gives the impression of having been rapidly constructed and perhaps even being unfinished. As seen in the plan map (fig. 7.10), the most substantial feature is a stacked cobble wall approximately 10 m long, oriented NNW–SSE, which then abuts to a three-sided room with exterior dimensions of 5 by 5 m, open to the WSW. An

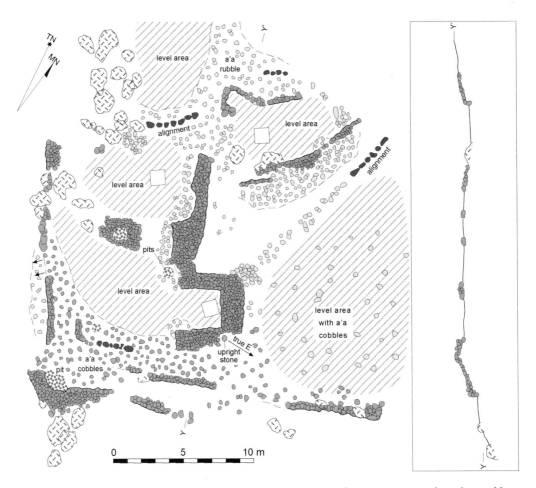

FIGURE 7.10. Plan of *heiau* KIP-115, showing three excavation units (*white squares*). (Based on plane-table survey by P. V. Kirch)

upright boulder is situated in the ESE corner of the wall defining this room. The configuration of the longer wall and adjacent room create a notched configuration, but there is no complete enclosure as in most notched *heiau*. Westward of these walls there is a roughly leveled terrace, about 9 by 15 m in extent, defined on the WSW side by a low retaining wall. There is a pit feature in the SSW corner of the terrace, and two other pit features on the terrace itself.

Eastward of the notched wall there is another roughly leveled area partly defined on the downslope (ESE) side by a crude boulder facing partly heaped on the underlying bedrock. Finally, to the NNW of the notched wall there is a slightly sunken, leveled soil area (4 x 7 m) defined by low, crudely stacked walls.

Excavation. Three test pits, 1 m² each, were excavated at KIP-115 in April 1999. Unit J21 revealed a charcoal-rich cultural deposit about 30–40 cm thick, containing 36 basalt flakes and 22 volcanic glass flakes, 10 pieces of marine shell, and 7 weathered coral fragments. Unit M11, with a similarly shallow cultural deposit, yielded 2 basalt flakes and 17 volcanic glass flakes, along with 2 pieces of marine shell and 1 piece of coral. Unit R23 exposed a rough pavement of ʻaʻā cobbles over a charcoal-flecked cultural deposit containing 99 basalt flakes, 18 volcanic glass flakes, 2 pieces of marine shell, and 8 pieces of coral.

Dating and chronology. A carbonized twig of the endemic, short-lived *Osteomeles* shrub (sample AA-38643) from excavation unit J21 yielded a radiocarbon date of 309 ± 34 BP, with a calibrated age range of AD 1482–1652. A second sample (AA-38644) of carbonized *Nototrichium* sp. wood from excavation unit R23 yielded a date of 379 ± 35 BP, with 2σ calibrated age ranges of AD 1442–1529 (57.9%) and 1552–1634 (37.5%). These relatively early dates indicate the use of this *heiau* between the fifteenth and seventeenth centuries.

Archaeological context. KIP-115 is part of the Heiau Ridge complex.

Viewshed. The viewshed is essentially the same as that at KIP-75, except that the eastward view is less restricted.

Orientation. This structure appears to be oriented toward the ENE, with its main axis—through the small notched room—facing an azimuth of ~70°. This is about 4° to the right of the rising point of the Pleiades around the sixteenth century.

KIP-77 (BISHOP MUSEUM SITE MA-A35-77)

Site KIP-77, a classic notched *heiau*, was first recorded during the 1966 Bishop Museum survey, when a compass-and-tape sketch plan was made by Peter Chapman. Kirch made a plane-table survey of the structure in June 1996 and remapped it in April 1999, at which time three test pits, 1 m² each, were excavated in the enclosure floor. Kirch and Ruggles jointly investigated the site on March 26, 2003, when both total station and GPS surveys were made, along with compass-clinometer observations.

Location and topographic setting. KIP-77 lies at an elevation of about 575 masl, on the spur of the relatively young Nawini ankaramite ‘a‘ā lava that forms Heiau Ridge. The *heiau* was built on the lip of the massive lava flow, so that the terrain falls away steeply below the high ENE wall of the structure.

Architecture. KIP-77 consists of a well-constructed enclosure with "notched" configuration (figs. 7.11 and 7.12). Its orientation is similar to that of KIP-75, deviating from cardinality by about 30° in a counterclockwise sense. The maximum exterior dimensions are 21 m WSW–ENE and 16 m NNW–SSE. The main walls are of the core-filled type, ranging between 0.6 and 1 m high on the interior faces, up to a maximum height of 2 m along the exterior facade of the ENE wall.

There are a number of noteworthy architectural features within the enclosure. The main court is divided into two parts by a low stone facing running N–S about 1/3 of the way from the WSW enclosure wall. Within the slightly lower, earthen space in this western third of the court, a rectangular cist-like feature is defined on three sides by low, upright slabs set on edge; the feature measures 2 by 1.5 m; it may be a burial (possibly postcontact). About 1 m to the S of this cist is a row of seven carefully placed ‘a‘ā slab uprights (the row is 3.25 m long, and the uprights about 0.5–0.6 m high), buttressed behind with a low core-filled wall. Farther to the NNE in the main court there is a shallow depression, 1.5–2 m in diameter, filled with ‘a‘ā cobbles; this may be a filled-in earth oven or offering pit.

In the top of the ENE wall of the southern court are two stone blocks placed side by side and supported by smaller boulders at each end so as to leave a gap underneath each. This arrangement, curiously, faces diagonally across the *heiau* toward a boulder outcrop, 1.2 m wide, within the main court; the form of the arrangement suggests the possibility that it functioned as a sighting device.

Excavation. Three units (P19, S17, and S22), 1 m² each, were excavated in the main court of

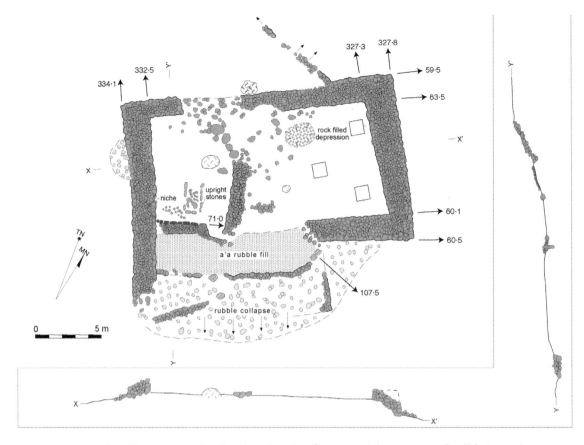

FIGURE 7.11. Plan of *heiau* KIP-77, showing the azimuths of intact straight segments of wall facing and a possible sighting device as determined by the total station survey. (Based on plane-table survey by P. V. Kirch)

FIGURE 7.12. Aerial view of *heiau* KIP-77 after excavation (three test pits, each 1 m², are evident in the main court). (Photo by P. V. Kirch)

KIP-77 in April 1999. Beneath a shallow layer of eolian overburden (3–5 cm thick), all three units contained a dark gray-brown cultural deposit between 40–60 cm thick, depending on the unit. The lower portions of this deposit contained dense ʻaʻā cobbles, which appeared to be fill laid down to level the floor of the court at the time of initial construction. Two small combustion features were encountered: feature 1, in unit S22, was a basin-shaped deposit of ash extending 20–40 cm below surface; feature 2, in unit P19, was a similar but slightly smaller ash deposit at the same depth. Both of these features appear to have been small hearths rather than earth ovens.

A diverse assemblage of cultural materials, including artifacts and faunal remains, was recovered from the three test excavations, as enumerated in table 7.2. Part of a bone fishhook, cut bone, two coral abraders, and an adze flake, along with numerous basalt flakes and volcanic glass flakes, all attest to various activities involving making or repairing a variety of items of material culture. As discussed in chapter 3, this is evidence that *heiau* such as KIP-77 were not only places of ritual activity and worship but also gathering places for men, most likely during the *kapu* days prescribed by the Hawaiian lunar calendar. The faunal remains included both pig and dog, along with rat (*Rattus exulans*), bird, and fish. Most of the bird was fragmentary and could not be identified but did include one chicken (*Gallus gallus*) bone, the tarsometatarsus of a flightless rail (Rallidae), and one bone of the wedge-tailed shearwater (*Puffinus pacificus*). The identifiable fish bones included a number of specimens of *Acanthurus* sp. and *Lutjanus* sp.; single specimens each of Balistidae, *Epinephelus* sp., and *Naso* sp.; and one shark tooth.

Dating and chronology. Two radiocarbon dates were obtained on samples excavated from the cultural deposits in the main court. The first sample (AA-38641), from level 4 of unit S17,

TABLE 7.2. Cultural content of KIP-77 excavations.

MATERIAL	UNIT P19	UNIT S17	UNIT S22
Fishhook	1	—	—
Coral abrader	1	1	—
Cut bone	—	—	1
Adze flake	—	—	1
Basalt core	—	—	1
Basalt flakes	83	170	131
Volcanic glass flakes	110	57	43
Coral	57	121	34
Charcoal (g)	145.3	252.9	146.4
Candlenut endocarps (g)	1.8	0.2	—
Pig bone (NISP)	6	23	1
Dog bone (NISP)	—	—	3
Medium mammal bone, pig or dog (NISP)	9	6	4
Rat bone (NISP)	8	14	43
Bird bone (NISP)	8	3	6
Fish bone (NISP)	42	67	31
Waterworn pebbles (ʻiliʻili)	20	22	23

consisting of wood charcoal from short-lived *Chamaesyce* sp., yielded a conventional age of 243 ± 34 BP, with calibrated age ranges of AD 1521–1575 (11.7%), 1585–1590 (0.4%), 1626–1684 (47.2%), 1736–1805 (28.9%), and 1935–1950 (7.2%), the last of which can be discounted. The second sample (AA-38642), from feature 1 in unit S22 and consisting of carbonized tuber (parenchyma), yielded a conventional age of 238 ± 33 BP, with calibrated age ranges of AD 1524–1559 (6.6%), 1631–1684 (46.7%), 1736–1806 (33.3%), and 1934–1950 (8.7%), the last of which can again be discounted. The virtually identical ages of these two samples lend confidence to an interpretation that the site was in use sometime during the seventeenth to the eighteenth century.

Archaeological context. KIP-77 is part of the Heiau Ridge complex, situated about 60 m downslope from site KIP-115. A small enclosure, KIP-114, lies about 25 m downslope from KIP-77; the function of KIP-114 has not been determined.

Viewshed. There is an uninterrupted view of the Haleakalā Ridge to the N and its eastern slopes sweeping down to Nuʻu and Kaupō, finishing at Ka Lae o Ka ʻIlio (az. 93°). The local *ʻaʻā* ridge, never more than ~150 m distant, limits the view from the S round to the WNW, between ~185° and ~305°. Open sea is visible to the SE, with the whole of Hawaiʻi Island in view given favorable weather conditions.

Orientation. The prominent ENE wall, whose outer face is 2 m high, together with the division of the main court with a lower part to the WSW, indicates that this structure faced ENE. The corner-to-corner orientation of the outer face of the NNW wall has an eastward azimuth of 59.5°, while the intact eastern part of the inner face yields 63.5°, so that our best estimate of the mean wall orientation is 61.5° (see fig. 7.11). That for the SSW wall east of the notch is 60.3°, so our overall best estimate for the mean axial orientation is 60.9°. The line of uprights has a somewhat different orientation, with an eastward azimuth of 71.0°. Our best estimates of the orientations of the ENE and WSW walls are 327.5° and 333.3°, respectively, for a mean northward orientation of 330.4°.

The *heiau* faces a stretch of the Haleakalā eastern slope between a conspicuous dip at azimuth 45°–46° and the rising position of the Pleiades (and the solstitial sun) around azimuth 66°–67°: the azimuth range 60.3°–61.5° corresponds to a declination range of +28.1° to +29.4°. The Hawaiian star Nānāhope (Johnson, Mahelona, and Ruggles 2015, 199)—that is, Pollux (β Gem)— rose here at the time of construction and use (dec. +28.9° in AD 1600, +28.5° in AD 1800), as did Alnath (β Tau, dec. +28.2° in AD 1600, +28.4° in AD 1800), although the latter star is not known to have been significant to the Hawaiians. The row of stone uprights is aligned farther down the slope (az. 71.0°, alt. 1.8°, dec. +18.3°), around which point rose Ka Nuku o Ka Puahi (ibid., 171), the V-shaped formation of the Hyades plus Aldebaran (dec. range +14.6° to +18.2° in AD 1600, +15.1° to +18.7° in AD 1800). This latter point is also the position of sunrise around May 12 and July 31 (Gregorian). The two positions also correspond roughly to the major and minor standstill limits of the moon (dec. +28.3° and +18.0°, respectively, which is the apparent declination after applying the lunar parallax correction for latitude 20.6° N; see chapter 4); however, the ethnohistory provides no reason to suppose that the lunar node cycle was of any significance to Hawaiians.

Toward the NNW, the WSW wall is oriented upon a wide cinder cone, but to the right of its peak, while the ENE wall is oriented to its left. The declination of the peak itself (+62.8°) has no obvious astronomical significance. To the WSW, the wall orientations are toward local ground, while to the SSE they are toward the low peak of Mauna Loa on Hawaiʻi Island, visible only in fine weather.

The azimuths of the "sighting-stones" as viewed from the center of the boulder outcrop are 105.7° (N stone, N side), 107.5° (where the stones touch), and 109.7° (S stone, S side). This means that the gaps between them have azimuths of 106.5°–106.7° and 108.3°–109.0°, respectively. Moving to either side of the boulder outcrop increases or decreases all these values by about 2.5°. While the interpretation of these two stone blocks and their supporting stones as a sighting device is speculative, our survey establishes that if one were to crouch down by the outcrop sufficiently as to bring the sighting-stones up to the level of the sea horizon (alt. −0.7°), then the declinations concerned would be −15.9° to −16.1° for the left (small) gap and −17.6° to −18.3° for the right (larger) one. Again, moving to either side of the boulder increases or decreases these declinations by about 2.5°.

The fact that the declination of the star Sirius (α CMa) ranged from −16.2° in AD 1600 to −16.4° in AD 1800 invites the suggestion that this curious structure was some sort of sighting device related to the rising of Hōkūhoʻokelewaʻa, the canoe-guiding star (also known by various other names); the brightest star in the sky, it is known to have been a star of considerable importance to Hawaiians (Johnson, Mahelona, and Ruggles 2015, 154, 211). This possibility merits further investigation and a more focused survey.

Additional remarks. As regards potential topographic and astronomical alignments of possible significance, this site includes a number of different possibilities. In particular, it preserves a possible device for observing the rising of Sirius, or Hōkūhoʻokelewaʻa, the navigators' star.

KIP-75, -76, AND -77

See "The 'Heiau Ridge' Complex."

KIP-80 (State Site 50-50-15-3847; Bishop Museum Site MA-A35-80)

A substantial terraced platform situated in the Kīpapa-Nakaohu uplands, site KIP-80 was first recorded during the 1966 Bishop Museum survey, at which time Kirch made a plane-table map. Kolb investigated the site in 1996; he reports excavating 2 m² of test pits (Kolb and Radewagen 1997, 70), from which a radiocarbon sample was obtained (Kolb 2006). Kirch mapped the structure with plane table and alidade in 1998 and visited the site again in March 2003, when a quick GPS survey was made along with observations on the viewshed. Although KIP-80 exhibits monumental architecture, for reasons outlined below we believe it is more likely to have been a *kahua* platform for the ritual performance of sacred *hula* dance than to have been a *heiau* proper.

Location and topographic setting. KIP-80 was built into the southwestern face of a small *puʻu* that was formed by a lobe of a massive, young ʻaʻā flow (the basanite of Kanahau-5, dating to 5000–10,000 BP). Situated at an elevation of 700 masl, the site is visible from much of the Kīpapa uplands; from the platform, one commands a sweeping view of this region. The terrain immediately below the structure drops away steeply into a natural swale formed of the much older basanite of Kahikinui lava (50,000–140,000 BP).

Architecture. KIP-80 consists of a platform surmounting stepped terraces (fig. 7.13). Unlike most southeastern Maui *heiau*, the site lacks any kind of enclosing wall. The main platform, elevated on all four sides, deviates from cardinality by about 10° in a clockwise sense. It measures 8 m E–W by 6.5 m N–S and is finely paved with small ʻaʻā clinkers; the platform surface slopes gently from N to S, as if designed to make whatever activity was performed on it clearly visible to viewers situated either in the swale downslope or on the adjacent ridges. Near the mid-

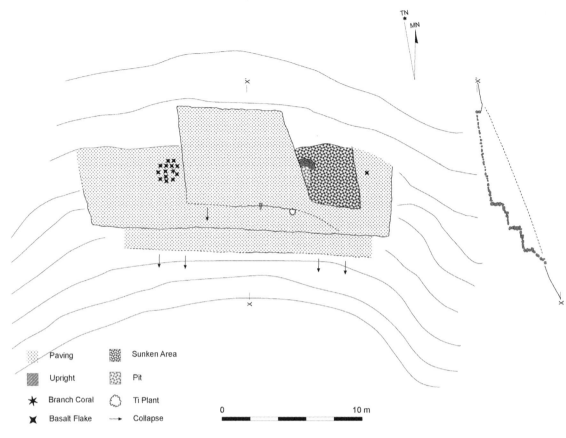

FIGURE 7.13. Plan of site KIP-80. (Based on plane-table survey by P. V. Kirch)

FIGURE 7.14. View of the main terrace facade at KIP-80, with phallic stone and *kī* plant (*Cordyline fruticosa*). (Photo by P. V. Kirch)

point in the prominent, 1 m high S face, a large, tapering ʻaʻā boulder has been set so as to project upward out of the face, in what almost certainly was a phallic representation (fig. 7.14). About 2 m to the E of the upright, a mature old kī plant (*Cordyline fruticosa*) with thick, gnarled stems grows out of the base stones; these plants are slow-growing and long-lived, and this one likely dates from the period of original site use. *Kī* was known to have ritual significance and was closely associated with *hula*: "Priests wore leaves about their necks as an indication of high rank or divine power, and it was among the plants customary on the altar of the *hālau hula*, representing Laka, the goddess of hula" (Abbott 1992, 115).

The upper platform is buttressed on the E, S, and W by a large terrace whose main facade, 1 m high, extends for 21 m. To the W of the platform, this terrace is roughly paved with ʻaʻā clinkers; several basalt flakes and a piece of *Porites* coral were observed on the surface. To the E of the platform, there is a rectangular sunken area or room in the terrace (about 4 by 4 m); this was the area test-excavated by Kolb in 1996. A second terrace face, also about 1 m high, parallels the main terrace downslope, extending for at least 18 m, although much of it has now collapsed. The three steplike faces of the terraces and platform have a combined height of more than 3 m and, when viewed from below, are quite impressive.

Excavation. Kolb excavated 2 m² in the sunken area to the E of the platform. Kolb and Radewagen report that "no features were located" and that "minimal traces of material culture were found including burnt organic material and bone" (1997, 70). No other details are available.

Dating and chronology. A sample of unidentified wood charcoal from a "basal" context in Kolb's test excavation was radiocarbon dated (Beta-122889), yielding a conventional age of 430 ± 70 BP, which gives 2σ calibrated age ranges of AD 1334–1336 (0.2%) and 1398–1643 (95.2%). Given the "basal" context, this sample likely dates land use or clearance activities predating the construction and use of the structure itself and therefore should be regarded as a *terminus post quem*.

Archaeological context. A small, stone-faced terrace (6 by 2 m) sits atop the hillock into which site KIP-80 is constructed. Otherwise there are no other archaeological sites in the immediate vicinity.

Viewshed. The location of KIP-80 on the steep slope of the ʻaʻā hillock completely restricts any views to the E or N, which are typical of most Kahikinui *heiau*. Rather, the structure looks out over the adjacent Kīpapa uplands to the S and SW. The temples of Heiau Ridge are visible below, as is much of the densely settled residential and agricultural region of Kīpapa. In the far distance, the Alena and Lualaʻilua coastal flats can be seen, including Wekea Point and the *pānānā* site at Hanamauloa (Kirch, Ruggles, and Sharp 2013). To the W, the Lualaʻilua Hills are also visible, as is the cinder cone of Manukani.

Orientation. The S facades of the upper platform and the two supporting terraces have an orientation of approximately azimuth 100° (corrected for mean magnetic declination).

Additional remarks. Although KIP-80 is a monumental site—and undoubtedly had some important ceremonial function—several lines of evidence suggest that it was not a *heiau* but rather a *kahua* (platform) designed for ritual *hula* performance. Architecturally, the site exhibits neither an enclosing wall (typical of most *heiau*) nor any kind of obvious altar. The absence of open views to the E or N (again typical of Kahikinui *heiau*) is likewise noteworthy. Furthermore, we did not observe any branch coral or ʻiliʻili stone offerings (two pieces of coral present were of the *Porites* variety).

Architecturally, the structure is best described as a stage—a platform prominently situated on the face of the steep hillock, sited such that it would have been clearly visible both from Heiau Ridge and from much of the residential zone of the Kīpapa uplands. The unusually fine paving of the sloped upper platform also suggests that it was designed as a stage for some kind of active performance.

In our view, the most likely functional interpretation of KIP-80 is as a *kahua* for the performance of sacred *hula* dance (see Kirch 2014, 184–186). We know that such *hula* platforms existed in Hawai'i, the most famous example being that at Kē'ē on Kaua'i Island (Kelly 1980); Emerson (1909) also describes the formal *hālau* (long house) in which *hula* dances were taught and performed. It seems likely that site KIP-80 was such a *kahua* for the performance of sacred *hula pahu*, dances accompanied by the beating of large temple drums (*pahu*), possibly on specified occasions in association with temple rites carried out at Heiau Ridge, as both sites are clearly intervisible.

KIP-115

See "The 'Heiau Ridge' Complex."

KIP-188 (State Site 50-50-15-3858; Bishop Museum Site MA-A35-188)

This midsize notched *heiau* in the Kīpapa uplands was recorded during the 1966 Bishop Museum survey, when it was plane-table mapped by Kirch. Kolb and Radewagen briefly report the site; Kolb (2006, table 1) gives a single radiocarbon date obtained from otherwise unpublished excavations. Ruggles visited this site on April 12, 2002, carrying out a compass-clinometer survey of the walls, and returned in better weather on April 16, 2002, to complete a survey of the horizons.

Location and topographic setting. The site is situated on gently sloping *'a'ā* terrain of the Alena basanite lava flow, at an elevation of about 620 masl.

Architecture. The structure consists of a classic notched enclosure whose orientation deviates from cardinality by about 7° in a clockwise sense, with the notch positioned in the NE corner. The maximum exterior dimensions of the enclosure are 20 m E–W by 17 m N–S; it has a basal area (calculated using aerial photo imagery of the site in ArcGIS) of 290 m². The walls are core-filled, and there is a secondary, buttressing terrace along the S (downslope) wall. The E wall is likewise buttressed with an external terrace.

Dating and chronology. Kolb (2006, table 1) reports a single radiocarbon date (Beta-122894) of 340 ± 50 BP from a "basal" context; no other excavation details are available. This corresponds to a calibrated age range of AD 1453–1645, which, given the basal context, must be taken as a *terminus post quem*.

Archaeological context. Site KIP-188 is part of a dense cluster of structures in this part of Kīpapa. Immediately upslope are a number of enclosures, shelters, and terraces that appear to be residential in function; another cluster of probable residential features lies a short distance to the W. The proximity of residential features suggests that this *heiau* may have been closely integrated with the local community, probably as an agricultural temple (*heiau ho'oulu 'ai*).

Viewshed. There are open views of the Haleakalā Ridge to the N and its upper slopes on each side, from WNW round to NE. Local ground around 90 m distant restricts the view of the lower slopes to the ENE and E, between ~65° and ~90°. The sea forms the horizon between ~90° and ~240°, except between ~125° and ~165°, where Hawai'i Island lies beyond, visible in good weather conditions. A ridgeline ~250m distant climbs from sea level in the WSW to form the remaining part of the horizon, with the peaks of the Luala'ilua Hills (2.4–3.0 km) visible beyond it; the S peak, at azimuth 244°, is seen just at the point where the local ridge would fall below the sea horizon, while the N peaks, one behind the other, are more prominent, rising to an altitude of ~3.5° at azimuth 260°.

Orientation. The architectural elaboration of the E wall suggests that the *heiau* may have had an easterly orientation, although a northerly orientation cannot be ruled out. Prismatic compass measurements were obtained with two different instruments and in both directions along all six wall segments. These yielded mean eastward azimuths for the S, central (short), and N walls of 89°, 96°, and 99°, respectively (in all cases corrected for mean magnetic declination), for a mean of 95°, and northward azimuths for the W, central (short), and E (short) walls of 358°, 9°, and 23°, respectively, for a mean of 10°.

The *heiau* faces, at least approximately, the point at which the ridge to the E meets the sea, which itself occurs close to azimuth 90° and the position of equinoctial sunrise. The S wall particularly appears to be oriented in this direction. To the N, the range of wall directions spans a portion of the Haleakalā rim, but there is no evident correlation with topographic features such as conspicuous cinder cones. To the W, the various walls are oriented toward a stretch of the slopes above the Luala'ilua peaks. The S wall is aligned approximately in the direction of equinoctial sunset; the southern Luala'ilua peak marks the winter solstice sunset, but no wall is oriented in this direction. To the S, the walls face open sea.

KIP-273 (Walker Site 177; State Site 50-50-15-177; Bishop Museum Site MA-A35-273A and -273B)

Site KIP-273, first recorded by Walker ([1930]), was also recorded during the 1966 Bishop Museum survey, at which time Kirch made a plane-table map of the structure. Kirch made a new plane-table map of the structure in March 1997. Kirch and Ruggles made both total station and GPS surveys of KIP-273 on March 22, 2008. Kolb and Radewagen (1997, 68) briefly describe the site.

Location and topographic setting. KIP-273 is about 240 m inland of the shoreline in the coastal sector of Kīpapa, at an elevation of about 30 masl. The site lies at the interface between the younger Alena basanite *'a'ā* flow, on which the eastern terrace was constructed, and the older Kīpapa ankaramite flow to the west, upon which the notched enclosure is situated. An escarpment ranging from 2 to 3 m in height separates the two flows, with the Alena flow being higher in elevation.

Architecture. The site is a compound structure with two components, designated KIP-273A and KIP-273B (fig. 7.15). Site KIP-273A, which lies to the W of the lava escarpment (and hence is lower in elevation), is a classic notched *heiau* enclosure, skewed from cardinality by about 8° in a counterclockwise sense, with the notch in the NW corner (fig. 7.16). The core-filled walls are unusually well constructed using *pāhoehoe* lava slabs in a kind of "brickwork" style, with wall faces up to 2 m high. The maximum exterior dimensions of the enclosure are 10.5 m E–W by 10 m N–S, with a basal area of approximately 150 m^2. The western room is set off by a low dividing wall; this room has a floor paved with *'ili'ili* gravel and contained numerous pieces of branch coral. The N and S walls of this western room have gabled peaks, indicating that they were meant to support a thatched roof. The floor of the eastern room or court is largely covered with collapsed rubble.

KIP-273B, situated atop the higher *'a'ā* escarpment, consists of a U-shaped enclosure

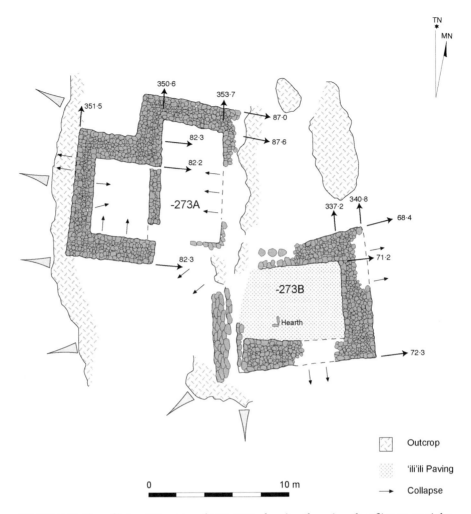

FIGURE 7.15. Plan of *heiau* KIP-273A and KIP-273B, showing the azimuths of intact straight segments of wall facing as determined by the total station survey. (Based on plane-table survey by P. V. Kirch)

skewed from cardinality by about 20° in a counterclockwise sense, open to the WSW, with dimensions of 9.5 m ENE–WSW and 7.5 m NNW–SSE and a basal area of 70 m². The walls are core-filled, about 50–70 cm high. The interior floor has scattered *'ili'ili* pebbles and some shell midden. Two *pāhoehoe* slabs set in the floor at right angles to each other appear to be two sides of a small hearth or fireplace (*kapuahi*).

Dating and chronology. Kirch and Sharp (2005) and Kirch, Mertz-Kraus, and Sharp (2015) report three ^{230}Th dates on branch corals from KIP-273. Two samples from KIP-273B yielded corrected ages of AD 1618 ± 7 and 1629 ± 4, which would correspond to the reign of the Maui *ali'i nui* Kamalālāwalu. A sample from an architecturally integral context (dating to the time of construction) from KIP-273A yielded a date of AD 1669 ± 3, around the end of the reign of king Kauhi-a-Kama or the beginning of the reign of Kalanikaumakaōwakea. These dates suggest that the eastern U-shaped enclosure may have been in use prior to the construction of the western notched *heiau*; both components clearly date to the seventeenth century.

FIGURE 7.16. Aerial photo of *heiau* KIP-273A, from the southeast. (Photo by P. V. Kirch)

Archaeological context. The only other feature near KIP-273 is a small platform *heiau*, KIP-275, situated about 80 m to the SSE.

Viewshed. KIP-273B has open views of the Haleakalā Ridge extending from the NW (~8 km distant) round to the ENE (~20 km distant) and of open sea from the E round to the SW. To the WSW, W, and WNW the view is restricted by lava ridges between 750 m and 2 km distant. KIP-273A is similar except that the view eastward is blocked by the lava escarpment.

Orientation. The eastern court of KIP-273A lacks both E and S walls: an E wall is probably lacking because the enclosure is built up against the lava escarpment, but the lack of a S wall suggests that the direction of orientation was northward. KIP-273B, in contrast, is open to the WSW and thus clearly oriented to the ENE.

Intact segments of the W, central, and E walls of KIP-273A yield best-estimate northward orientations of 351.5°, 350.6°, and 353.7°, respectively, for a mean of 351.9°. The eastward orientations of the S and central walls, 82.3° in both cases, are close to perpendicular to this, but the N wall is significantly skewed, with a mean eastward azimuth of 87.3°. Intact segments of the outer faces of the NNW and SSE walls of the E enclosure, KIP-273B, yield 68.4° and 72.3°, respectively, for a mean of 70.4°. However, these two walls widen toward opposite ends, and the long, intact inner face of the NNW wall may provide the best estimate of the intended orientation—namely, 71.2°. Our best estimate of the northward orientation of the ENE wall is 339.0°.

KIP-273B faces a prominent lava outcrop on a nearby ridge about 100 m away, which just cuts the more distant eastern horizon between azimuths of ~69.0° (the precise azimuth depends on the height of the observer) and 70.2°, as viewed from a position toward the SSW corner; there is a small but conspicuous notch in its center (az. 69.5°, alt. 2.0°, dec. +19.5°, to the nearest half degree). This is close to the rising point of the vertex of the conspicuous V shape of Ka Nuku o Ka Puahi, or Hyades-Aldebaran (Johnson, Mahelona, and Ruggles 2015, 171), which had a declination of +18.5° around AD 1700 (see table 4.2). However, we conjecture that this feature might have been used instead from a position roughly 5 m farther S, outside the enclosure itself (which would reduce the azimuth of the nearby foresight to about 66°), as a sighting device marking the rising of the Pleiades (which, at the time of construction in the early or mid-seventeenth century, would have occurred at around az. 66°). This would parallel the situation at ALE-140, where a sighting device for the Pleiades appears to have functioned from outside the *heiau* itself.

To the N, KIP-273A faces a relatively featureless Haleakalā rim, toward the left end of the highest part, within 5° of the red cinder cone Kanahau.

Additional remarks. The two components of this site had distinct orientations and seem to have faced different directions. It is therefore possible that they had different functions.

KIP-275 (BISHOP MUSEUM SITE MA-A35-275)

This platform temple was first recorded during the 1966 Bishop Museum survey. Kirch prepared a plane-table map in March 1997. Kirch and Ruggles carried out a total station survey on March 22, 2008, and mapped the site with GPS at that time as well.

Location and topographic setting. The site lies in the coastal zone of Kīpapa, about 170 m inland of the coastline and about 80 m SSE of KIP-273. It was constructed on the western edge of the young Alena basanite ʻaʻā flow, overlooking a *kīpuka* (patch of older lava) of the Kīpapa ankaramite lava flow.

Architecture. The site consists of a platform or terrace originally rectangular in plan view, its orientation deviating from cardinality by about 11° in the counterclockwise sense, but with the SSW corner somewhat collapsed (fig. 7.17). The main platform surface, paved with fine ʻaʻā clinkers and covered in scattered ʻiliʻili pebbles and branch coral, measures 10 m NNW–SSE by 9 m WSW–ENE. One larger waterworn cobble was also noted on the platform surface. A second, lower terrace abuts the main platform on the WSW. From the surface of the main platform to the base of the lower terrace, there is a drop in elevation of 3.5 m, giving the western facade a prominent aspect when viewed from the level ground immediately WSW of the site. On the ENE, the platform is bounded by a core-filled wall 50–60 cm high, with a distinct notched jog.

Dating and chronology. Kirch, Mertz-Kraus, and Sharp (2015, table 1) report a ^{230}Th date of AD 1625 ± 6 on a piece of branch coral collected from the platform surface. This corresponds to the reign of Maui *aliʻi nui* Kamalālāwalu.

Archaeological context. The closest feature is the KIP-273 *heiau* complex to the N. There are no residential features in the immediate vicinity of KIP-275.

Viewshed. There are open views of the Haleakalā Ridge from the NW round to the NE and of open sea from the ESE round to the SW. To the WSW, W, and WNW, the view is restricted by lava ridges between 800 m and 2 km away. Some 15 m to the E of the platform is a jagged lava ridge that blocks the view to the ENE and E.

Orientation. The lower terrace to the WSW and the higher wall to the ENE indicate that this *heiau* faced eastward. A well-preserved segment of the outer face of the NNW wall has an

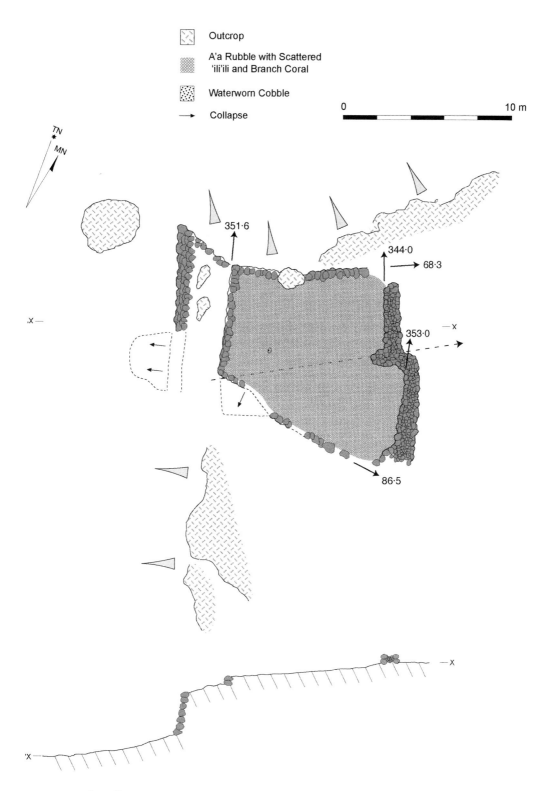

FIGURE 7.17. Plan of *heiau* KIP-275, showing the azimuths of intact straight segments of wall facing and platform edges as determined by the total station survey. The dashed line shows points from close to which the Pleiades would be seen to rise up the left side of the outcrop around the time of construction. (Based on plane-table survey by P. V. Kirch)

eastward orientation of 68.3°, while the skewed boulder facing on the southern side has an eastward azimuth of 86.5°. The mean direction is 77.4°. The best-preserved NNW–SSE structural segments, on the WSW platform edge, and the inner edges of the northern and southern halves of the ENE wall, yield 351.6°, 344.0°, and 353.0°, respectively, for an average of 349.5° (perpendicular to 79.5°).

During our survey, we noticed that from a seated position on a cobble and coral concentration on the western side of the platform, a large stacked ʻaʻā boulder feature 8 m ENE of the platform (comprising a hefty stone evidently placed atop a ground-fast stone and chocked in place) aligns with a broad natural outcrop 7 m farther to the ENE, which forms a prominent feature breaking the skyline to the ENE (fig. 7.18). From the center of the coral area, the left-hand end of the outcrop has an azimuth of 72°, and the notch high up at its center an azimuth of 74°. From the WNW corner of the platform these values increase to ~81.5° and ~83.5°, respectively, whereas from the southern end of the facing on the WSW side they decrease to 67° and 69°, respectively.

The *heiau* would face the bottom of the Haleakalā slope to the ENE and E, where it meets the sea, if the outcrop did not intervene. The alignment of features (coral concentration to stacked boulders to prominent outcrop) is broadly oriented upon the rising position of Ka Nuku o Ka Puahi, the conspicuous V-shaped asterism comprising the Hyades together with Aldebaran (Johnson, Mahelona, and Ruggles 2015, 171). Around AD 1625 this asterism spanned a declination range of +14.7° to +18.3°, while the declinations of the left side and notched highest part of the outcrop are around +18° and +16°, respectively. On the other hand, from a point outside the platform at the southern end of the facing on the WSW platform edge, viewing along a line passing almost exactly through the center of the platform (shown as a dashed line in fig. 7.17), the Pleiades (dec. +22.7° to +23.2° in AD 1625) would have risen closely up the left side of the outcrop (see fig. 7.18). (With a foresight this close, a lateral shift of only 30 cm from a position on the WSW side of the platform, and even less from positions farther ENE, would have shifted the azimuth of the outcrop by ~1°.) This suggests the strong possibility, at least, that this temple was related to observations of the Pleiades.

In other directions the walls and boulder facings are oriented upon a stretch of the Haleakalā rim to the NNW and N, including the prominent red cinder cone Kanahau (noted at various other sites) at azimuth 346°, a featureless stretch of the ridgeline to the W, and open sea to the S and SSE.

KIP-306 (Bishop Museum site MA-A35-306)

This small fishing shrine (*koʻa*) was first recorded during the 1966 Bishop Museum survey. Kirch and Ruggles visited it briefly on March 21, 2008.

Location and topographic setting. The site lies just 30 m from the shoreline, at an elevation of about 5 masl.

Architecture. The shrine consists of a rectangular platform of stacked cobbles, defined by low walls around the perimeter, which have largely collapsed. It is broadly cardinally oriented; there is the hint of an entryway on the W. There are two distinct concentrations of branch coral (and some waterworn *Porites* coral as well) and larger waterworn basalt cobbles—a larger concentration in the northern part of the platform and a smaller concentration near the SE corner.

Dating and chronology. Kirch, Mertz-Kraus, and Sharp (2015) report two ^{230}Th dates on branch corals from KIP-306 with ages of AD 1579 ± 3 and 1595 ± 4. The close agreement between

FIGURE 7.18. Horizon to the ENE at *heiau* KIP-275, from a point outside the platform at the southern end of the facing on the WSW platform edge. DTM-generated profile superimposed on a photograph of the local ʻaʻā outcrop. The solid shaded line indicates the rising path of the sun at the June solstice and the dotted line the rising path of the Pleiades in AD 1625. (Composite of photo by Clive Ruggles and graphic by Andrew Smith)

these two dates indicates deposition of coral offerings at the *koʻa* during the late sixteenth century, a period that spanned the reigns of Maui *aliʻi nui* Piʻilani and his son Kiha-a-Piʻilani. Use of the shrine, however, is likely to have continued into the late precontact period.

Archaeological context. *Heiau* site KIP-307 lies 50 m to the ENE, and a large habitation enclosure (KIP-308) lies 50 m to the NNE. The KIP-308 site was occupied from precontact times into the mid-nineteenth century.

Viewshed. The site commands an excellent view of the ridgeline of Haleakalā, as well as views out to the ocean.

Orientation. The larger concentration of branch coral and waterworn cobble offerings in the N sector of the platform is suggestive of a northerly orientation. Prismatic compass measurements of the wall orientations, corrected for mean magnetic declination, indicate a northerly azimuth of 358° and an easterly azimuth of 88°; in other words, the shrine is cardinally oriented to within 2°–3°, whereas an orientation perpendicular to the coastline at this point would have resulted in an orientation displaced counterclockwise from this by about ~20°.

KIP-307 (Walker Site 176 [Possible]; Bishop Museum Site MA-A35-307)

Site KIP-307 was first recorded during the 1966 Bishop Museum survey, when it was thought to be a postcontact-period house site. This structure is probably also that recorded by Winslow Walker as his site 176 (Walker [1930], 246), which he described as being located in the "village site of Ka lae o Ka Pulou on the edge of the shore." Walker did not make a plan, and his description of this structure is so minimal that it is difficult to be certain, although his note that "this is a large site with a wall 100 feet long on the east and 80 feet long on the north" is consistent with the configuration of KIP-307. Kolb (2006, table 1) reports a radiocarbon date from Walker's site 176.

Kirch mapped KIP-307 with plane table and alidade in July 1996. Kirch and Ruggles visited the site in March 2008, making both total station and GPS surveys, as well as taking detailed observations on the viewshed.

Location and topographic setting. KIP-307 is situated a mere 15 m from the shoreline, atop a low cliff with an elevation of about 10 masl. The substrate consists of the Kīpapa-1 ankaramite ʻaʻā lava flow.

Architecture. The architecture at KIP-307 is complicated by the fact that the structure was clearly rebuilt for use as a residential structure in the postcontact period, probably around the middle of the nineteenth century (as with site KIP-728 in the Kīpapa uplands). Nonetheless, it is clear that the original structure was a classic notched *heiau*, its orientation deviating from cardinality by about 15° in a counterclockwise sense, although with an unusual addition of a long NNW–SSE-trending wall (fig. 7.19). The original notched enclosure had exterior dimensions of 16 m ENE–WSW by 12 m NNW–SSE, with the notch situated in the WNW corner of the enclosure. The western interior shows traces of an original *pāhoehoe* slab and ʻiliʻili pavement. Whereas the NNW and ENE walls are still intact (and incorporate massive ʻaʻā boulders in their basal course), the WSW and particularly SSE walls were robbed of all but the base course, presumably in order to construct a wall (nearly 2 m high) that bisects the original *heiau* court to form a smaller room on the east. This room, along with a freestanding three-sided enclosure with doorway immediately to the W of KIP-307, presumably functioned as the nineteenth-century dwelling structures.

What sets KIP-307 off from other notched *heiau* in Kahikinui is the long NNW–SSE wall, which is an extension of the ENE wall of the *heiau*. This wall has a total length of 31 m, with a well-faced gap or entryway (3.5 m wide), which begins at 14.5 m from the wall's NNE corner. At present this gap is filled with a crudely constructed, low wall of roughly stacked cobbles, probably added in recent times by fishermen who have camped out at the structure. To the west of the seaward end of the long wall lies a terrace, roughly paved with ʻaʻā clinkers, buttressed on the SSE and WSW with low facings; the terrace measures approximately 8 by 10 m.

The long wall, with its entryway, appears to have been a part of the original *heiau* construction. Given the location of this site on the immediate coastline, the entryway may have been associated with the precontact foot trail (the later Hoapili Trail of the early nineteenth-century lies a short distance inland). KIP-307 occupies a low promontory that could have been the boundary between Kīpapa and Nakaohu. It is interesting to speculate that KIP-307 may have functioned as a Lono *heiau* associated with annual Makahiki rites, especially a clockwise circuit of the Lono priests moving from *ahupuaʻa* to *ahupuaʻa*.

Excavation. We did not excavate at KIP-307, although Kolb evidently did excavate a test pit in Walker's site 176, which is probably this site. No details of Kolb's excavation are available.

Dating and chronology. Kolb (2006, table 1) reports a radiocarbon date (Beta-122903) of 260 ± 60 BP from a "pavement" context in Walker's site 176, yielding 2σ calibrated age ranges of AD 1461–1691 (70.3%), 1729–1811 (18.9%), and 1924–1950 (6.2%). Kirch, Mertz-Kraus, and Sharp (2015, table 1) obtained two specimens of branch coral, both in architecturally integral contexts wedged under the basal stones of the original, western part of the notched *heiau* wall; these

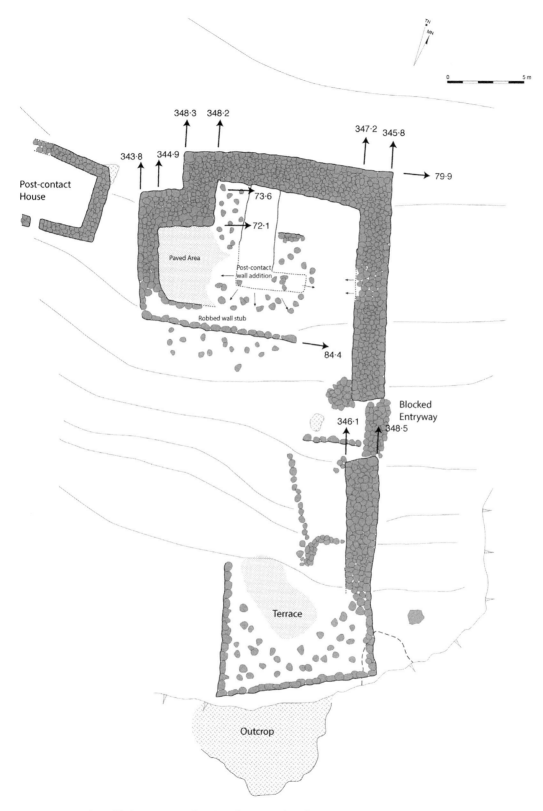

FIGURE 7.19. Plan of *heiau* KIP-307, showing the azimuths of intact straight segments of wall facing as determined by the total station survey. (Based on plane-table survey by P. V. Kirch)

yielded ^{230}Th ages of AD 1647 ± 3 and 1643 ± 3. The high degree of consistency in the coral dates lends confidence that the structure was constructed in the middle of the seventeenth century, at a time corresponding to the reign of Maui ruler Kauhi-a-Kama.

Archaeological context. Site KIP-308 (also known as site M11), a postcontact *pā hale* (house) enclosure with stratigraphically underlying precontact pavement and midden deposit, lies about 25 m NW of KIP-307 (see Chapman and Kirch 1979). Fishing shrine site KIP-306 lies 50 m to the WSW of KIP-307.

Viewshed. There are good all-around views. The Haleakalā ridgeline arches over from the NW (where it is ~8 km distant) to the ENE, where its slopes (~20 km distant) run down behind Ka Lae o Ka 'Ilio at azimuth 80°. Ridgelines between 1 and 3 km form the horizon to the WSW and W, running down to the sea at azimuth 240°. The Luala'ilua Hills, 4.5–5 km away, protrude above the closer horizon to the WNW, peaking at azimuths 294.5° and 304.5°. There is open sea from the E round to the SW.

Orientation. The lack of any entranceway in the WSW wall suggests that the original entrance into the main notched enclosure would have been toward or at the eastern end of the SSE wall, which would imply that the *heiau* was NNW-facing.

The eastward orientation of the NNW wall (measured corner-to-corner, as the outer face is somewhat convex) is at azimuth 79.9°, while that of the remaining base stones of the SSE wall is 84.4°, giving a mean estimated orientation of 82.1°. The shorter ENE–WSW-oriented wall segment forming the SSE side of the notch is skewed with respect to this, with a mean azimuth of 72.8°. The straight outer face of the ENE wall north of the entryway has a northward orientation of 345.8° (the best-fit orientation for the outer face of the wall as a whole is 346.8°), while the outer face of the WSW wall has an orientation of 343.8°, for a mean of 344.8°. Shorter NNW–SSE segments of wall facing yield values up to 348.5°.

To the NNW, the *heiau* is directly oriented upon the prominent red cinder cone Kanahau noted at a number of other sites: to the nearest half degree, its peak is at azimuth 344°, altitude 14°, and declination +73.5°. This has no obvious astronomical significance; it is likely that the alignment, if deliberate, was topographic rather than astronomical in nature. The NNW wall is directly aligned upon Ka Lae o Ka 'Ilio (fig. 7.20), which is at azimuth 80°, altitude 0°, and declination +9°. While the star Humu (Johnson, Mahelona, and Ruggles 2015, 161)—that is, Altair (α Aql, dec. +8.0° in AD 1645)—would have risen just to the right of this point, it may be that this wall was deliberately built at a skew, rather than perpendicular to the ENE wall, in order to incorporate a second topographic alignment.

The alignments in the opposite directions, toward the open sea in the SSE and the lower part of the slope down from Haleakalā to the WSW, have no obvious significance, either topographic or astronomical.

KIP-330 (Bishop Museum Site MA-A35-330)

This *koʻa* (fishing shrine) near the coast in Kīpapa was first recorded during the 1966 Bishop Museum survey. Kirch mapped the site with plane table and alidade on July 17, 1996. Kirch and Ruggles visited the site briefly on June 16, 2002, and took some prismatic compass measurements. In January 2008, Kirch and Sharp visited the site to collect coral samples for ^{230}Th dating.

Location and topographic setting. The site lies about 100 m inland of the rocky shoreline in Kīpapa, at an elevation of about 20 masl. The substrate is Nawini ankaramite *ʻaʻā* lava.

FIGURE 7.20. View eastward along the NNW wall of *heiau* KIP-307, showing its alignment upon Ka Lae o Ka 'Īlio in the distance. (Photo by Clive Ruggles)

Architecture. The main part of the shrine consists of a rectangular enclosure whose orientation deviates from cardinality by about 17° in a counterclockwise sense. It measures 4.5 m ENE–WSW by 5.5 m NNW–SSE and has stacked-stone walls 50–75 cm thick and 50–60 cm high (figs. 7.21 and 7.22). The ESE corner of the enclosure comes to a distinct point or acute angle, while the NNE corner incorporates a large lava boulder about 1 m in diameter; the boulder is reddish and was probably regarded as a ritual attractor. There is a formal entryway on the WSW side, leading to a court with a carefully set pavement of *pāhoehoe* lava slabs. To the ENE of the paved court, a single-course alignment of basalt cobbles forms the facade of a slightly raised bench that runs the length of the ENE interior of the enclosure. Two whole *Pocillopora* coral heads, several coral branches, and three waterworn basalt cobbles are sitting atop this bench, which probably served as an altar. Other pieces of branch coral are scattered about the paved court, while a large coral head rests on the WSW wall of the enclosure. Another whole coral head lies on the ground just outside the ESE corner of the enclosure.

Attached to the enclosure on the NNW (inland) side are two low, stone-filled rectangular platforms, each about 2.5 by 3 m. The platforms are filled with *'a'ā* clinkers. These are almost certainly burials that were added to the fishing shrine in the early to mid-nineteenth century.

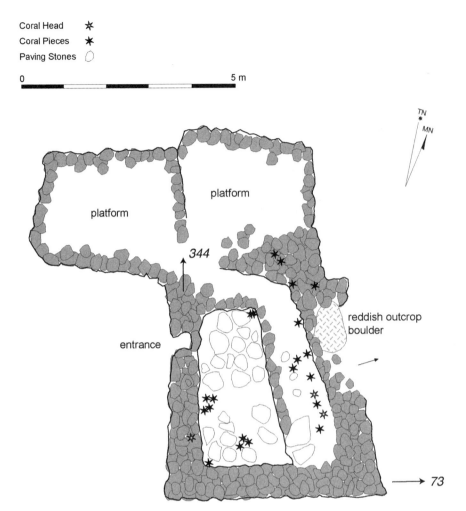

FIGURE 7.21. Plan of *ko'a* KIP-330, showing the azimuths of wall segments as determined by prismatic compass corrected for mean magnetic declination. (Based on plane-table survey by P. V. Kirch)

Dating and chronology. Two pieces of branch coral were dated with the ^{230}Th method (Kirch, Mertz-Kraus, and Sharp 2015, table 1). The first sample (KIP-330-CS-1) was a multilobed branch that had been wedged deeply between wall stones in the NNE corner of the enclosure, near the large natural boulder; this yielded an age of AD 1658 ± 3. The second sample (KIP-330-CS-3) was a multilobed branch wedged between stones on the altar bench; this yielded an age of AD 1733 ± 2. These dates indicate use of this *ko'a* from the mid-seventeenth to the mid-eighteenth century, a period spanning the reigns of Maui *ali'i nui* Kalanikaumakaōwakea and his successor, Lonohonuakini.

Archaeological context. This *ko'a* is part of a small coastal residential cluster (*kauhale*) that sits on the lava slope just below the shrine. The closest structure is KIP-331, an L-shaped shelter that excavation indicated had probably functioned as a men's house, or *mua* (Van Gilder 2005, 234–236). Farther down the ridgeline is KIP-335, which was interpreted as a *hale noa*, or dwelling house (232–234).

FIGURE 7.22. Main part of *koʻa* KIP-330, viewed from the northwest. (Photo by Clive Ruggles)

Viewshed. There is an unobstructed view out to sea from this location—which would be appropriate for a fishing shrine—stretching from the E round to the SW. There are also clear views northward, but the views are more restricted (closer than ~150 m) between WSW and NW and to the ENE.

Orientation. The orientation of this shrine is most simply explained by its being placed perpendicular to the shoreline. The entryway is on the WSW side, and the bench that functioned as an altar runs along the inner ENE wall, suggesting that the site was ritually oriented to the ENE. The position of the large red boulder in the NNE corner may also be ritually significant. While there is a broad view of the sea (from az. ~90° to az. ~240°), the *heiau* faces local ground ~150 m away around azimuth 73°. With an altitude of ~2.5° and a declination of ~+16.5°, it is broadly aligned upon the rising of Ka Nuku o Ka Puahi, the Hyades plus Aldebaran (dec. ~+15.0° to ~+18.5° around AD 1700), and in the opposite direction (az. 253°, alt. 1°, dec. ~–15.5°) upon the setting of Hōkūhoʻokelewaʻa, or Sirius (α CMa, dec. ~–16.5° around AD 1700) (see table 4.2), although the latter is probably fortuitous. Our measurements indicate that the WSW wall faces, at least approximately, the red cinder cone Kanahau on the Haleakalā Ridge noted at several other sites (az. 343° from here), while the skewed ENE wall, whose orientation is about 330°/150° as judged from the plane-table survey, is approximately aligned southward upon the low summit of Mauna Loa on Hawaiʻi Island.

KIP-366 (BISHOP MUSEUM SITE MA-A35-366)

This probable *heiau* site was first identified during the 1966 Bishop Museum survey, when it was initially identified as a house site. During resurvey in 1996 by Kirch the site's function was revised to "ritual/ceremonial," based on its architecture and location within the intermediate zone inland of the coast. Kirch and Ruggles visited the site briefly on June 16, 2002, taking prismatic compass measurements, and returned to undertake a fuller compass-clinometer and GPS survey on March 21, 2008.

Location and topographic setting. Site KIP-366 is situated atop a prominent ridge spur at an elevation of about 125 masl. It sits on an older basanite of Kahikinui lava flow.

Architecture. The site consists of a large enclosure whose orientation deviates from cardinality by about 8° in a counterclockwise sense. It measures 24 m N–S by 18 m E–W, with stacked cobble and boulder walls averaging about 50 cm high; the E wall is more prominent than those on the S and W. The space enclosed within these walls is mostly level and consists of earth. In the NE corner of the larger enclosure is a smaller, square enclosure or room, 7 m on each side, whose E and N walls are those of the larger enclosure. This smaller enclosure was presumably covered by a thatched roof when the structure was in use.

Excavation. A test pit, 1 m², was excavated by James Coil in July 1998. Under a 5 cm layer of overburden, a cultural deposit containing occasional charcoal pieces extended to a depth of about 25 cm. No features were encountered. In addition to charcoal, the cultural deposit yielded 69 basalt flakes, 259 NISP of marine shell, 2 NISP of unidentified bone, and 2 pieces of coral.

Dating and chronology. A sample of carbonized candlenut endocarp (*Aleurites moluccana*) was recovered from level 3 of the excavated test pit and radiocarbon dated (AA-38651). This yielded a conventional radiocarbon age of 226 ± 25 BP, with 2σ calibrated age ranges of AD 1642–1680 (45.7%), 1763–1802 (39.3%), and 1938–1950 (10.5%). This date indicates use of the site sometime during the seventeenth to eighteenth centuries.

Archaeological context. KIP-366 is not part of a site complex, and there are no other structures in the immediate vicinity. The absence of other structures nearby is one possible indication of its ritual function.

Viewshed. The site commands an excellent view out over the Kīpapa coastline as far as Alena to the WSW. There is also a clear view of the Haleakalā Ridge rising behind the more local ridgeline (~ 1 km) to the NW and sweeping over toward the ENE. Only the lowest part of the eastern slopes, around azimuths 75° to 80°, is obscured by closer ground (~200 m distant).

Orientation. The more substantial nature of the E wall suggests that this structure faced eastward. The eastward orientation of the N wall is about 82°; this wall is oriented, at least approximately, upon the point where the ridge that forms the skyline to the E reaches the sea (alt. −0.3°, dec. +7.1°). A possible explanation of the skewed orientation of the S wall (~63°) is that it is oriented approximately upon the point where the western slopes reach the sea in the opposite direction.

The point where the horizon descends to sea level in the E coincides with the rising point of 'Aua (Johnson, Mahelona, and Ruggles 2015, 146), or Betelgeuse (α Ori, dec. +7.3°)—as well as being close to that of Humu (ibid., 161), or Altair (α Aql, dec. +8.0°)—around AD 1660. There is no reason to suppose that this was of significance to those who built and used this temple. Added to this, the eastward orientation of the skewed S wall (az. 63°, alt. +3°, dec. +26°) is appreciably to the left of the rising position of the Pleiades around AD 1660 and the June solstice sunrise, which occurred around azimuth 66°.

To the W, on the other hand, the S wall (az. ~243°, alt. 0°, dec. −25°) is oriented more closely to the setting position of the December solstice sun. Furthermore, the June solstice sun (and also the Pleiades around AD 1660) set directly into the most prominent of the Luala'ilua Hills, upslope at azimuth 292°. This suggests the possibility, at least, that the seasonal changes in the position of the setting sun in the west were tracked from this temple. The westward orientation of the N wall (az. 262.0°, alt. 2.5°, dec. −6.5, all to the nearest half degree) is upon a point within the solar setting arc, but not one of any obvious significance in itself.

The northward orientation of the E wall is approximately 350°, while that of the W wall is ~353°, for a mean of ~351°. This is toward an unremarkable point on the Haleakalā ridgeline and toward open sea in the opposite direction.

KIP-394 (Walker Site 175; State Site 50-50-15-175; Bishop Museum Site MA-A35-394)

This structure was first recorded by Walker ([1930], 245), who described it as "an open platform 45 feet square built up in two step terraces to a height of 15 feet on the front." Walker noted that "in the northeast corner is a higher platform walled to a height of 3 feet." Walker did not make a plan map of the structure. In 1966 the site was resurveyed by the Bishop Museum team, with Kirch making a plane-table survey of the structure. Kolb surveyed the structure and excavated two test pits (total of 3 m²) in 1996 (Kolb and Radewagen 1997, 70, fig. 5.6). Kirch and Ruggles visited the site on March 27, 2003, making both total station and plane-table surveys. For reasons indicated under "Additional Remarks," we believe that this structure is more likely to have functioned as a chiefly residence than as a *heiau*.

Location and topographic setting. Site KIP-394 is situated at an elevation of about 350 masl, immediately adjacent (N) to the highway and about 0.5 km ENE of the ruins of St. Ynez Church. The structure sits in a low swale within the relatively young, massive Nawini ankaramite lava flow.

Architecture. The site consists of a large terraced platform, broadly cardinally oriented, rising on the E and S in a double-step facing while merging into the lava flow slope on the N and W. The main platform measures about 13 m E–W by 10 m N–S. The prominent double facing on the S side rises to a maximum height of 3 m above the natural ground, giving the structure an impressive aspect when viewed from below. The faces are constructed of stacked *'a'ā* cobbles and boulders, while the platform surface is roughly paved with *'a'ā* clinkers.

As Walker initially observed, the NE corner of the platform is taken up by a smaller space cell, measuring 4 m N–S by 3 m E–W, defined on the E and N by stacked cobble walls and on the S and W by a low stone facing; the interior of this room is slightly elevated above the level of the main platform. A small wall stub partly defined the edge of the room on the W side as well. This partly walled, slightly elevated area was likely to have been roofed over with a thatched superstructure when the site was in use.

Excavation. Kolb and Radewagen report that the two test pits excavated in KIP-394 (one in the smaller room in the NE corner and one in the main platform surface) yielded "small traces of midden" with "charcoal[,] . . . coral, land snails, shellfish, fish bone, and nut shell" (1997, 70).

Dating and chronology. Kolb (2006) reports a single radiocarbon date on unidentified wood charcoal, obtained from the test unit in the smaller space cell in the NE corner of the structure. This sample (Beta-95905) yielded a conventional age of 170 ± 60 BP, with calibrated ages of AD 1648–1894 (78.3%) and 1905–1950 (17.1%), the second of which can be discounted. This puts the use of the site sometime during the seventeenth to the eighteenth century.

Archaeological context. Although the area immediately north and upslope of the site has not been surveyed, several enclosures and terraces were noted to the east and south of KIP-394, indicating that it appears to have been part of a larger complex.

Viewshed. The position of this structure within a deep swale in the ʻaʻā flow greatly limits its viewshed. To the E a massive ʻaʻā lava ridge about 400 m distant blocks the view toward Kaupō. Similarly, to the W another massive ʻaʻā ridge blocks the view of the Lualaʻilua Hills. The only direction in which the larger Kahikinui landscape is visible is thus northward, where the higher part of the Haleakalā rim is visible. The sea view to the S, between azimuths ~70° and ~230°, includes the whole of Hawaiʻi Island, whose peaks are visible on clear days.

Orientation. A noteworthy aspect of the main platform is that the retaining terraces on the E and S join, not at a right angle, but at about 75°. If one stands on the platform, the point formed by the intersection of the two faces (the line bisecting the angle) is oriented directly toward the peak of Mauna Kea on Hawaiʻi Island, which can be readily observed across the ʻAlenuihāhā Channel on a clear day.

This observation is borne out by the survey data, which show the southern terrace to be oriented with an azimuth of 99.9° toward the E, while the eastern platform edge has an azimuth of 174.6° toward the S. The direction "faced" by the corner—that is, of the line bisecting the angle—therefore has an azimuth of 137.3°. The profiles of both Mauna Kea and Mauna Loa were clearly visible on the day of our survey; the summit of Mauna Kea has an azimuth of 136.2°.

Additional remarks. Two lines of evidence suggest that this structure—in spite of its impressive construction—functioned not as a *heiau* but, rather, as a chiefly or elite residential platform. First is the close association of the structure with several nearby enclosures and terraces, suggesting that it was part of a larger *kauhale* (domestic cluster). Second is the absence of any branch coral offerings, as well as the absence of an altar platform; the midden material recovered by Kolb in his excavation also is suggestive of domestic use.

The architecture of site KIP-394 closely resembles that of a large platform (with a smaller enclosure in the NE corner) at the coastal village site of Auwahi (AUW-20), which was identified by Walker ([1930], 95) as a chief's house platform (see also Sterling [1998, 210], site 67, "Makee Village"). We believe that the most likely interpretation of KIP-394 is similarly as an elite residence, and not a *heiau*. Nonetheless, the orientation of the rather pointed corner to the peak of Mauna Kea may well have been significant.

KIP-405 AND -410

See "The ʻHeiau Ridgeʼ Complex."

KIP-414 (State Site 50-50-15-4247; Bishop Museum Site MA-A35-414)

Site KIP-414 was originally recorded during the 1966 Bishop Museum survey. Kirch revisited the site in August 1995, mapping it with plane table and alidade. Kolb excavated a single test pit in the structure in 1996 (Kolb and Radewagen 1997, 70–71). Ruggles visited this site on April 17, 2002, carrying out a compass-clinometer survey of the walls and horizons.

Location and topographic setting. Site KIP-414 lies at an elevation of about 725 masl, on a long ridge formed by a tongue of the young, massive basanite of Kanahau-5 ʻaʻā lava flow.

Architecture. Although Kolb and Radewagen describe the structure as a "notched enclosure" (1997, 70), the form of this compound structure is quite irregular (fig. 7.23). The main feature is an enclosure, built of stacked ʻaʻā cobbles and boulders partly incorporating natural

FIGURE 7.23. Plan of *heiau* KIP-414, showing the azimuths of intact straight segments of wall facing as determined by prismatic compass corrected for mean magnetic declination. (Based on plane-table survey by P. V. Kirch)

outcrops. However, although this has an overall notched configuration and broadly follows a cardinal orientation in the western part, its eastern extension does not follow neat, orthogonal lines as found in most Kahikinui notched *heiau*. The S wall in particular is not straight, instead having several bends. The dimensions of this main enclosure are about 21 m E–W by 14 m N–S. Abutting the E wall of the main enclosure on the exterior is a small, square enclosure or room (3 by 4 m) with an earthen floor; there seems to have been a formal passage or entryway connecting the main enclosure with this smaller room. As noted on the plan map, several artifacts were found situated on the E and N walls of the main enclosure: two grindstone slabs, two basalt flakes, and a quadrangular basalt adze (illustrated in Kirch 2014, 203, fig. 57). Also present were branch coral, *Cellana* shell, and a waterworn pebble. The SE corner, though somewhat irregular, may be another example of a canoe-prow corner.

On the W side of the main enclosure, an L-shaped wall stub abutting the exterior wall at the NW corner defines another small room or space cell, also with an earthen floor. To the NE of the main enclosure is a secondary structure, a double enclosure of stacked cobbles and boulders with two interior rooms. This secondary structure measures 11 m E–W by 8 m N–S.

Excavation. Kolb and Radewagen report finding "coral, pig bone, charcoal, and volcanic glass" in their test excavation in KIP-414 (1997, 71).

Dating and chronology. A single piece of branch coral from the surface of the enclosure wall was ^{230}Th dated, yielding an age of AD 1574 ± 3 (Kirch, Mertz-Kraus, and Sharp 2015, table 1). This date would place the use of the *heiau* within the reign of Maui *aliʻi nui* Piʻilani.

Archaeological context. A small stone platform that may be a burial (KIP-500) lies about 15 m S of KIP-414. About 20 m to the NE, there is also a rectangular enclosure (~9 by 12 m), which may be a residential site.

Viewshed. Site KIP-414 has good, broad views of the Haleakalā ridgeline stretching from the WNW round to the ENE, toward Kaupō to the E, and of the open sea from the ESE round to the SW. The Lualaʻilua Hills are prominent to the WSW, at the foot of the western ridgeline; the northern peaks break the skyline, while the southern peaks are below the sea horizon.

Orientation. The concentration of artifacts on the eastern side and the longer E–W axis suggest that this *heiau* faced E rather than N. Various E–W-oriented wall segments yield orientations between 77° and 97°, corrected for mean magnetic declination, while the N–S-oriented segments yield 359° and 5°. The general axis of orientation appears to be about 95°, with the western part skewed counterclockwise by some 15°–20°, although the westernmost wall segment is broadly perpendicular to the longer main axis.

The principal orientation is upon open sea to the E (az. 95°, alt. −0.8°, dec. −5°), directly above the promontories of Apole Point and Ka Lae o Ka ʻIlio in Kaupō, but not in the direction of equinoctial sunrise. To the N, the W walls are oriented upon a cluster of cinder cones at the eastern end of the Haleakalā rim. Although the December solstice sun would have set directly into the NE Lualaʻilua peak (az. 245.1°, alt. +0.8°, dec. −23.5°), which is the most prominent of the three peaks as seen from here, the westward orientation of KIP-414 is a long way upslope (az. 275°, alt. 4°, dec. +6°), close to the setting point of Ka ʻŌnohi Aliʻi (Johnson, Mahelona, and Ruggles 2015, 172)—that is, Procyon (α CMi, dec. +6.2° around AD 1600)—but of no other obvious astronomical significance. The southward wall orientations are out to sea.

The variant canoe-prow SE corner is oriented with an azimuth of ~140°, corresponding to the direction of Mauna Kea on Hawaiʻi Island.

Additional remarks. The fact that this structure appears to be oriented upon noteworthy topographic features both to the E and N, together with the lack of any obvious astronomical correlates, suggests that the significant visual influences on its orientation, if any, were topographic rather than astronomical in nature.

KIP-424 (State Site 50-50-15-4366)

Site KIP-424 is a small, square enclosure first recorded in 1995 during an intensive survey of the Kīpapa uplands. Kolb subsequently investigated the site and excavated a 1 m² test pit (Kolb and Radewagen 1997, 69–70).

Location and topographic setting. The enclosure lies at an elevation of about 440 masl, a short distance inland of the main road. The geological substrate consists of the relatively young and unweathered Kīpapa-1 ankaramite ʻaʻā lava flow.

Architecture. Although Kolb and Radewagen state that KIP-424 is a "notched enclosure," in fact it is a square enclosure, its orientation deviating from cardinality by about 10° in a counterclockwise sense, measuring 11.5 m both N–S and E–W (exterior dimensions). The core-filled walls have a maximum height of about 1 m; there is an entryway in the middle of the S wall, with several flat slabs. The 1995 survey team noted the presence of branch coral.

Excavation. Kolb and Radewagen record that the excavated test pit contained a "burn episode," along with "shell, coral, volcanic glass, an adz fragment, ʻiliʻili, land snails, bone, basalt flakes, and charcoal" (1997, 70). Apparently, no radiocarbon dates were obtained.

Archaeological context. KIP-424 is part of a cluster of structures, including other enclosures and C-shaped shelters, that probably constituted a *kauhale* (residential cluster).

Viewshed. No detailed information on the viewshed is available, although it is known that from this location there are neither good easterly nor good westerly views. The rim of Haleakalā is visible.

Orientation. No precise compass readings are available, but the general orientation seems to be northward, around azimuth 350°, with the entryway on the S. The *heiau* thus faces broadly in the direction of the red cinder cone Kanahau (az. 351.9°).

Additional remarks. Kolb and Radewagen (1997, 70) opined that this was a "small *hale mua* style *heiau*," implying that it was a men's house within a residential complex. The available evidence, including the presence of branch coral, support this interpretation.

KIP-567 (Bishop Museum Site MA-A35-567)

Site KIP-567 was initially recorded during the 1966 Bishop Museum survey. Kirch mapped the structure with plane table and alidade in 1995, and Ruggles and Kirch carried out GPS and total station surveys on March 27, 2003.

Location and topographic setting. The site sits at an elevation of about 425 masl on a gently sloping surface of Nawini ankaramite lava.

Architecture. The structure consists of a nearly square platform, 11 by 11.5 m, with a basal area of 125 m^2 (fig. 7.24). Its orientation deviates from cardinality by 29° in a counterclockwise sense. Low walls (~60–80 cm high) bound the platform on the NNW and ENE, which is otherwise defined by well-faced edges about 0.5–1 m high. The part of the platform on the ENE side is well paved with cobbles; there is a faced pit (1.5 m diameter) in the southeastern part of this pavement. Two pieces of branch coral were found in and adjacent to the pit, one of which was ^{230}Th dated. A second, smaller pit (0.5 m diameter) is situated in the southern part of the platform.

Dating and chronology. The branch coral from the pit was ^{230}Th dated with an age of AD 1574 ± 3 (Kirch, Mertz-Kraus, and Sharp 2015, table 1). We believe that this coral was a dedicatory offering, in which case the *heiau* was constructed during the reign of Maui *aliʻi nui* Piʻilani.

Archaeological context. There is a buffer zone of about 70 m surrounding the site, with no other structures. About 100 m to the west of the platform is a cluster of features of presumed residential function.

Viewshed. There is a clear view of the Haleakalā Ridge sweeping over from the NW to the E, where its eastern ridgeline descends down toward Ka Lae o Ka ʻIlio in Kaupō. From the WSW round to the NW, between azimuths ~240° and ~320°, the view is restricted to a lava ridge 100–200 m distant. The view out over the open sea extends from E round to SW (az. 80° to ~240°), including the entire profile of Hawaiʻi Island between ~125° and ~165°, the peaks of Mauna Kea and Mauna Loa being visible during clear weather.

Orientation. The cobble pavement on the ENE side of the platform, together with the fact that the ENE wall is higher than the NNW wall, implies that this structure faced ENE. The outer faces of the NNW and ENE walls and the platform edges on the SSE and WSW sides are in sufficiently good condition for corner-to-corner orientations to be determinable on all four

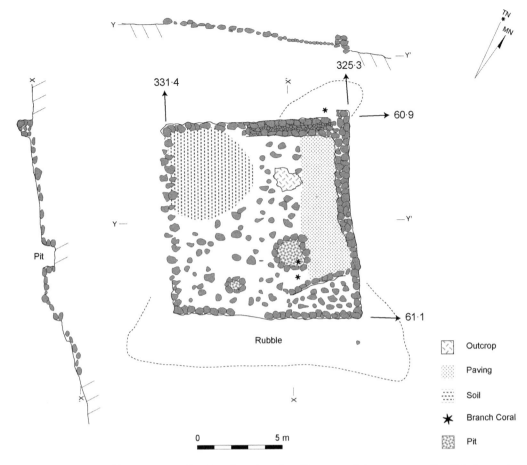

FIGURE 7.24. Plan of *heiau* KIP-567, showing the azimuths of intact straight segments of wall facing and platform edges as determined by the total station survey. (Based on plane-table survey by P. V. Kirch)

sides. The eastward azimuths of the NNW wall and SSE side are 60.9° and 61.1°, respectively, for a mean of 61.0°, while the northward azimuths of the WSW and ENE sides are 331.4° and 325.3°, respectively. Thus, the WSW side is perpendicular to the NNE and SSW sides to within half a degree; the anomalous value for the ENE wall reflects that it is built out somewhat at the SSE end around a triangular terraced step in the ESE corner.

The platform faces an unexceptional point on the lower eastern slopes of Haleakalā (az. 61.0°, alt. 4.3°, dec. +28.5°). This declination corresponds almost exactly to the upper limb of the moon rising at major standstill: the geocentric lunar declination is +28.9°, which is almost exactly the same as the geocentric declination of the major standstill moon for around AD 1600 (+28.65°), plus the semidiameter of the moon (0.25°). This is a theoretical target much discussed by archaeoastronomers (Ruggles 1999a; González-García 2014) but unlikely to have been precisely observable in practice (Ruggles 1999a, 60–62; 2014b, 466–468) and is one for which there is no evidence whatsoever of an interest among Hawaiians. However, the Hawaiian star Nānāhope (Johnson, Mahelona, and Ruggles 2015, 189)—that is, Pollux (β Gem)—did rise close to here around AD 1575 (dec. +29.0), as did Alnath (β Tau, dec. +28.2°), although Alnath is not known to have been significant to the Hawaiians. The platform alignments to the NNW (upper slopes of

Haleakalā to the W of the main rim) and WSW (part of the nearby lava ridge, just above where it cuts the sea horizon) are similarly unexceptional, but to the SSE the WNW wall aligns quite closely upon the summit of Mauna Loa (az. 150.5°, alt. 1.0°, dec. −54.5°, all to the nearest half degree), suggesting a possible topographic motive for the orientation. Gacrux (γ Cru)—the topmost star of the Southern Cross, or Newe (ibid., 200)—would have been seen to rise here around the time of construction (dec. −54.7° in AD 1575).

KIP-728

Site KIP-728 was first recorded during the 1995 intensive survey of the Kīpapa uplands, at which time it was initially interpreted as a postcontact house site with adjacent enclosed yard (i.e., a classic *pā hale*; see Kirch and Sahlins 1992). In July 1996 the site was thoroughly cleared of obscuring vegetation and mapped with plane table and alidade, and subsurface excavations were conducted within the putative historic house enclosure. As described in detail elsewhere (Kirch 2014, 229–241), it became apparent during the course of this intensive study that although in its latest phase the structure had indeed been occupied as a house site in the mid-nineteenth century, this represented a reuse of an older *heiau*.

Location and topographic setting.   The structure straddles a low ridge of the Kīpapa-1 ankaramite *'a'ā* lava flow at an elevation of 540 masl.

Architecture.   As seen in the plan map (fig. 7.25), the main part of this complex structure lies on the E, consisting of two parallel walls oriented approximately N–S. The more westerly of these two walls is longer (36 m long), breached at one point with an opening or entryway. This entrance gives access to a broad, nearly level terrace on the gentle slope, defined on three sides by a low but well-constructed stone retaining wall. The more easterly of the two main walls (about 21 m long) is almost 1.5 m high at maximum; it is against this wall that the rectangular, postcontact house was abutted (fig. 7.26). The elongated rectangular space defined by the E and W walls is enclosed on the S by a fairly low, irregular wall breached by a gap or entryway, and on the N by another crudely constructed, low wall of smaller cobbles that traces a sort of irregular arc between the ends of the main E and W walls. This rectangular space (about 10 by 30 m in interior area) has two distinct levels, separated by a low, partly collapsed retaining terrace wall; thus, the space is divided into upper and lower courts. The entire structure has a surface area of about 620 m².

Examination of the walls, including the size of *niho*, or base, stones used in construction and the techniques of stone stacking, revealed strikingly discrete patterns. The main E and W walls are very straight, nearly true N in orientation, and incorporate massive *'a'ā* lava boulders in their basal courses. They were built using the core-filled construction technique, in which inner and outer facings of larger boulders and first laid down, and the space between them then filled in with smaller pebbles and cobbles. The core-filled method is typical of precontact Hawaiian architecture. In contrast, the N and S walls that enclose and define the two courts are low and crudely constructed, using only cobbles and smaller boulders, and do not follow straight lines. The walls of the house proper (see fig. 7.26), about 1 m high, are similarly made of stacked (as opposed to core-filled) cobbles and smaller boulders and are clearly abutted to the main E wall; hence, they were constructed at a later time than the main wall itself. Given the evident care taken to construct the main E and W (i.e., N–S-trending) walls—using impressively large base stones and keeping the alignments to true north—the crudeness and irregularity of the E–W-trending cross walls are striking.

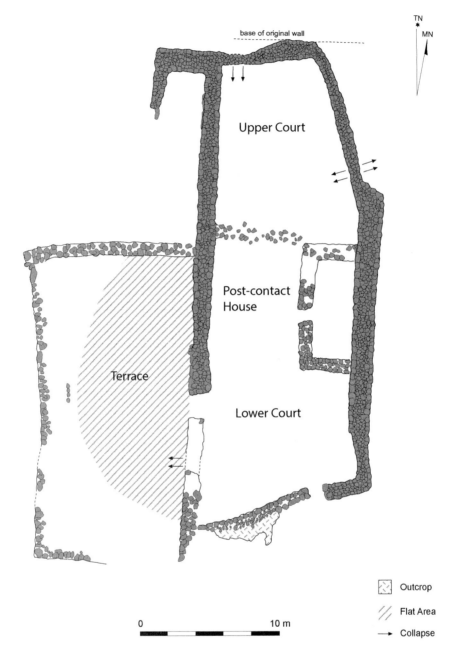

FIGURE 7.25. Plan of *heiau* KIP-728, modified as a house site in the postcontact period. (Based on plane-table survey by P. V. Kirch)

Based exclusively on the evidence of surface (i.e., visible) architecture, several tentative conclusions could be drawn. First, two distinct construction stages were evident, the earlier involving the main E and W (N–S-trending) core-filled walls and presumably also the large level terrace to the W that abuts the main W wall. This first construction stage made use of massive boulders, and considerable effort was made to keep the walls in a nearly true N–S alignment. The second construction stage involved building the rectangular house, as well as the low, crude walls on the N and S of the rectangular courts. It also appeared that most of the stones used in

FIGURE 7.26. View of the postcontact house enclosure constructed within *heiau* KIP-728. (Photo by P. V. Kirch)

this second construction stage had been taken, or robbed, from the tops of the main walls of the first stage.

Excavation. Five units, each 1 m², were excavated within the interior floor area of the postcontact house, four of these forming an L-shaped trench (units Q24–26 and R26) and the fifth unit (Q28) positioned against the N wall. Beneath a thin layer of eolian overburden, we encountered a compact cultural deposit containing substantial numbers of artifacts of Euro-American manufacture. These included an iron chisel, an iron hoe, pieces of iron barrel hoop, assorted buttons, glass beads, a thimble, a metal belt buckle, fragments of slate and flint, and fragments of glass. This artifact assemblage was comparable to those excavated in Anahulu Valley house sites of the mid-nineteenth century (Kirch and Sahlins 1992, 2:179–182). Thus, our original interpretation of the rectangular structure with doorway as a mid-nineteenth-century house site was fully supported by the artifact assemblage recovered from the house interior.

The stratigraphic picture in the trench, however, was more complicated than a simple, one-phase postcontact occupation. At about 17 cm below surface, charcoal became more frequent, and we encountered traces of a disturbed pavement of flat *pāhoehoe* lava slabs. A hearth or burn feature and a possible post mold were associated with this pavement. Euro-American artifacts were no longer present, but pig bone (16 pieces representing one subadult individual), dog bone (3 pieces representing one individual), and pieces of branch coral (*Pocillopora meandrina*) were dispersed among the disturbed paving stones. Pig, of course, was an important sacrificial offering at Hawaiian *heiau* (Valeri 1985, 119). Likewise, branch coral was typically placed as an offering both at coastal fishing shrines and on the altars of upland agricultural temples throughout Kahikinui, as in other parts of the islands (Kirch and Sharp 2005). These findings confirmed that prior to being used as a house in the mid-nineteenth century, site KIP-728 had functioned as a *heiau*.

In order to ascertain whether the original NE corner of the *heiau* wall had been taken down and robbed for stones when the postcontact house enclosure was constructed, we laid out a trench, 1 by 3 m, where triangulation indicated that the original wall should have stood (see fig. 7.25). The original wall stub indeed appeared just a few centimeters below surface. This wall was 1.2 m thick, with two courses of large *pāhoehoe* base slabs still in position. Between the inner and outer faces was a fill of *'a'ā* rubble and cobbles, displaying the classic precontact core-filled method of wall construction. This excavation demonstrated that the KIP-728 walls originally extended as indicated by the dashed lines in figure 7.25. We conclude that both the NE and SE corners of the original temple walls had been purposively destroyed—their visible courses removed and taken down to ground level. The stones thus removed had then been used to construct the dwelling house and the new, low, curved walls that now define the N and S ends of the larger enclosure, converting the *heiau* to a simple *pā hale*.

Dating and chronology. Two pieces of branch coral recovered from the excavations (in units Q25 and Q26) were ^{230}Th dated (Kirch, Mertz-Kraus, and Sharp 2015, table 1). These corals, which were presumably deposited as ritual offerings when the *heiau* was in use, returned identical ages of AD 1572 ± 4, indicating use of the temple in the late sixteenth century, during the reign of Maui *ali'i nui* Pi'ilani.

Archaeological context. A small terrace (to the N), a C-shaped shelter (to the E), and a small enclosure (to the S) were recorded within a 20 m radius of KIP-728; these features are all probably associated with the postcontact use of the site as a habitation.

Viewshed. We did not make detailed observations of the viewshed from KIP-728, but it commands an excellent view of the Luala'ilua Hills to the W, as well as over much of the middle-elevation terrain of the Kīpapa area. The Haleakalā ridgeline is also clearly visible.

Orientation. We did not conduct total station or compass-clinometer surveys at KIP-728. Based on the 1966 plane-table map, however, the main N–S walls of the structure are oriented with azimuth 4°/184°.

Additional remarks. Site KIP-728 was originally a substantial *heiau*, with two courts dividing the main N–S-oriented enclosure and with an adjoining terrace on the W. The presence of the higher court on the N indicates that this *heiau* had a N orientation. Branch coral recovered from the excavations indicate that the temple was in use during the later half of the sixteenth century.

KIP-1010 (STATE SITE 50-50-15-4279)

This large *heiau* was first reported during an aerial helicopter survey of the Kīpapa-Nakaohu uplands performed by Hammatt and Folk (1994) for the Department of Hawaiian Home Lands; Hammat and Folk assigned site number CS-1010 to the structure. Kolb and Radewagen (1997, 71) conducted a series of test excavations totaling 21.75 m². Kirch mapped the site with plane table and alidade in July 1995. Ruggles visited the site on April 15, 2002, and carried out a compass-clinometer survey. Kirch and Ruggles first visited the site together on June 16, 2002, then returned on March 25, 2003, to undertake both total station and GPS surveys, as well as taking detailed observations on the viewshed and orientations.

Location and topographic setting. KIP-1010 is located on a tongue of the relatively young basanite of Kanahau-5 *'a'ā* lava flow, at an elevation of about 660 masl.

Architecture. This structure exhibits some of the most complex architecture of all Kahikinui *heiau*, resulting from its having had two stages of construction. In essence, the site consists of two notched *heiau*, both cardinally oriented to within about 10°; the one on the E seemingly represents a larger version of the smaller one on the W and mimics many of the latter's details (fig. 7.27; see also fig. 3.9 for an aerial view). With a total basal area of approximately 1,385 m², KIP-1010 is also the largest *heiau* within Kahikinui *moku*.

As seen in the plan map, the W enclosure measures about 24 m N–S by 16 m E–W on the exterior, with the notch on the NE. The enclosure's SE corner—unlike the other corners—is not orthogonal and instead forms a distinct acute angle, or canoe prow (a feature noted at other sites, such as LUA-29, KIP-75, KIP-330, and KIP-1156). A gap in the W wall appears to be a formal entryway, leading into the enclosure from a rough terrace to the W of the structure. Within the enclosure, the earthen ground has two distinct courts on the S and N, with the N court elevated about 1.2 m higher. There is a small niche in the inner face of the N wall, near the NE corner, possibly constructed to hold offerings or a sacred object such as an image.

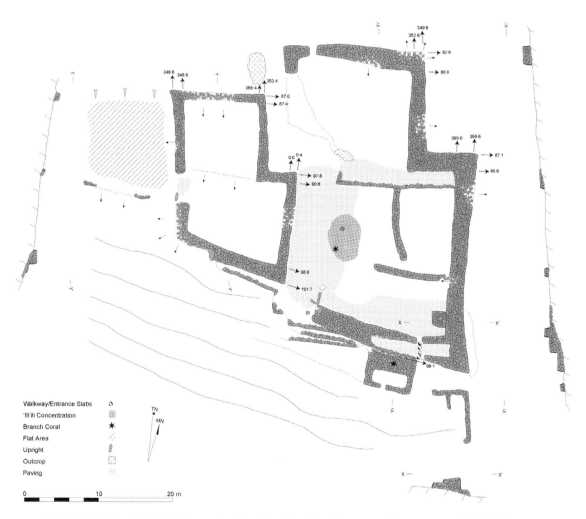

FIGURE 7.27. Plan of *heiau* KIP-1010, showing the azimuths of intact straight segments of wall facing as determined by the total station survey. (Based on plane-table survey by P. V. Kirch)

The much larger E enclosure has exterior dimensions of approximately 40 m N–S by 22 m E–W. Its walls are more massive and higher (up to 2 m wide and 1.8 m high in places), especially along the E side. Once again, the SE corner has a pronounced obtuse angle, making it resemble a canoe prow. The S wall of the enclosure is furthermore buttressed by up to three stepped terrace faces. The interior of this notched enclosure (with the notch again being in the NE corner) is also divided into a higher court on the N (about 1.6 m higher) and a lower court on the S, with a low, paved terrace demarcating the division between the courts. A small niche is again situated in the N wall interior face, near the NE corner. The main S court has a number of features, including a concentration of *ʻiliʻili* gravel with a small posthole feature, two low walls that subdivide the court into discrete rooms, and a small raised platform (~2 by 2 m) near the SE corner of the court.

There is a well-defined entryway into the S court of the main enclosure, marked by a line of flat *pāhoehoe* stepping-stones and a gap into the S wall of the structure. Adjacent to the main S wall immediately W of the entryway is a small rectangular enclosure, about 6 by 4 m, which was probably the foundation for a thatched structure, possibly a house where the priest could prepare himself prior to entering the temple.

Excavation. Kolb and Radewagen (1997, 71–72, fig. 5.9) report excavating 16 different test units within KIP-1010, which yielded "pig bone, coral, *kukui* nut, shellfish, adz fragments, basalt flakes, bird bone, sea urchin, land snails, *ʻiliʻili*, a bone awl, a worked shell piece, and a stone abrader." They also note that an earth oven feature was exposed "under a stone-paved floor" in the W enclosure, and that "the *imu* contained large amounts of shellfish including limpet shell, and sea urchin, as well as some pig and bird bone" (72).

Dating and chronology. Kolb (2006, table 1) reports three radiocarbon dates from his excavations in KIP-1010, all on unidentified wood charcoal from "basal" contexts. Two samples (Beta-095907 and -097500) were collected from test pits in the larger, E enclosure and have similar ages of 340 ± 60 and 350 ± 70 BP (with 2σ calibrated age ranges of AD 1445–1654 and 1432–1664, respectively); since these are reported to come from "basal" contexts, they presumably date burning of the original land surface under the *heiau* architecture. We believe that both of these dates should be regarded as *terminus post quem*. (Kirch visited during Kolb's excavations at KIP-1010 and was present when one of the "basal" contexts in the E enclosure was sampled for radiocarbon dating. The charcoal appeared to come from an agricultural soil underlying the temple's pavement, thus dating a period of land use before construction of the temple.) It is clear from remarks in Kolb and Radewagen (1997, 72) that a third sample (Beta-095906) was recovered from the earth oven in the smaller, W notched enclosure. This sample yielded an age of 200 ± 70 BP, with calibrated age ranges of AD 1522–1575 (6.1%), 1585–1590 (0.4%), and 1625–1950 (88.9%). Because this latter sample derived from an earth-oven context, it can be taken as indicative of the period of use of the W *heiau*. The highest-probability calibration interval suggests that the W enclosure was in use sometime during the seventeenth to the eighteenth century.

Kirch and Sharp (2005) dated a single piece of branch coral that was found in a surface context adjacent to the formal entryway to the large E enclosure. This branch coral, which may have been a dedicatory offering for the larger enclosure, yielded a ^{230}Th age of AD 1580 ± 10, falling within the reign of Maui *aliʻi nui* Piʻilani, who is credited with consolidating the island into a single unified polity.

Kolb and Radewagen maintained that the larger, E enclosure was constructed between AD 1445 and 1665 and that "the smaller enclosure to the east [*sic*; it is to the W] was added sometime after AD 1660" (1997, 72). Their interpretation depends on accepting the "basal" radiocarbon dates from the E enclosure as accurately indicating the age of construction; we reject this interpretation because these samples clearly came from stratigraphic contexts underlying the *heiau* structure and thus predating its construction. We believe that it is more likely that the original structure was the smaller notched *heiau* on the W, with the larger, E enclosure being built later, mimicking while enlarging all of the essential architectural features of the smaller *heiau*. Such an enlargement and replication of the features of a successful *heiau* would be in keeping with Hawaiian ritual practices.

Archaeological context. Site KIP-1010 is relatively isolated. A small enclosure on the same ridgeline a few meters downslope might have been a residence associated with the priest in charge of the temple.

Viewshed. There is a clear view of the Haleakalā ridgeline extending up from Manukani (2.8 km, az. 293°) in the WNW round to the E, where its eastern slopes reach down to the Kaupō Peninsula and Ka Lae o Ka ʻIlio. To the NE, Puʻu Pane (3.4 km, az. 37°) dominates the foreground, some 8° to the left of, and on a level with, a prominent rounded notch in the distant skyline; it stands out particularly prominently when backlit by the rising sun. To the WSW and W, between azimuths ~243° and ~283°, the view is restricted by the Nawini ʻaʻā flow in the direction of Heiau Ridge, 400–800 m away. Just above where the Nawini flow meets the sea horizon, the northeastern Lualaʻilua peak just breaks the skyline (3.9 km, az. 251°). The sea horizon extends from azimuths ~82° to ~243°, within which the entire profile of Hawaiʻi Island is visible in clear weather conditions (between az. ~125° and 165°).

Orientation. Although the W enclosure has a formal entrance on the W side, the two enclosures are clearly N-facing: both have elevated northern courts and niches for offerings in their N walls. The terrace faces on the S side of the E enclosure reinforce this conclusion.

We were able to obtain good estimates of the orientation (either from corner-to-corner alignments or from intact straight sections of wall facing) of both the inner and outer faces of nearly all wall segments in both enclosures (see fig. 7.27). The mean wall orientations can be tabulated as follows.

WALL SEGMENT	W HEIAU	E HEIAU
Northward orientations of N–S-oriented wall segments		
W wall	349.3°	—
E wall N of notch	354.4°	351.0°
E wall S of notch	0.2°	355.8°
Eastward orientations of E–W-oriented wall segments		
N wall	87.2°	84.9°
Central wall (within notch)	94.3°	86.8°
S wall	100.3°	99.1° (outer face only)

Taking the mean of the azimuths of the available N–S-oriented wall segments in each case, we obtain our best estimates of the northward orientation: 354.6° for the W enclosure and 353.4°

for the E enclosure. Since the E enclosure does not have a W wall, a fairer comparison might be the mean of the E wall segments alone, which yields 357.3° and 353.4°, respectively for the W and E enclosures. The N wall orientations (87.2° [W], 84.9° [E]) are close to the orthogonal directions (87.3° [W], 83.4° [E]). The skewed S walls (100.3° [W], 99.1° [E]) are similarly oriented.

From this location, the Haleakalā ridgeline is crowned by a prominent cluster of cinder cones visible between azimuths 350° and 357°. The two enclosures are clearly oriented upon these (e.g., fig. 7.28). In the eastward direction, the N and central walls align upon the sea horizon above the Kaupō Peninsula, spanning azimuths 82° to 97° (fig. 7.29). Although the position of sunrise at the equinoxes falls within this range (az. ~89.5°)—as does that of the rising of Orion's Belt, part or all of Nā Kao (Johnson, Mahelona, and Ruggles 2015, 198), around AD 1600 (az. ~90.5° to ~92.0°, alt. −0.7°, dec. ~−1.0° down to ~−2.5°)—there is no reason to suppose that it was of any particular significance. The wall alignments to the W are upon an unremarkable stretch of the Nawini flow ridgeline, while those to the S are toward open sea. Apart from the fact that the E and W ranges include the equinoctial directions and the rising and setting of Orion's Belt, there is no obvious astronomical significance to any of these alignments.

As noted earlier, KIP-1010 exhibits two of the most prominent examples of canoe-prow corners. For the W enclosure, the direction in which this acute corner "points"—that is, the direction of the line that bisects the corner—has an azimuth of 140°. For the E enclosure, the azimuth would be about 137.5° if the E wall were straight, but it is decreased by several degrees because the wall curves out toward the corner. While the summit of Mauna Kea, the highest point on Hawai'i Island, has an azimuth of 137° and could therefore have been the intended reference point, it is also noteworthy that the great war temple of Mo'okini situated at 'Upolu Point on Hawai'i Island (Stokes 1991, 173–178) has an azimuth of 135°. Although Mo'okini Heiau itself would not have been visible from across the 'Alenuihāhā Channel, its general direction would have been known to the Maui chiefs, and it is therefore entirely possible that Mo'okini was the intended reference point.

Additional remarks. Given the similarity in form and orientation of the adjacent enclosures, this double *heiau* may give us a unique insight into the level of precision with which a structural orientation was, in practice, aligned upon an intended visual (topographic or astronomical) target.

There can be little doubt that the orientations of various features at this complex structure were influenced by various visible topographic features, including the prominent cluster of cinder cones crowning the Haleakalā ridgeline to the N, the Kaupō Peninsula to the E, and the summit of Mauna Kea to the SE.

KIP-1025

This rather unusual structure was discovered and recorded during an intensive survey in 1997. Kirch mapped the site with plane table and alidade on February 10, 1997. Kirch and Ruggles visited KIP-1025 on March 21, 2008, to make total station and GPS surveys.

Location and topographic setting. KIP-1025 is situated at an elevation of about 90 masl, slightly less than 0.5 km inland of the coastline. The structure was built on a spur of the relatively young Nawini ankaramite lava, overlooking a *kīpuka* (patch of older lava) of basanite of Kahikinui lava.

Architecture. The layout of this structure is unusual (fig. 7.30), giving us the distinct impression that it may never have been finished, as suggested especially by the low, discon-

FIGURE 7.28. View from the SE corner of the E enclosure at *heiau* KIP-1010, showing the alignment along the E wall toward a prominent cluster of cinder cones on the Haleakalā ridgeline to the N. (Photo by Clive Ruggles)

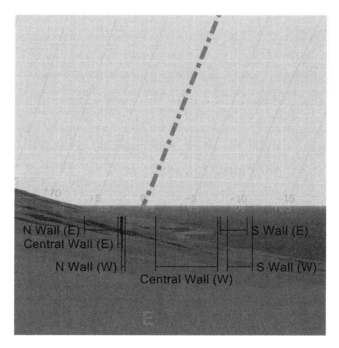

FIGURE 7.29. DTM-generated profile of the horizon to the E as seen from *heiau* KIP-1010. The horizontal bars indicate the direction of orientation of the N, central, and S walls of each of the two enclosures. The dash-and-dot line indicates the rising path of the sun at the equinoxes. (Graphic by Andrew Smith)

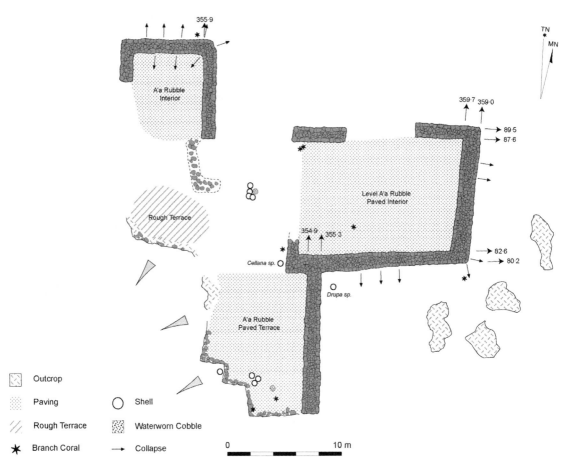

FIGURE 7.30. Plan of *heiau* KIP-1025, showing the azimuths of intact straight segments of wall facing as determined by the total station survey. (Based on plane-table survey by P. V. Kirch)

nected wall stub on the N and the incomplete W wall of the larger enclosed space. To within about 5°, it is cardinally oriented. The main feature is a rectangular, leveled court roughly paved with ʻaʻā clinkers, bounded by core-filled walls on the E and S and having an incomplete wall on the N. The W wall is mostly missing. The exterior dimensions of the enclosed space are 16 m E–W by 12 m N–S. The E wall is especially well constructed, rising to heights of 1.5–2 m, while the S wall is about 1.2 m high.

To the SW of the main enclosure there is a broad terrace, also roughly paved with ʻaʻā clinkers, defined on the E by a core-filled wall about 1 m high that abuts perpendicularly to the S wall of the main enclosure. The terrace edge is faced on the S and W with a stone retaining wall 0.8–1 m high; the face makes two right-angle turns.

In the NW part of the complex a rectangular space about 6 by 7 m is bounded on the W, N, and E by a low stone wall rising at most to a height of 0.8 m.

Several pieces of branch coral and some waterworn basalt cobbles were noted on the site surface, especially on the SW terrace. A cluster of shell fragments and a waterworn cobble were also noted slightly W of the main enclosure.

Archaeological context. There are no other structures in the immediate vicinity of KIP-

1025. The site lies within an intermediate zone between the coast and more densely settled uplands, an area with a low density of sites.

Viewshed. There is an unobstructed view of the Haleakalā ridge extending from the WNW to the ENE, where its eastern slopes reach down to Ka Lae o Ka ʻIlio at azimuth 82°. There is also a clear view out to sea from here round to the WSW, at azimuth ~245°. The western horizon between this azimuth and ~305° is formed by ridges 1.5–3 km distant, except where the Lualaʻilua Hills, 4.5–5 km away, appear conspicuously on the skyline, with their peaks at azimuths 285.8° and 296.5°.

Orientation. The prominent E wall suggests that this site has an easterly orientation. The longest E–W-oriented wall segment, that on the S side of the main enclosure, has an eastward orientation of 81.4° (the orientations of individual wall faces, which never differ by more than ~1° on either side of the mean orientation, are shown in fig. 7.30). The shorter N wall segment on the NE side of the enclosure has an azimuth of 88.6°, so that our best estimate of the mean axial orientation is 85.0°. The E wall, with a northward azimuth of 359.3°, is more closely orthogonal to the N wall (orthogonal direction 358.6°) than to the S wall (orthogonal direction 351.4°).

We were also able to obtain reliable orientation estimates for the inner face of the E wall of the separate rectangular space to the NW of the main enclosure (az. 355.9°) and for the orientation of the E wall of the SW terrace (355.1°).

The S wall of the main enclosure is directly aligned upon Ka Lae o Ka ʻIlio, while the shorter N wall is oriented well to the right, toward (but not exactly upon) the position of equinoctial sunrise. The S wall alignment (az. ~+81.5°, alt. ~−0.5°, dec. +8°) is also close to the rising point of the Hawaiian stars Humu (Altair, α Aql, dec. +8°) and ʻAua (Betelgeuse, α Ori, dec. ~+7.5°) around the likely time of construction (see table 4.2), and the span of alignments also corresponds broadly to the rising position of the northern half of Nā Kao (Orion), but the belt (dec. −2.0° to −0.5°, to the nearest half degree) rose somewhat to the right of the N wall alignment.

To the N, the walls are aligned upon part of the section of the Haleakalā ridgeline, but it is a part that lacks prominent topographical features such as cinder cones and is of little obvious interest astronomically (dec. ~+80.5° to ~+81.5°), except that Hōkūpaʻa (Polaris, α UMi) would hang in the sky some 7° above the horizon close to due N. The western profile is interesting in that the June solstice sunset (as well as the setting path of the Pleiades around the likely time of construction) coincided with the most prominent of the Lualaʻilua Hills. Furthermore, the N wall of the main enclosure is aligned almost exactly on the position of equinoctial sunset (az. 268.6°, alt. 3.3°, dec. −0.2°), while December solstice sunset coincided with the junction of the foot of the ridge and the sea horizon. Although the evidence is not conclusive, it is at least possible that this combination of natural and constructed markers could have provided a means of tracking the seasonal variation in the position of the setting sun.

Additional remarks. Among the possible factors that could have influenced the orientation are a topographic alignment upon Ka Lae o Ka ʻIlio in Kaupō and a desire to track the changing position of the setting sun over the seasons.

KIP-1146

This unusual temple was discovered and mapped by Kirch with plane table and alidade in February 1997. On March 23, 2008, Kirch and Ruggles revisited the site to make total station and GPS surveys.

Location and topographic setting. At an elevation of about 290 masl, KIP-1146 sits astride a low ridge spur of the Kīpapa-1 ankaramite ʻaʻā lava flow.

Architecture. The architecture at KIP-1146 is unusual, and at first we were uncertain how to classify it, although with detailed study we are convinced of its ritual function. The most prominent features are two stone-faced and stone-paved terraces constructed on the downslope (S) side of a lava ridge. The lower and larger terrace runs W–E and measures 9 by 4 m, with a front facade height of 1.9 m. The upper and lower terraces are separated by a single-course facing just 30 cm high; this runs WNW–ESE and measures 6.5 by 2 m (fig. 7.31). Adjoining the lower terrace on the W is a small stone-walled enclosure with interior dimensions of 3.5 by 3 m; this enclosure was probably the foundation for a thatched structure. Immediately N (upslope) of the enclosure is a stone-filled terraced platform, measuring 5 by 3 m, with a sunken depression in the center; this is almost certainly a burial platform, probably postcontact in age. A fragment of a poi pounder was found on the E end of the lower terrace.

Immediately E of the two terraces and at the same elevation on the lava ridge is a small, leveled space about 2 m in diameter (B in fig. 7.31), defined on the NW and SE by low stacked walls of lava rock. About 5 m to the NE of this leveled space, on the NE side of the lava ridge, a well-constructed, core-filled rock wall runs for 16 m in a NW–SE direction, terminating in a large slablike upright, 1 m wide and 1.2 m high (fig. 7.32, and A in fig. 7.31). The upright is not natural and was purposefully erected in this position at the end of the wall. Situated near the bedrock outcrop that is the terminus of the lava ridge, the upright occupies a commanding position over the landscape, which slopes away to the east and south.

Archaeological context. Two small enclosures about 20 m NW of KIP-1146 might have been temporary habitation features. Otherwise, the site density in this area is quite low, with a few minor terraces that may have had an agricultural function. This elevation is at the lower margin for growing sweet potato in terms of the annual average rainfall, and there are no major habitation complexes.

Viewshed. There are open views of the Haleakalā rim extending from the NW round to the E, where its eastern slopes run down to Ka Lae o Ka ʻIlio in Kaupō. The Lualaʻilua Hills (peaks at 279.8° and 297.2°) are conspicuous features on the horizon to the W and WNW, protruding above a lava ridgeline some 200 m distant. There is open sea from azimuth 78.5° round to ~238°.

Orientation. The lower and upper terraces have eastward orientations of 95.5° and 111.3°, respectively, while that of the 16 m wall is 132.1°. More significant, however, may be the orientation of the upright when viewed from the small cleared area. This view is directly toward the Kaupō Peninsula and Ka Lae o Ka ʻIlio (see fig. 7.32).

All of these orientations, apart from that of the wall, fall within the solar rising arc (fig. 7.33). The declinations for the sea horizon (alt. −0.5°) and the approximate Gregorian dates when the sun would rise at the relevant position are given in table 7.3.

The sun would rise over the Kaupō Peninsula, between the point where the slope down from Haleakalā meets the sea horizon to the left and the promontory of Ka Lae o Ka ʻIlio to the right, between August 26 and September 14 (Gregorian) moving from left to right and between March 28 and April 16 moving from right to left. This latter period coincides with the end of the Makahiki season, raising the possibility that this temple was laid out so as to keep track of the timing of the close of the Makahiki, when the tribute (*hoʻokupu*) would be collected by the chiefs in the name of Lono, during the *akua loa* circuit (see chapter 2). This time of late March to April is

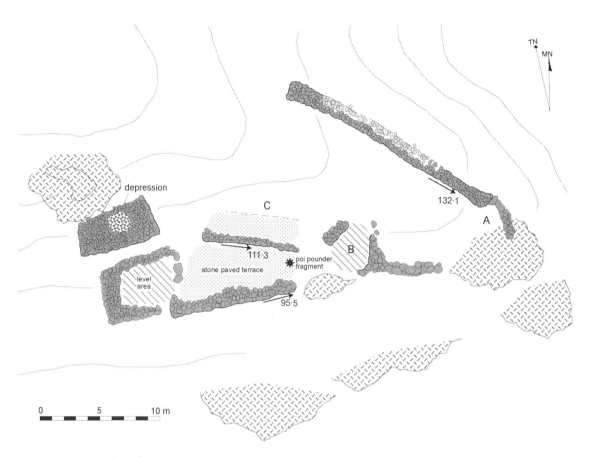

FIGURE 7.31. Plan of site KIP-1146, showing the azimuths of the terraces and the 16 m wall as determined by the total station survey. A is a 1.2 m high slablike upright, and B is a small leveled space. (Based on plane-table survey by P. V. Kirch)

FIGURE 7.32. Upright stone at KIP-1146 (A in Figure 7.31) as viewed from the site's leveled area (B in Figure 7.31), with Ka Lae o Ka 'Ilio directly behind it. (Photo by Clive Ruggles)

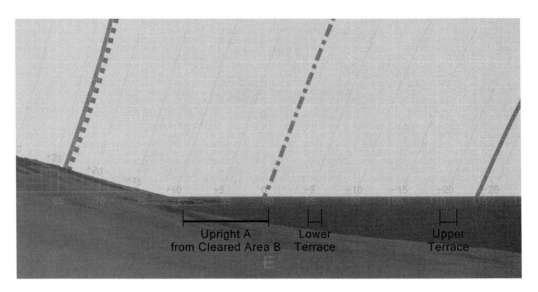

FIGURE 7.33. DTM-generated profile of the horizon to the east as seen from site KIP-1146. The horizontal bars indicate the direction of orientation of the upright A when viewed from the small cleared area B (see fig. 7.31) and of the terraces. The solid shaded lines indicate the rising paths of the sun at the June solstice (*left*) and the December solstice (*right*); the dash-and-dot line indicates the rising path of the sun at the equinoxes; the dotted line indicates the rising path of the Pleiades in AD 1600. (Graphic by Andrew Smith)

TABLE 7.3. Dates of sunrise in directions of possible significance at site KIP-1146.

POINT ON HORIZON	AZIMUTH	DECLINATION	APPROXIMATE GREGORIAN DATES
Slope meets sea horizon to left of Kaupō Peninsula	78.5°	+10.3°	Aug 26, Apr 16
Above Ka Lae o Ka 'Īlio	86.2°	+3.1°	Sep 14, Mar 28
Lower terrace direction	95.5°	−5.6°	Oct 7, Mar 6
Upper terrace direction	111.3°	−20.3°	Nov 22, Jan 19

also the flowering season for the *wiliwili* tree, which is so characteristic of Kahikinui and other dryland leeward regions. A Hawaiian proverb states: *Pua ka wiliwili nanahu na manō* (When the *wiliwili* tree blooms, the sharks bite) (Pukui 1983, 295). The shark was a well-known metaphor for the ruling chiefs (Kirch 2010, 2012).

KIP-1151

KIP-1151 was discovered and recorded during the 1997 field season; Kirch mapped the site with plane table and alidade on February 19, 1997. Kirch and Ruggles returned to the site on March 23, 2008, to undertake total station and GPS surveys.

Location and topographic setting. KIP-1151 is situated at an elevation of 310 masl. It lies a short distance to the west of the long cattle wall between Kīpapa and Alena, although we have included the site within the numbering sequence for Kīpapa. The substrate here is the Alena-2 basanite ʻaʻā lava flow, less than 10,000 years old and hence very rocky and barely weathered. The site was constructed on the west side of a massive lava pressure ridge running north–south. The most prominent topographic feature in the vicinity of the temple is a prominent *puʻu* 60 m to the E of the structure. This *puʻu* rises about 10 m above the surrounding terrain and dominates the view to the E of the site.

Architecture. KIP-1151 consists of a nearly square terraced platform, skewed from cardinality by about 11° in a counterclockwise sense. Its exterior dimensions are 11 m N–S by 12 m E–W; its surface is well paved with fine ʻaʻā clinkers (figs. 7.34 and 7.35). On the N and E, the platform is bounded by well-faced retaining walls rising to a height of 1.6 m above the paved floor of the court. On the S and W, the terrace itself is retained by facades that drop off as much as 2 m (on the S). Because these S and W facades were constructed on steeply sloping ʻaʻā bedrock formations, the terrace is elevated as much as 8 m above the surrounding terrain in these directions. On the N and E, however, the retaining walls are built up against outcrops that rise up in those directions; thus, the site is nestled into the harsh lava terrain.

A large natural lava boulder 2.6 by 1 m stands up 0.5 m through the paved floor near the W end of the terrace. Adjoining this boulder on its E side is a small stone-lined pit, possibly for the disposal of offerings. A few fragments of shell and *Porites* coral were noted on the terrace pavement. A basalt hammerstone was situated on top of the retaining wall in the NE corner.

At the SE corner of the main terrace a well-constructed ramp about 2 m wide and 5 m long (partly incorporating natural outcrop) leads upslope to a cleared, leveled space about 4 m in diameter, defined on the E by partly modified natural outcrop. At the juncture of the ramp and the leveled space we noted a cluster of four small elongate ʻiliʻili pebbles (the elongate shape giving the pebbles a distinct phallic appearance). On the floor of the leveled space was a flaked basalt cobble. This small leveled space or room sits at an elevation of about 2 m above the main terrace and hence overlooks the latter. It would have provided a natural space for a *kahuna* or other ritual practitioner to perform ceremonies while others observed the rites from the main terrace.

Archaeological context. No intensive survey was conducted around KIP-1151, but a cursory examination suggests that there are no major sites close to the temple.

Viewshed. There is a clear view of the Haleakalā ridgeline from the NW (az. ~310°) round to the ENE, beyond which, as already noted, the view is dominated by a lava hillock about 60 m to the E that obscures everything in the farther distance down to the sea horizon. Where this hillock intersects the more distant Haleakalā slope depends on the position of the observer, but it is around azimuth 72° for an observer in the upper leveled area or in the center of the main platform. (Although moving from the former position to the latter shifts the hillock to the right relative to the more distant horizon, it also shifts it upward, with the net result that the junction between the two profiles changes little in azimuth.) The Lualaʻilua Hills, 2–3 km distant, dominate the western skyline between ~275° and ~305°, below which a closer ridgeline ~1 km distant runs down to meet the sea horizon at ~250°.

Orientation. The S platform edge has an eastward orientation of 76.3°. This is similar to that of the convex exterior face in the eastern segment of the N wall (az. 76.5°), but the high inner face of the N wall has an azimuth of 81.0°, suggesting that our best estimate of the main axial orientation should be 78.7°. The W platform edge has a northward orientation of 344.5°, and the high inner face of the E wall has an azimuth of 346.9°, so that our best estimate of the N–S axial

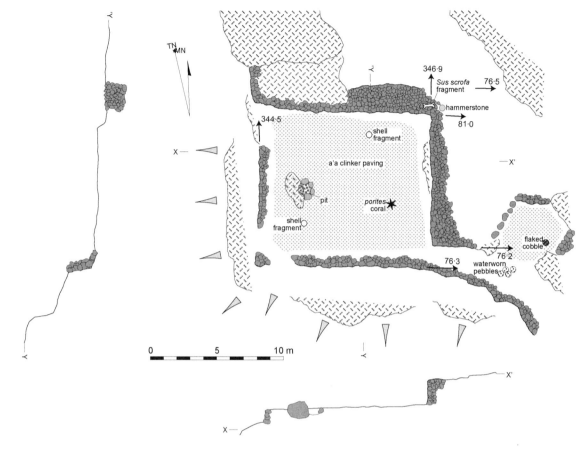

FIGURE 7.34. Plan of *heiau* KIP-1151, showing the azimuths of intact straight segments of wall facing and platform edges as determined by the total station survey. (Based on plane-table survey by P. V. Kirch)

FIGURE 7.35. Aerial photo of *heiau* KIP-1151, from the southeast. (Photo by P. V. Kirch)

orientation is 345.7°. A straight segment of inner wall facing on the N side of the ramp has an orientation of 76.2°.

The view from the center of the upper leveled space is combined with the digitally generated horizon profile in figure 7.36. As viewed from the platform, the foreground hillock would be displaced about 4° to the right and 2° upward relative to the more distant Haleakalā slope (and the astronomical rising lines). The hillock would be ideal for picking out the rising points of a number of Hawaiian navigation stars, but the rising point of Makaliʻi (the Pleiades) is always farther up the Haleakalā ridgeline to the left. The presence of a partly collapsed wall built up the northern slope of the hillock, which may or may not have been present at the time of use, means that we cannot be certain as to whether any more specific markers originally existed.

It is also possible that the hillock was used to track the changing position of the rising sun, although not close to the solstitial extremes, where the sun spends a disproportionate amount of time. Broadly speaking, the sunrise position would spend a quarter of the year beyond the hillock at each end, and a quarter crossing it in each direction.

To the W, the June solstice sun set just above the notch formed by the junction between the Lualaʻilua Hills, although the wall orientations are well downslope of this. It is at least possible that the prominent hills on the western horizon could have been used additionally from this site to track the changing position of the setting sun, especially in the spring and summer months. The western alignment (az. ~258.5°, alt. 1°, dec. ~−10.5°) is also close to the setting point of the star Hikianalia (Johnson, Mahelona, and Ruggles 2015, 149)—that is, Spica (α Vir, dec. ~−9.5° around AD 1700). To the N, the walls are aligned upon part of the Haleakalā ridgeline between conspicuous cinder cones at 342° and 352°, while to the S they are aligned upon open sea just off to the right of Hawaiʻi Island. It seems unlikely that either of these alignments was of any significance.

FIGURE 7.36. View from the center of the upper leveled space at site KIP-1151, overlaid on the digitally generated horizon profile (which cannot reproduce the nearby landform). The solid shaded line indicates the rising path of the sun at the June solstice; the dash-and-dot line indicates the rising path of the sun at the equinoxes; and the dotted line indicates the rising path of the Pleiades in AD 1600. An additional azimuth scale is provided at the bottom of the figure. (Composite of photo by Clive Ruggles and graphic by Andrew Smith)

The existence of the higher, leveled space on the E side of the main terrace suggests that the focus of attention was to the E rather than to the N and that the platform was intended to face this direction. At the same time, the positioning of the higher, leveled space to one side of, rather than directly in front of, the view to the E suggests that this view could have been important both to a *kahuna* or other ritual practitioner in the higher space and to other participants on the main terrace.

Additional remarks. We believe that KIP-1151 was oriented upon a prominent landscape feature—the *puʻu* to the E of the site. This could have served a practical purpose related to tracking the changing position of sunrise or observing the rising of navigation stars.

KIP-1156

Site KIP-1156 was first recorded by Kirch during an archaeological survey in Kīpapa in February 1997; at that time the structure was mapped with plane table and alidade. Test excavations were also made during the 1997 field season. On March 23, 2008, Kirch and Ruggles jointly visited the site to conduct a total station survey; a GPS survey was also made at that time.

Location and topographic setting. KIP-1156 is situated at an elevation of about 350 masl, on a gentle slope of the Kīpapa-1 ankaramite ʻaʻā lava flow. The structure lies on the western lip of a prominent intermittent drainage channel that carries water during *kona* (southerly) storms; this channel may have had more frequent flow in precontact times when the forest was lower and the water table higher due to fog drip precipitation (Stock, Coil, and Kirch 2003). We believe that the positioning of this temple adjacent to the watercourse was intentional and significant.

Architecture. The structure is a nearly square enclosure, whose orientation deviates from cardinality by about 23° in a counterclockwise sense, but with a pronounced acute angle at the intersection of the NNW and ENE walls, creating a distinct canoe-prow point in the NNE corner (fig. 7.37; see also fig. 3.7). The WSW and SSE walls are each 14 m long, while the ENE wall measures 13 m and the NNW wall 17 m, respectively. The walls range in height from about 0.5 to 1 m and are 1.5 m thick; the construction technique is classic core-filled. A piece of branch coral was noted on the distinct point of the NNE corner wall junction, presumably an offering. The prominence accorded to the NNE corner in the overall layout of the structure is one indication that the *heiau* may have been ritually oriented in this direction, although orientations to NNW and ENE are also possible.

A gap in the SSE wall near the ESE corner may have been a formal entryway into the enclosure; a piece of branch coral and a *Cellana exarata* shell were observed next to this entryway feature. The interior of the structure is subdivided into three distinct rooms or space cells. The ENE half of the interior comprises the largest of these rooms, with a level floor of earth and scattered stones. To the SSW, one steps down about 0.5 m into a space cell that measures 6 by 7 m and also has an earthen floor. The WNW part of the interior makes up a third space cell, elevated about 0.5 m above the adjacent SSW room and entirely paved over with ʻaʻā clinkers. Near the NNW end of this WNW paved space cell is a low, single-course alignment of cobbles defining what may have been a low altar about 1 m from the interior face of the NNW wall.

Excavation. Three test excavation units (G10, I14, and L11), each 1 m², were dug in the interior of KIP-1156. In all units a cultural deposit of gray-brown loam was exposed underlying a thin (3–5 cm) layer of overburden; the cultural deposit extended to a depth of about 40–45 cm

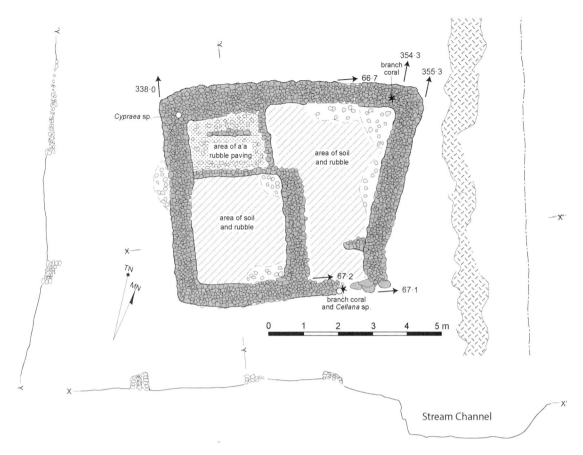

FIGURE 7.37. Plan of *heiau* KIP-1156, showing the azimuths of intact straight segments of wall facing as determined by the total station survey. (Based on plane-table survey by P. V. Kirch)

below surface. In unit L11 a concentration of whitish ash about 40 cm in diameter and 10–15 cm thick, designated as feature 1, probably represents a single-use hearth feature.

The cultural content of the KIP-1156 excavation units is detailed in table 7.4. There were a significant number of basalt flakes, indicative of tool production or adze maintenance within the *heiau*. Most of the faunal material consisted of marine shell, with the identifiable taxa being 'opihi (*Cellana exarata*), *pipipi* (*Nerita picea*), snake's head cowrie (*Cypraea caputserpentis*), and some larger Thaididae species. Of the small quantity of bone present, one specimen of parrotfish (Scaridae) and one of bird (otherwise not identifiable to taxon) were recognizable.

Dating and chronology. A single radiocarbon date was obtained from wood charcoal excavated from the hearth feature in unit L11, with an age of 150 ± 50 BP, which yields calibrated age ranges of AD 1665–1894 (78.2%) and 1905–1950 (17.2%). Most likely, the site was in use during the late seventeenth or the eighteenth century.

Archaeological context. A small scatter of basalt flakes and a possible adze preform were noted about 20 m downslope (S) from KIP-1156. No other sites are in the immediate vicinity, but at a farther distance there are various probable residential features surrounding the site within a radius of 200 m. At this elevation within Kīpapa the site density becomes relatively low as the lower annual rainfall limits for sweet potato cultivation are approached.

TABLE 7.4. Cultural content of KIP-1156 excavations.

MATERIAL	UNIT G10	UNIT I14	UNIT L11	FEATURE 1
Basalt flakes	62	6	27	3
Coral	12	8	19	—
Charcoal (g)	58.4	61.2	121.1	4.0
Marine shell (NISP)	53	11	88	17
Sea urchin (NISP)	—	—	1	8
Bone (NISP)	8	—	2	—

Viewshed. There is a clear vista of the Haleakalā Ridge from the NW (az. ~315°) round to the ENE, where its eastern slopes sweep down to the Kaupō Peninsula and Ka Lae o Ka ʻIlio. Open sea is seen from azimuths ~80° (above the Kaupō Peninsula) round to ~230°. Between SW and NW, the view is restricted by the nearby Alena flow between 100 and 200 m away.

Orientation. As already noted, the main orientation is probably to the ENE. The straight and largely intact outer face of the SSE wall is oriented with an eastward azimuth of 67.1°. The outer face of the NNW wall, most of which is also intact, yields 66.7°, and the inner face of the SSE wall within the SSW room yields 67.2°, so we obtain a reliable estimate of the intended orientation: 67.0°. The WSW wall is reasonably orthogonal to this, with a northward inner face orientation of 338.0°. The skewed ENE wall has good straight inner and outer faces oriented at 354.3° and 355.3°, respectively, for a mean of 354.8°.

The *heiau* faces a specific part of the eastern Haleakalā slope (az. 66.7°, alt. 2.6°, dec. +22.6°, to az. 67.2°, alt. 2.5°, dec. +22.1°) that is unremarkable topographically but is very close to the rising position of the Pleiades (dec. +22.6° to +23.1° in AD 1600, +23.0° to +23.5° in AD 1700). Furthermore, in the foreground, just under the more distant slope, is a slight promontory on which the ruins of St. Ynez Church now sit. This promontory is reputably the site of a former *heiau* that was destroyed when the church was built in the 1830s.

These factors could certainly account for the main orientation, but they do not explain the prominent point in the NNE corner or, alternatively, the orientation of the skewed ENE wall. By analogy with the canoe-prow corners at sites such as KIP-1010, it might be suggested that the diagonal alignment out through the pointed corner was significant, but in this case the direction concerned (az. 31°, alt. 12°, dec. +59°) is an unremarkable part of the upper eastern Haleakalā slope. This part of the slope corresponded around AD 1700 to the rising position of the stars Tsih (γ Cas), possibly one of the three stars forming Mūlehu (Johnson, Mahelona, and Ruggles 2015, 195), and Megrez (δ UMa), one of the seven stars forming Nā Hiku (ibid., 196), but otherwise has no obvious astronomical significance. The E wall might have been skewed so as to incorporate a secondary orientation of significance. The strongest possibility is to the N, where the wall is aligned upon the prominent red cinder cone of Kanahau, which extends from 350° to 355° and forms the highest point on the northern skyline. To the S, the wall is aligned upon open sea. The easterly direction perpendicular to this skewed wall (az. 84.8°) is toward the visually prominent Kaupō Peninsula but well to the left of Ka Lae o Ka ʻIlio (az. 87.3°).

The axial alignment in the opposite direction (az. 247°, alt. 2°, dec. −21°) has no obvious astronomical significance, although the southeastern peak of the Lualaʻilua Hills, which just protrudes above the closer lava ridge to the WNW, coincides with the setting point of the Pleiades as well as that of the June solstice sun. To the NNW, the WSW wall is oriented 2–3° to the left of one of the larger cinder cones on the upper western Haleakalā slope; the corresponding declination (+68°) also has no obvious significance. To the SSE, the WSW wall is aligned upon open sea.

KIP-1306

Site KIP-1306 is a small agricultural shrine in the Kīpapa uplands, first identified during an archaeological survey in 1995. The structure was plane-table mapped by James Coil in March 1999, as part of an investigation of adjacent features (KIP-1304) believed to be a ritual garden complex (Coil 2004, fig. 6.66).

Location and topographic setting. The shrine is situated in the northeastern corner of a large natural swale or depression in the undulating ʻaʻā topography, at an elevation of 475 masl. The swale floor appears to be a *kīpuka*, or patch of older lava (probably a basanite flow dating to >50,000 BP), surrounded by the younger Kīpapa ankaramite dating to 10,000–30,000 BP. The swale was intensively cultivated, as indicated by terraced retaining walls running across the floor.

Architecture. The KIP-1306 shrine consists of a well-defined platform constructed against the sloping face of the massive ʻaʻā flow defining the edge of the swale. The platform is approximately cardinally oriented and measures 4 by 4 m, with well-defined faces on the S and W and to a lesser extent on the N. On its eastern side the platform grades into the surface of the ʻaʻā lava flow. The prominent western face, which rises 1.2 m up from the earthen floor of the swale, consists of four to five courses of carefully stacked boulders. This face is not straight but has a slight concave arc; in the center of this arc near the base of the stacked facade is a small niche that might have served as a place to deposit offerings. In the center of the platform there is also a shallow pit, about 1 m in diameter.

Dating and chronology. The KIP-1306 shrine itself was not directly dated. However, a test excavation in the immediately adjacent garden terrace revealed an earth oven with charcoal that was radiocarbon dated (AA-38678) to 147 ± 32 BP, which yields 2σ calibrated age ranges of AD 1667–1783 (46.1%), 1796–1891 (32.7%), and 1909–1950 (16.6%).

Archaeological context. The shrine is part of a complex of adjacent features that includes two well-defined rectangular terraces (each about 5 by 7 m), separated by a low stone platform. Excavations in these terraces revealed that they were used for planting, with small waterworn ʻiliʻili stones placed in the bottoms of individual planting depressions. One of the terraces also contained a small earth oven lined with basalt cobbles, probably a ritual oven (*imu hoʻomaʻalili*) of the kind described by Kamakau (1976, 30) for cooking offerings when sweet potato fields were first planted. (Kamakau writes that this was the "*imu* from which the planter ate, and from which he called to the gods.") The evidence suggests that these terraces, situated in the northeastern (i.e., sacred) corner of the larger garden swale, functioned as a ritual garden. The KIP-1306 shrine would have been a part of this ritual agricultural complex.

Viewshed. The viewshed from this shrine, situated at the base of a steeply rising ʻaʻā lava slope, is very restricted. However, standing on the platform, one would have an excellent view out over the intensively gardened swale.

Orientation. No precise measurements were made of the platform orientation. The plane-table map, however, shows that the platform is laid out more or less along cardinal directions, with the principal western face running N–S.

Additional remarks. KIP-1306 is a good example of a small agricultural shrine, associated with several other features that appear to be part of a ritual garden. This ritual complex occupies the northeastern corner of a large swale that was intensively cultivated, probably primarily in sweet potato but possibly with dryland taro and other crops as well.

KIP-1307

KIP-1307 was discovered and recorded in March 1999 during investigation of a large agricultural swale and associated residential complexes (sites KIP-725 and -726) in the Kīpapa uplands. At this time, Kirch mapped the structure with plane table and alidade; two test pits were also excavated in the feature under the direction of S. Millerstrom.

Location and topographic setting. The structure sits on the gently rising slope of a massive ʻaʻā flow to the east of a large swale used for intensive cultivation (see the description of site KIP-1306), at an elevation of 475 masl. The substrate consists of the Kīpapa ankaramite lava flow, with considerable surface rock.

Architecture. The main part of the structure consists of a rectangular terrace, broadly cardinally oriented, measuring 9 m E–W by 7.5 m N–S, defined by stone-faced retaining walls on the W, S, and E. The well-defined S facade of the terrace rises to a height of 1 m (fig. 7.38). The interior floor of this terrace consists of earthen fill and was tested with two excavation units. In the NE corner of the terrace there is a well-defined niche, separated from the main floor area by a low dividing wall. The niche may have been a location for offerings; about 1 m N of the niche is a large flat stone that also could have been used for offerings. A basalt core was found next to this stone. Several small waterworn cobbles (ʻiliʻili) were noted along the S edge of the terrace.

To the E of the main terrace there is a second, low terrace roughly paved with ʻaʻā clinkers, defined on the S and E with low retaining walls. The terrace grades into the hillslope on the N. The NE part of this low terrace is taken up with a shallow depression 2.5 m in diameter, which is probably a filled-in pit, possibly for the disposal of offerings.

Excavation. Two adjacent units (G10 and G11), forming a 1 by 2 m trench, were excavated in the floor of the main terrace. Beneath a thin (5–8 cm) layer of eolian overburden, the excavation revealed a cultural deposit, very dark brown to black, with much included charcoal. The cultural deposit extended to a depth of 26 cm below surface. A small concentration of charcoal and ash in the southeastern corner of unit G10, at a depth of 20–26 cm, probably represents a small hearth.

The cultural deposit in site KIP-1307 was unusually rich in artifacts and faunal remains (table 7.5). One basalt adze came from unit G11, and the substantial quantity of flaked basalt (totaling 508 flakes) demonstrates that significant stone artifact production took place within this small *heiau*. Numerous pieces of coral and small waterworn pebbles (ʻiliʻili) were probably offerings. The faunal assemblage includes small quantities of both dog and pig (the medium mammal bone is also of one or both of these animals), along with rat (*Rattus exulans*), chicken, and unidentifiable bird (a small passerine species). The fish bone included convict tangs (*Acanthurus* sp.), jacks (*Caranx* sp.), groupers (*Epinephelus* sp., Serranidae), wrasses (Labridae), and parrotfish (Scaridae).

FIGURE 7.38. Plan of *heiau* KIP-1307. (Based on plane-table survey by P. V. Kirch)

TABLE 7.5. Cultural content of KIP-1307 excavations.

MATERIAL	UNIT G10	UNIT G11
Basalt adze	—	1
Basalt flakes	214	294
Volcanic glass flakes	2	7
Waterworn pebbles (*'ili'ili*)	11	16
Coral	42	73
Charcoal (g)	175.1	193.8
Candlenut (g)	3.8	5.6
Marine shell (NISP)	735	834
Pig bone (NISP)	1	1
Dog bone (NISP)	1	—
Medium mammal bone (NISP)	11	8
Rat bone (NISP)	3	2
Chicken bone (NISP)	—	2
Bird bone (NISP)	1	1
Fish bone (NISP)	30	43

Dating and chronology. Two radiocarbon dates were obtained from the test excavations, one (AA-38680) from the feature 1 hearth and the other (AA-38681) from level 4 of unit G11. The feature 1 sample yielded an age of 82 ± 32 BP, with calibrated 2σ age ranges of AD 1686–1731 (25.6%) and 1808–1927 (69.8%), while the second sample dated to 157 ± 32 BP, with calibrated age ranges of AD 1665–1709 (16.6%), 1718–1786 (33.6%), 1793–1889 (27.0%), and 1911–1950 (18.3%). These dates indicate that this site was in use in the eighteenth century.

Archaeological context. There are no other structures in the immediate vicinity of KIP-1307, an indication of its probable ritual function. The site appears to be part of a large complex associated with the large agricultural swale, including two residential structures to the W of the swale. KIP-1307 is likely to have served as a local *heiau ho'oulu 'ai* for the occupants of these residential structures. In this regard the position of KIP-1307 to the E of the swale is noteworthy.

Viewshed. The view to the E is limited, owing to the rising slope of the massive 'a'ā lava flow. From the main terrace, one has a good view down over the adjacent agricultural swale, as well as to the looming Luala'ilua cinder cones about 2 km distant. The rim of Haleakalā can also be seen to the N, but the view to the S is limited.

Orientation. Based on the plane-table survey, the main terrace is laid out with the facing walls close to true cardinal directions. The position of the niche suggests some importance to the NE corner, as does the large pit in the NE corner of the upper terrace.

Additional remarks. KIP-1307 is a small *heiau* associated with a nearby group of residential structures and a large swale used for intensive gardening. Radiocarbon dates indicate that it was in use in the eighteenth century.

KIP-1317

Site KIP-1317 was discovered, mapped, and excavated during the March 1999 field season. Kirch made a plane-table map of this small shrine while S. Millerstrom excavated a test pit.

Location and topographic setting. KIP-1317 lies in the uplands of Nakaohu at an elevation of 480 masl. The substrate in this area is the Nawini ankaramite 'a'ā lava, a relatively young flow (10,000–30,000 years old) with many outcrops and surface rocks.

Architecture. The feature consists of a simple, semicircular stacked-stone wall built up on the E side of a prominent natural outcrop of jagged 'a'ā lava, creating an enclosed space about 1.5 m in diameter (fig. 7.39). The low enclosing wall incorporates an upright stone (80 cm tall) at the W end of the arc. The outcrop—which was presumably the ritual focus of this feature—stands about 1.6 m above the natural slope.

Excavation. A single 1 m² test pit was excavated in the floor area circumscribed by the low enclosing wall. Aside from some recent rat and unidentified mammal (goat?) bones in the thin overburden, the thin cultural deposit yielded only a small amount of charcoal, including a charred seed of the endemic *wiliwili* tree (*Erythrina sandwicensis*) and some carbonized parenchyma from a tuber.

Dating and chronology. The carbonized tuber fragment from the test pit was submitted for radiocarbon dating (AA-38689), yielding an age of 967 ± 41 BP, with a calibrated 2σ age range of AD 996–1161. This is the oldest radiocarbon date from any archaeological site within the entire Kahikinui District and indeed one of the earliest dates from any cultural context in the Hawaiian Islands.

Archaeological context. A small C-shaped shelter lies about 15 m S of KIP-1317, and a resi-

FIGURE 7.39. View of shrine KIP-1317, a large natural outcrop with a small enclosure on the W side. (Photo by P. V. Kirch)

dential complex, or *kauhale* (including sites KIP-1341, -1342, and -1343) lies about 50 m to the S. In light of the early radiocarbon date from KIP-1317, however, it is doubtful that any of these later residential features were directly related to the shrine.

Viewshed. No detailed observations of the viewshed were made during the 1999 field season, but the general setting would lend itself to good views of the Haleakalā ridgeline as well as the Lualaʻilua Hills to the W.

Orientation. The low enclosure adjoins the prominent natural outcrop on its W side, so that a ritual officiate seated within the enclosure would face the outcrop to the E. It is also noteworthy, however, that there is a placed upright in the W side of the wall. Thus, both E and W orientations are indicated in this shrine.

Additional remarks. KIP-1317 is a classic example of a *pōhaku o Kāne*—a stone representing the male procreator god Kāne (Kamakau 1976, 130). Kāne was associated with the sun and its path across the sky, which is consistent with the E–W orientation of this shrine, including the positioning of an upright stone in the W wall. What is unusual about this site is the early radiocarbon date, indicating the burning of a tuber and its placement as an offering within the first century or so following the discovery of the Hawaiian Islands by Polynesians. This ritual event occurred well before the main occupation of Kahikinui, which began in the fifteenth century. There are, however, hints in Hawaiian oral traditions that Kahikinui was an important location with respect to early voyagers between southern Polynesia and Hawaiʻi (see Kirch 2014, 88–97; Kirch, Ruggles, and Sharp 2013). It seems probable that this prominent outcrop was identified as having some sacred significance during an early exploration of the Kahikinui uplands.

KIP-1398

Site KIP-1398, an agricultural shrine, was mapped with plane table and alidade during August 1999, when agricultural and residential features in the adjacent swale (KIP-1400) were investigated (see Coil 2004, 206, fig. 6.35; Coil and Kirch 2005).

Location and topographic setting. The site is situated on the E side of a rocky swale within the massive Kīpapa-1 ankaramite ʻaʻā lava flow, at an elevation of 325 masl. The structure is built up against a prominent, linear lava pressure ridge running roughly N–S up and down the hillslope.

Architecture. The main architectural feature is a stone-faced terrace constructed to the W of the natural lava pressure ridge (fig. 7.40). The main facade of this terrace is about 1 m high. The terrace itself is about 10 m N–S) by 4 m E–W and was positioned directly W of a prominent high point along the pressure ridge, where a natural three-pointed upright stands about 2.2 m high. This natural upright was presumably the ritual focus of the shrine. Scattered branch coral pieces on the level floor of the swale in front of the terrace are an indication of its ritual function.

To the N of the shrine itself a second, well-defined rectangular level area appears to have been a temporary residential site for persons cultivating the adjacent swale during the winter growing season (the swale is on the lower-elevation margin of the upland cultivation zone in Kīpapa).

Excavation. A single test pit (TP-1) was excavated in the floor of the terrace. A cultural deposit of dark gray-brown sediment extended to a depth of about 26 cm below surface. Feature 1, partly exposed in the northern half of the unit, was a combustion feature with a diameter in excess of 1 m and a depth of 30 cm, filled with large chunks of charcoal and fire-altered rocks (2–6 cm in size), and was interpreted as an earth oven (*imu*). The excavation yielded 71.3 g of charcoal (much of it from feature 1), 34 NISP of bone, 782 NISP of marine shell, 2 basalt flakes, 18 volcanic glass flakes, and 52 pieces of coral. The bone and shell from this excavation have not been further analyzed to taxa.

Dating and chronology. *Chenopodium* charcoal from feature 1 in the test pit was radiocarbon dated (AA-38690) to 200 ± 39 BP, yielding 2σ calibrated age ranges of AD 1642–1697 (24.9%), 1725–1815 (48.0%), 1835–1878 (4.1%), and 1917–1950 (18.4%).

Archaeological context. The KIP-1398 shrine is part of a complex of agricultural and temporary residential features clustered in a rocky swale showing evidence of intensive cultivation. The swale floor is covered with dozens of small planting mounds and is divided by a low embankment possibly indicating a social division within the garden area.

Viewshed. No precise observations of the viewshed were made.

Orientation. Lying directly to the W of the prominent natural lava ridge and outcrop, the terrace is clearly oriented to this landscape feature and thus has a general easterly orientation.

Additional remarks. This site is a good example of a small agricultural shrine directly associated with an intensively cultivated swale. The presence of numerous branch coral offerings attests to its ritual significance. The primary ritual attractor here was a prominent outcrop along the lava pressure ridge; this outcrop may be another instance of a *pōhaku o Kāne*.

The *Ahupuaʻa* of Nakaohu

The land section name of "Nakaaha," like that of "Nakaohu," first appears on the 1882 map of Kahikinui by Hawaiian Government Survey member W. R. Lawrence, where it is written in fine pencil, presumably added after the original map was drafted. It is uncertain whether Nakaaha was actually an *ahupuaʻa* (or possibly an *ʻili*, a land unit smaller than an *ahupuaʻa*), but it was certainly a land section of some kind. Geologically, Nakaaha straddles the interface between the much older Kula volcanic series to the east and the younger Hāna volcanic series to the west. We use the abbreviation "NAK" for archaeological sites within Nakaaha.

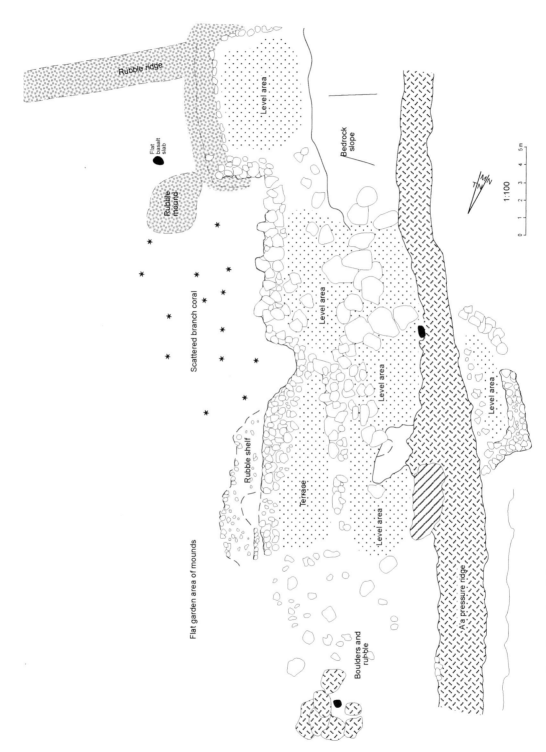

FIGURE 7.40. Plan of agricultural shrine KIP-1398. (Based on plane-table survey by P. V. Kirch)

Archaeological survey within Nakaaha has been far less extensive than in Kīpapa and Nakaohu. The coastal sector was intensively surveyed, and Kirch and his students made several transect forays across the Nakaaha uplands, resulting in the discovery of the *heiau* reported in this section of our catalog. However, there is no reason to think that all *heiau* within Nakaaha have been recorded; future survey work is likely to reveal the presence of additional sites.

NAK-27

NAK-27 was first recorded during a reconnaissance transect across the Nakaaha uplands in June 2003; the notched plan strongly suggested a ritual function for the enclosure. Kirch and Ruggles visited the site on March 26, 2008, making total station and GPS surveys. The site has not been mapped in detail with plane table and alidade.

Location and topographic setting. NAK-27 lies at an elevation of 610 masl in the Nakaaha uplands, on fairly steeply sloping terrain. The geological substrate is the older basanite of Kahikinui ʻaʻā lava flow.

Architecture. In plan view, NAK-27 is a classic notched enclosure whose orientation deviates from cardinality by about 32° in a counterclockwise sense, with the notch in the SSW corner. The core-filled walls are well constructed and in some places are more than 1.1 m high. The exterior dimensions of the enclosure are 11.8 m WSW–ENE by 8 m NNW–SSE. Because there was no time to clear the enclosure interior of the thick, obscuring vegetation, it is uncertain whether there are any internal architectural features. No branch coral was observed.

Archaeological context. A small C-shaped shelter was noted about 12 m to the NW of NAK-27, and a small enclosure about 30 m to the NW. Both of these structures appear to be residential sites.

Viewshed. A massive ʻaʻā outcrop about 60 m from the site blocks the view to the NE. On the enclosure's S side the outcrop falls away sharply to reveal the more distant Haleakalā slopes beyond. The view of the Kaupō Peninsula to the E is also blocked by a nearby ridge, which is currently covered in thick vegetation; thus, the extent to which the more distant view would be blocked in the absence of this vegetation is unclear. In other directions there are open views. The peak of Manukani (3.2 km, az. 290°) stands out on the WNW horizon.

Orientation. The fact that the WSW–ENE axis is the longer one suggests that this small enclosure faced ENE rather than NNW, but in the absence of a detailed survey we cannot be certain. Long, straight intact segments of wall facing yielded best-fit orientations for the NNW wall of 58.4° (outer face) and 59.2° (inner face), for a mean eastward azimuth of 58.8°. The orientation of the SSE wall, as determined by prismatic compass, is 60° (corrected for mean magnetic declination), which is consistent with this. The outer face of the ENE wall yielded 328.8°, exactly orthogonal to the NNW wall, but the inner face yielded 326.0°, for a mean of 327.4°. A single determination on the inner face of the WSW wall, north of the notch, yielded 319.4°, suggesting a mean axial orientation of 323.4°.

As viewed from the SSW corner of the *heiau*, the SSE wall of the *heiau* is seen to be aligned upon a small rock with a conspicuous vertical face marking the foot of the steep side of the large ʻaʻā outcrop, at azimuth ~60.5°. However, owing to the proximity of the outcrop, this situation alters significantly from observing positions farther to the N. In particular, moving to the WNW corner shifts the outcrop significantly to the right, increasing its azimuth by approximately 7°, so that from the vicinity of the WNW corner the Pleiades (and the June solstice sun) would have

FIGURE 7.41. View to the ENE as seen from the WNW corner of enclosure NAK-27, overlaid on the digitally generated horizon profile (which cannot reproduce the nearby landform). The solid shaded line indicates the rising path of the sun at the June solstice, while the dotted line indicates the rising path of the Pleaides in AD 1600. (Composite of photo by Clive Ruggles and graphic by Andrew Smith)

been seen to rise close to the foot of the outcrop (fig. 7.41). Because of the large errors involved, it is impossible to say on the basis of the current data (which did not include EDM fixes on the outcrop because of its difficulty of access) whether the ideal locations from which the Pleiades' rising position would have coincided exactly with the foot of the outcrop would have been from inside or outside the *heiau* on the NNW side. In any case, the configuration of this *heiau* in relation to the outcrop and horizon to the ENE suggests two possibilities: first, that it incorporated a device for anticipating the rising position of the Pleiades, thus facilitating their observation for calendrical purposes; and second, that it was placed so that the first light of the rising sun, even at its farthest N at the June solstice, would fall across the entire enclosure rather than just part of it. Further dating evidence and more extensive survey data may help clarify these possibilities.

To the WNW, the Pleiades and the June solstice sun set behind the ridgeline some 2° upslope from the peak of Manukani, around azimuth 292°. Meanwhile to the WSW, the NNW and SSE walls are orientated quite closely upon the point where the western slopes fall below the sea horizon, but the declination (~−29.5°), well beyond the solar setting range, has no evident astronomical significance.

The northerly orientation is upon an unremarkable stretch of the upper Haleakalā slopes.

The horizon between the directions of orientation of the WSW and ENE walls spans declinations +51.0° to +58.5° (to the nearest half degree), which includes the setting point of the asterism Mūlehu (assuming it to be α, β, and γ Cas) around AD 1600, as well as part of Nā Hiku (Ursa Major) (see table 4.2). To the SSE, the wall orientations fall between the peaks of Mauna Kea (az. 137°) and Mauna Loa (az. 151°) on Hawai'i Island. The orientation of the ENE wall (az. 147.5°, alt. 0.5°, dec. 52.0°, all to the nearest half degree) is close to the rising point of Canopus (α Car)—known to Hawaiians as Ke Ali'i o Kona i Ka Lewa, "chief of the southern heavens" (Johnson, Mahelona, and Ruggles 2015, 180)—with a declination of ~−52.5° in the sixteenth through the eighteenth centuries.

NAK-29

Site NAK-29, along with its close neighbors NAK-30 and NAK-35, were discovered during the same reconnaissance transect survey on June 13, 2003, that found site NAK-27. Kirch returned two days later to make a plane-table and alidade survey of the structure. On March 26, 2008, Kirch and Ruggles visited the site, making both total station and GPS surveys as well as detailed observations on the viewshed.

Location and topographic setting. NAK-29, at an elevation of 590 masl, sits on a discrete geological feature—a small, narrow *kīpuka* (patch) of older *'a'ā* lava (basanite of Kahikinui, 50,000–140,000 years old), sandwiched between much older Kula volcanics to the east and the very young Kamole Gulch basanite *'a'ā* flow (3,000–5,000 years old) just a few meters to the west.

Architecture. The architecture at NAK-29 consists of two main components: a small but well-constructed notched enclosure overlooking an adjacent leveled terrace on the downslope side (figs. 7.42 and 7.43). The orientation of the notched enclosure deviates from cardinality by about 27° in a counterclockwise sense, with the notch situated in the WNW corner. It has exterior dimensions of 13 m WSW–ENE by 9.75 m NNW–SSE. The walls are well constructed, core-filled, 0.6–1.0 m high, and, on the ENE side, up to 1.5 m thick. The main court on the ENE is subdivided by a low, single-course stone alignment; within the space cell set off to the NNW by this dividing alignment is a small cist-like feature defined by several slabs set on edge. In the lower part of the court to the SSE there is a large, flat *pāhoehoe* slab that appears to have been a seating stone, presumably for the *kahuna* or other officiant during rituals.

The WSW part of the enclosure is also divided into two space cells, with a small room or chamber on the upslope side, measuring about 2 by 1.5 m (interior dimensions), and a small depression on the downslope side. The depression may be an oven or a pit for the disposal of offerings.

Adjoining the notched enclosure on the downslope side is a large, leveled terrace, roughly paved with *'a'ā* clinkers, retained on three sides by stacked-stone facings. The terrace has exterior dimensions of 15 by 9 m. This terrace gives the impression of an assembly or seating area for persons witnessing rites in the adjacent notched *heiau* immediately upslope.

Dating and chronology. A piece of branch coral exposed by a partial collapse of the WSW corner of the lower terrace, and hence originally in an architecturally integral context, was ^{230}Th dated, yielding an age of AD 1603 ± 3. This date falls within the reign of Maui *ali'i nui* Kiha-a-Pi'ilani, according to the genealogical chronology (Kirch 2014, table 2).

Archaeological context. NAK-29 is a part of a complex together with NAK-30 and NAK-35, which are situated only about 20 m downslope to the SSW.

Viewshed. There are clear views of the Haleakalā Ridge stretching from the NW over to

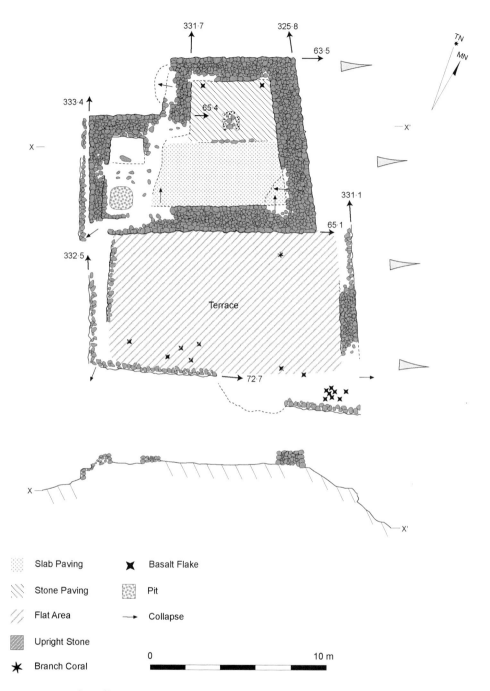

FIGURE 7.42. Plan of *heiau* NAK-29, showing the azimuths of intact straight segments of wall and terrace facing as determined by the total station survey. (Based on plane-table survey by P. V. Kirch)

FIGURE 7.43. *Heiau* NAK-29, viewed from the NNW, showing its landscape setting. (Photo by Clive Ruggles)

the E where its eastern slopes sweep down toward Ka Lae o Ka ʻIlio in Kaupō. The slope meets the sea horizon at azimuth 83°, and the Kaupō Peninsula appears below the sea horizon between azimuths 89° and 97°. The site looks out over open sea from the ESE round to the WSW. The profile of Hawaiʻi Island extends between ~125° and ~165°, the peaks of Mauna Kea and Mauna Loa being visible during favorable weather conditions. Between SW and NW (az. ~214° to ~311°) the view is restricted by ridges less than 150 m distant.

Orientation. The longer WSW–ENE axis and the more substantial ENE wall imply that the direction of orientation was to the ENE. The outer face of the NNW wall has an eastward orientation of 63.5°, while that of the outer face of the SSE enclosure wall is 65.1°, so that the mean axial orientation is 64.3°. The facing on the SSE side of the southern terrace is skewed by several degrees from this direction, the largest intact segment having an azimuth of 72.7°. The enclosure narrows toward the NNW, its three NNW–SSE-oriented wall segments having northward orientations of 333.4°, 331.7° and 325.8°, respectively (from west to east), for a mean of 330.3°. The NNW–SSE orientation of the southern terrace—its WSW and ENE sides having azimuths of 332.5° and 331.1°, respectively—is broadly consistent with this.

The *heiau* faces the eastern Haleakalā slope, near a point where two small *puʻu* at azimuths 67° and 70°, form conspicuous bumps on the horizon. In AD 1600 the Pleiades would have risen directly behind the left side of the leftmost of these *puʻu*. While this *puʻu* would have marked the rising position of the Pleiades, the actual alignment of the enclosure walls is still farther to the left, closer to the rising position of the June solstice sun; around the equinoxes the sun would have risen over the Kaupō Peninsula. To the WSW, the wall alignments fall just to the right of the setting position of the December solstice sun.

The northerly wall orientations center upon an isolated *puʻu* on the Haleakalā Ridge (peak at az. 330.5°, alt. 16.0°, dec. +61.5°, all to the nearest half degree). While no bright stars set here directly, this is roughly midway between the setting positions of Dubhe (α UMa) and Merak (β

UMa)—or Hiku Kahi and Hiku Alua, the first and second of the seven stars in Nā Hiku, Ursa Major (Johnson, Mahelona, and Ruggles 2015, 150)—whose declinations were +63.8° and +58.5°, respectively in AD 1600. The setting positions of these stars (and other stars in Nā Hiku—see table 4.2) also correspond roughly to the horizon points upon which the W and E enclosure walls are aligned (corresponding respectively to declinations +64.5° and +57.0°, to the nearest half degree). However, these topographic and astronomical associations may well have been coincidental, and it may be more significant that in the southern direction the enclosure was broadly oriented upon the rounded peak of Mauna Loa on Hawai'i Island (az. ~151.5°, alt. ~0.5°, dec. ~−55.5°), over which Gacrux (γ Cru)—the top star of Newe, the Southern Cross (ibid., 200)—would have risen around AD 1600 (dec. −54.9°).

Additional remarks. There seems little doubt that this *heiau* was intended to face either the rising position of the Pleiades or that of the June solstitial sun. To the SSE, the enclosure walls were roughly oriented upon the peak of Mauna Loa.

NAK-30

This small platform *heiau* with adjacent terrace was discovered during the same reconnaissance transect survey on June 13, 2003, that encountered sites NAK-27 and NAK-29. Kirch returned on June 15 to make a plane-table and alidade survey of the structure. On March 26, 2008, Kirch and Ruggles visited the site, making total station and GPS surveys as well as detailed observations on the viewshed.

Location and topographic setting. NAK-30 lies about 20 m downslope (SSW) from NAK-29 at an elevation of 580 masl, on the same narrow *kīpuka* of Kahikinui basanite *'a'ā*.

Architecture. The site has two components: a small well-constructed platform on the upslope side, with an adjacent *'a'ā*-clinker-paved terrace immediately downslope (fig. 7.44). The platform, whose orientation deviates from cardinality by 23° in a counterclockwise sense, measures 6 m NNW–SSE by 7.5 m ENE–WSW. It is elevated 1–1.25 m on the WSW and SSE but only about 0.5 m on the ENE and NNW. A low core-filled wall runs along the N side of the platform, whose surface is paved with *'a'ā* cobbles. Two prominent upright slabs are situated in the NNE and ESE corners of the platform, standing 0.45 and 0.65 m high, respectively. A single *'ili'ili* cobble lay on the pavement midway between the two uprights.

The lower terrace measures 7.5 by 8 m and has a low facing (0.4 m high) on the downslope side. A basalt awl, basalt flakes, and a single *'ili'ili* cobble were found on the terrace surface. The terrace has the appearance of being an assembly area where persons may have been seated to observe rites performed on the adjacent higher platform.

Archaeological context. NAK-35, a U-shaped, core-filled wall lies immediately NW of the NAK-30 platform. The notched *heiau* NAK-29 is situated about 20 m farther upslope along the same ridge.

Viewshed. There are clear views northward to the Haleakalā Ridge, as at NAK-29, and its eastern slopes are visible down to the sea horizon, but the eastern view appears very different because of the presence of an *'a'ā* ridge with several distinctive outcropping rocks a mere 5 m E of the platform, just beneath the more distant profile. Figure 4.10 shows the view from a seated position on the WSW side of the platform, as might well have been adopted by a priest while performing a ritual. The two uprights at the NNE and ESE corners demarcate a stretch of horizon running between azimuths ~55° and ~80°, with a large pointed boulder on the nearby ridge marking the

FIGURE 7.44. Plan of *heiau* NAK-30, showing the azimuths of intact straight segments of platform edges and wall facing as determined by the total station survey. (Based on plane-table survey by P. V. Kirch)

approximate center of this range, at about azimuth 67°. The Kaupō Peninsula is not visible from this position. As at NAK-29, the westward view is restricted by nearby ridges; from here they are up to 200 m distant and extend from azimuth ~235° to azimuth ~300° before the more distant Haleakalā ridgeline (~6 km away) becomes visible. The profile of Hawai'i Island again extends between azimuths ~125° and ~165°.

Orientation. The presence of the two uprights on the ENE side and an entrance feature at the SSE end of the WSW side indicates that the *heiau* faced ENE, a judgment reinforced by the striking view in that direction, as already described. The outer face of the NNW wall and the SSE platform edge are oriented with eastward azimuths of 71.3° and 62.3°, respectively, for a mean axial orientation of 66.8°. The inner face of the NNW wall has orientation 73.6°; if this is taken into account, the mean NNW wall orientation is 72.5° and the mean axial orientation is 67.4°. The WSW and ENE sides of the platform are oriented with northward azimuths of 336.9° and 337.4°, respectively, for a mean axial orientation of 337.2°, which is almost exactly perpendicular to the ENE–WSW axis.

The axial orientation coincides almost exactly with the rising position of the Pleiades around AD 1600. The orientation of the pointed boulder from midway along the WSW side of the platform is ~67°, implying that the *heiau* was carefully placed so that its main central axis would align upon the pointed boulder and also the rising position of the Pleiades. In the opposite direction, the axial orientation is over 2° to the right of the position of the setting December solstice sun, implying that this was not the intended target.

To the NNW, the platform is aligned upon an unexceptional part of the Haleakalā profile, with a declination around +68°; to the SSE, it is aligned upon the northern slopes of Hualālai Volcano on Hawai'i Island, which would have been visible only in exceptional conditions, with a declination of −60°. The latter corresponds to the rising point of Rigil Kentaurus (α Cen)—possibly the Hawaiians' Ka-maile-hope (Johnson, Mahelona, and Ruggles 2015, 168)—with a declination of −59.4° in AD 1600, falling to −60.2° in AD 1800. However, both of these alignments arise as simple consequences of the perpendicularity of the NNW–SSE axis to the principal ENE–WSW one and are unlikely to have had any significance.

Additional remarks. It seems that this *heiau* was placed so as to align with a prominent pointed boulder in the ridge ~5 m to the east, which marks the point on the more distant horizon, just above, where the Pleiades would have risen at the time of construction.

NAK-34 (State Site 50-50-15-1156 [Probable])

Site NAK-34, a small notched enclosure, was discovered and briefly recorded during a reconnaissance survey in the Nakaaha uplands in June 2003, when a GPS survey and field notes were taken. Ruggles had visited this site on April 17, 2002, and undertaken a compass-clinometer survey.

NAK-34 is probably the same site as Kolb's "Nakaaha Heiau," site 50-50-15-1156 in his extension of the state numbering system (Kolb and Radewagen 1997, 71, fig. 5.7). Kolb describes site 50-50-15-1156 as a small notched *heiau* with the notch in the NE corner, which is consistent with our observations.

Location and topographic setting. NAK-34 is situated at an elevation of about 650 masl, on a substrate of relatively young 'a'ā lava (basanite of Kanahau).

Architecture. The structure consists of a notched enclosure whose orientation deviates from cardinality by 9° in a counterclockwise sense. Its exterior dimensions are 10.4 m N–S by 10 m E–W; the notch is in the NE corner. The walls are of core-filled construction, averaging about 1.25 m high. Kolb's map of his site 50-50-15-1156 shows a small terrace attached to the notched enclosure on the W side.

Excavation. Kolb placed two test pits within his site 50-50-15-1156 and noted that nothing other than charcoal was recovered.

Dating and chronology. Kolb (2006, table 1) reports a single radiocarbon date (Beta-122890), on unidentified wood charcoal from a "basal" context: 340 ± 70 BP, yielding 2σ calibrated age ranges of AD 1437–1666 (94.4%) and 1785–1794 (1.0%). As with other samples recovered by Kolb from "basal" contexts, this date is probably best considered as a *terminus post quem*.

Archaeological context. The archaeological context is unknown because no intensive survey surrounding NAK-34 has yet been conducted.

Viewshed. There is an unobstructed view of the Haleakalā ridgeline from the NW round to the NE and of the open sea from the ESE round to the SW. To the WSW, W, and WNW (az. ~235° to ~305°) the view is restricted by ʻaʻā ridges 0.5–1 km distant, while to the NE, ENE, and E (az. ~40° to ~105°) the view is blocked by rising ground within 50 m.

Orientation. The site is most likely oriented northward, given the substantial architecture of the N wall. Prismatic compass determinations, corrected for mean magnetic declination, give the northward orientation of the W and E walls as 350° and 352°, respectively, for a mean azimuth of 351°, while the N and S walls have eastward orientations of 80° and 82°, respectively, for a mean azimuth of 81°. Thus, the enclosure is impressively rectangular, although the short E–W-oriented wall segment, forming the S side of the notch in the NE corner, has a slightly different bearing of 87°.

The *heiau* is oriented toward the right-hand end of a large cinder cone that, from this location, crowns the Haleakalā Ridge between azimuths 344° and 351°. There is no apparent astronomical connection, and the referent appears to be purely topographic. To the S the axial orientation is upon open sea, while to the E and W it is upon unremarkable parts of the local ridges with no obvious astronomical significance.

The *Ahupuaʻa* of Mahamenui

Mahamenui (or Mehamenui in some sources) is a land section to the east of Nakaaha. In Royal Patent Grant 2824 issued to Helekunihi in 1861, "Mahamenui" is given as an *ahupuaʻa* name (Kirch 2014, table 6). Geologically, Mahamenui spans two substrates of rather different ages: a section of old Kula volcanics (<225,000 years old) with several incised intermittent drainage channels on the W; and a single massive ʻaʻā lava flow about 90,000 years old that emanated from the Puʻu Pane cinder cone, perched at an elevation of about 1,000 masl.

Approximately 3 km^2 of Mahamenui has been archaeologically surveyed, primarily by Holm (2006) for her doctoral research at the University of California, Berkeley. A total of 359 sites have been recorded in this surveyed area (Kirch 2014, table 1), but only 2 definite *heiau* sites are known.

MAH-231 (Walker Site 170; State Site 50-50-15-170)

Site MAH-231 was discovered by Holm in 2002 during her intensive survey of parts of Mahamenui and was plane-table mapped by Kirch and Holm on June 30, 2002. Kirch and Ruggles visited the site on March 22, 2008, making total station and GPS surveys, as well as viewshed observations.

MAH-231 is presumably the same structure recorded by Walker in 1929 as his site 170 (Walker [1930], 240), given that it fits his description of a "small heiau 40 feet square," consisting of "an open platform." Walker gave the location of site 170 as "East of Kepuni Gulch above the Kula Trail about a quarter of a mile."

Location and topographic setting. MAH-231 lies an elevation of 490 masl on the old Kula volcanics substrate, just a few meters west of the massive Puʻu Pane ʻaʻā lava flow. The old jeep road that runs up through Mahamenui passes next to the structure on the east and north.

Architecture. The structure consists of a substantial stone platform (fig. 7.45) whose orientation deviates from cardinality by 3°–5° in a clockwise sense, built up primarily on the W, S,

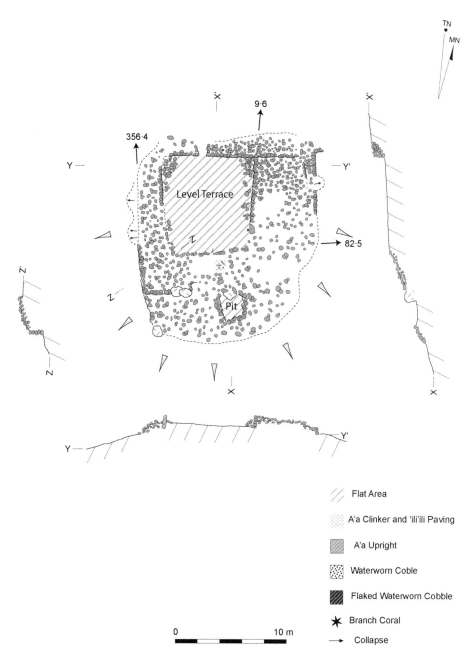

FIGURE 7.45. Plan of *heiau* MAH-231, showing the azimuths of intact straight segments of platform facing as determined by the total station survey. (Based on plane-table survey by P. V. Kirch)

Kīpapa to Manawainui • 279

and E with ʻaʻā cobbles and boulders and having overall dimensions of 18 m N–S by 16 m E–W. The outer parts of the platform have mostly collapsed, although a faced corner still remains in the SW; if this is indeed Walker's site 170, then it originally had "three stepped terraces" in the seaward face. There is a small enclosed space or pit near the bottom of the S slope of rubble.

The top of the platform consists of a leveled area 8.5 m N–S by 7.5 m E–W, defined on all four sides by low facings. The low wall along the N is gapped by an apparent entryway about 1 m wide. An upright ʻaʻā boulder stands in the NW corner with several ʻiliʻili cobbles near it. More ʻiliʻili cobbles and some branch coral were noted near the seaward (S) edge of the leveled area.

Excavation. Holm (2006) excavated two units, each 1 m^2, in MAH-231; TP-1 was situated in the main court against the N wall, just to the E of the entryway, while TP-2 was located in a smaller space within the S terrace slope. In TP-1 a cultural deposit 25–30 cm thick contained 41 basalt flakes and a single volcanic glass flake, along with charcoal, candlenut, and coral (Holm 2006, 312–313). Feature 1, a small circular depression about 10 cm in diameter, was interpreted as a posthole that may have been associated with a thatched superstructure. Identifiable faunal material included 1 dog femur, 37 NISP of medium mammal (much of this probably from dogs), and 1 parrotfish (*Scarus* sp.) pharyngeal grinder. TP-2 yielded waterworn pebbles (ʻiliʻili), charcoal, candlenut, and 17 basalt flakes.

Dating and chronology. Holm (2006) obtained two samples of charcoal from her test excavations. The first sample from TP-1 (Beta-179405), identified as a short-lived *Chenopodium* sp., was radiocarbon dated to 380 ± 30 BP, yielding 2σ calibrated age ranges of AD 1445–1524 (61.6%) and 1558–1632 (33.8%). The second sample from TP-2 (Beta-179406), of short-lived *Chamaesyce* charcoal, was dated to 160 ± 30 BP, yielding calibrated age ranges of AD 1664–1707 (16.7%), 1719–1826 (47.4%), 1832–1884 (12.6%), and 1914–1950 (18.6%), the last two of which can be rejected.

Archaeological context. There are a number of probable residential features (terraces, enclosures) within a 100 m radius of the site.

Viewshed. There are clear views of the Haleakalā ridgeline to the N and of its western slopes extending down to Alena Point in the SW at azimuth ~218° and meeting the sea horizon around azimuth 243°. The Lualaʻilua Hills are prominent at the foot of these slopes, peaking at azimuths 251° and 258.5°, and Manukani also stands out upslope at azimuth 284°. A number of small cinder cones around the summit of Puʻu Pane, about 2.5 km from the site, are visible below the Haleakalā skyline to the N. Farther round to the E, the distant view is blocked by the massive Puʻu Pane lava flow merely ~50 m away. The exact point at which the local ground rises to obscure the Haleakalā ridgeline depends on the exact position and height of the observer.

Orientation. The platform widens markedly both toward the N and, less so, toward the W. The E and W facings of the platform are oriented with northward azimuths of 356.4° and 9.6°, respectively, yielding a mean axial orientation of 3.0°. We were unable to obtain an accurate estimate of the orientation of the N facing, but the S facing yields 82.5° and the plane-table plan suggests that the N facing is oriented with an azimuth of ~87.5°, for a mean E–W orientation of about 85°.

Although the spatial disposition of the structural elements offers little to confirm this, the restricted view to the E and the broadening out toward the N suggest that the *heiau* was oriented toward the N rather than the E. In the northerly direction is the highest part of the Haleakalā ridgeline, itself devoid of distinctive features such as cinder cones but below which stand the various cones around the summit of Puʻu Pane. The highest of these (as it seems from this

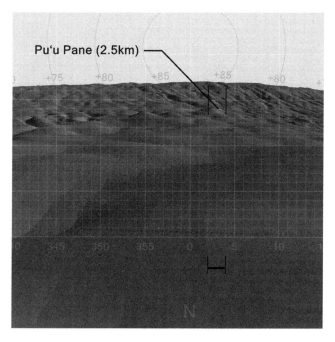

FIGURE 7.46. DTM-generated profile of the northern horizon as seen from *heiau* MAH-231. The horizontal bar indicates the main axial orientation of the *heiau*. Note that the nearby ridge, which is mentioned in the text and obscures the distant horizon farther to the E, is too close to be accurately reproduced in the DTM. (Graphic by Andrew Smith)

vantage) is seen at azimuth 3°, in exactly the axial direction (fig. 7.46). In the opposite direction the platform is oriented upon open sea.

To the W, the axial orientation falls several degrees upslope of the Luala'ilua Hills and midway between the positions of sunset at the December solstice and the equinoxes. It therefore seems unlikely that this direction was of significance.

Additional remarks. Site MAH-231 directly faces the highest visible cinder cone at the source of the lava flow beside which it was constructed.

MAH-363, Kamoamoa (Walker Site 172; State Site 50-50-15-172)

MAH-363, also known as Kamoamoa Heiau, was first recorded by Walker in 1929, who gave it site number 172 in his sequence (Walker [1930], 242). Kirch visited the site and made a plane-table and alidade map on June 9, 2002. No excavations or further survey of the structure have been made.

Location and topographic setting. MAH-363 lies about 120 m inland of the shoreline, at an elevation of about 20 masl. The geological substrate consists of the basanite of Pu'u Pane.

Architecture. As seen in the plan view (fig. 7.47), the architecture of MAH-363 is highly disturbed, and it is impossible to discern with confidence the original configuration of the structure. Walker ([1930], 242) described it as "an irregular shaped platform 94 feet across the front and extending back 80 feet." Today most of the site consists of a massive heap of 'a'ā cobbles and boulders about 20 m E–W by 18 m N–S. The only clearly defined features are found in the SW, where

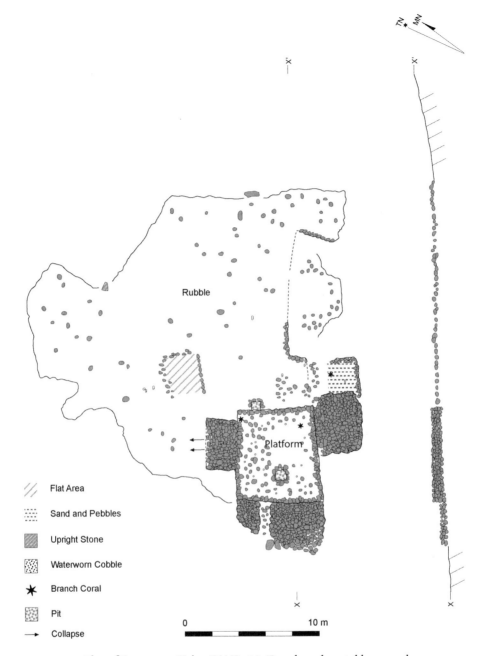

FIGURE 7.47. Plan of Kamoamoa Heiau (MAH-363). (Based on plane-table survey by P. V. Kirch)

there is a well-faced platform (6.5 by 6 m) paved with 'a'ā clinkers and with scattered 'ili'ili on its surface. A small pit about 1 m in diameter penetrates the platform fill. Lower, irregular platforms abut this main platform on the N, W, and S; these may be postcontact burial platforms.

Archaeological context. MAH-363 is surrounded by a number of features within a 150 m radius, many of which may be postcontact residential structures. Walker ([1930], 242) refers to this complex as the "village of Kamoamoa."

Viewshed. No observations were made of the viewshed from this site.

Orientation. The well-preserved platform has an orientation of approximately 67°, which suggests that the structure might have been associated with the Pleiades.

The *Ahupua'a* of Manawainui

This land section (alternatively called Manuwainui) may or may not have been an *ahupua'a* (see Kirch 2014, appendix B). The area consists entirely of old lava flows of the Kula volcanic series, bounded on the east by the deeply incised Manawainui Gulch. Holm (2006) surveyed approximately 3.3 km² of the Manawainui landscape for her doctoral research project at the University of California, Berkeley, and recorded 351 archaeological sites (Kirch 2014, table 1), only one of which can be considered a formal *heiau*.

MAW-2

Site MAW-2 was first recorded in May 2001 during Holm's intensive survey of Manawainui (Holm 2006). Kirch and Ruggles visited the site on March 27, 2003, to make total station and GPS surveys.

Location and topographic setting. MAW-2 lies at an elevation of 150 masl, on the western lip of Manawainui Gulch. The terrain consists of gently sloping, deeply weathered old *'a'ā*.

Architecture. The structure consists of a notched enclosure whose orientation deviates from cardinality by about 28° in a counterclockwise sense. Its exterior dimensions are 12 m ENE–WSW by 10 m NNW–SSE; the notch is situated in the WNW corner. The walls have largely collapsed, so that original facings are visible only in a few areas. There seems to have been a low stone platform in the ESE corner.

Excavation. Holm excavated a 1 m² unit (TP-1) in MAW-2 (Holm 2006, 343–345, fig. 6.56); the unit was located adjacent to the inner face of the E wall. The cultural deposit, designated layer II, was only 3–8 cm thick but contained three waterworn *'ili'ili* stones, "a large fine-grained basalt flake, and a formal reworked adze made from non-local material," all found during excavation. In addition, charcoal, marine shell, and 84 basalt flakes were recovered during screening.

Dating and chronology. A sample of wood charcoal (Beta-179667) from Holm's TP-1 excavation, identified as *Nototrichium* sp., was radiocarbon dated to 70 ± 30 BP, yielding 2σ calibrated age ranges of AD 1691–1730 (24.3%) and 1810–1924 (71.1%). This suggests that MAW-2 was used in the eighteenth century to possibly the early nineteenth century, which is consistent with the relatively late occupation of the Manawainui area as demonstrated by other radiocarbon dates from residential features (Kirch 2014, 111).

Archaeological context. A U-shaped enclosure 20 m SSE (downslope) is probably a residential site associated with MAW-2.

Viewshed. Passing from the NE across to the NW, one looks up the narrow channel of Manawainui Gulch, with the Haleakalā rim looming behind. The eastern slopes of Haleakalā extend down to meet the sea horizon at azimuth 88°. To the E there is a good view of Nu'u Bay and landing, with the Kaupō Peninsula beyond, extending out to Ka Lae o Ka 'Ilio at azimuth ~95.5°. Hawai'i Island extends from azimuths 126° to 166°, with the peaks of Mauna Kea, Mauna Loa, and Hualālai at azimuths 138°, 153°, and 160°, respectively. From the SW round to the WNW (az. ~230° to ~300°), the view is restricted by ridges ~0.5–1 km from the site.

Orientation. On the basis of the archaeological evidence, it is difficult to say whether this structure faced ENE or NNW. The wall orientations in these directions, as judged from extant fragments of intact facing, are 58.3° for the NNW wall outer face, 60.0° for the NNW wall inner face, 64.4° for the SSE wall inner face, 332.9° for the WSW wall outer face, 333.0° for the WSW wall inner face, and 328.0° for the ENE wall inner face. Thus, we obtain an azimuth of 59.6° as our best estimate of the NNW wall orientation, 62.0° for the mean ENE–WSW axial orientation, and 330.5° for the mean NNW–SSE axial orientation.

The wall orientations to the NNW bracket an unremarkable stretch of the Haleakalā ridgeline a few degrees to the right of the Puʻu Pane cinder cone, 4.3 km distant and visible in the foreground just below the Manawainui Gulch at azimuth 323°. The declinations, +64° for the WSW wall and +59° for the ENE wall, correspond quite closely to the setting points of Dubhe (α UMa) and Merak (β UMa)—or Hiku Kahi and Hiku Alua, the first and second of the seven stars in Nā Hiku, or Ursa Major (Johnson, Mahelona, and Ruggles 2015, 150)—in the sixteenth century (see table 4.2) but less so around AD 1800, when their respective declinations had fallen to around +63.0° and +57.5°. In the southward direction, the WSW wall is oriented upon the summit of Mauna Loa, while the ENE wall and axial orientation fall within its long, gradual slope on the northern side.

To the ENE, the wall orientations fall to the left of the rising position of the June solstice sun and the Pleiades (which occur between az. 66° and 67°); the mean direction (az. 62°, alt. 5°, dec. +28°) is close to the rising position of Pollux (β Gem)—that is, Nānāhope (Johnson, Mahelona, and Ruggles 2015, 199)—around AD 1800 (dec. +28.5°). Directly to the E there is a low *puʻu* on the horizon, just where it meets the sea (peak at az. 85.0°, alt. −0.1°, dec. +4.4°), but this is several degrees to the left of the rising position of the equinoctial sun, which is over the center of the Kaupō Peninsula, and does not correspond to any bright or known Hawaiian stars; the closest around AD 1800 is Procyon (α CMi, dec. +5.7°), or Ka ʻŌnohi Aliʻi (ibid., 172). In any case, there is no obvious architectural relationship to this *puʻu*. Thus, there is little to suggest or confirm that this *heiau* was of importance in relation to the rising of the sun or stars.

To the WSW, the wall orientations bracket an unremarkable stretch of the ridgeline a few hundred meters distant, although the SSE wall is aligned directly upon the setting position of the December solstice sun (az. 244.5°, alt. 1.5°, dec. −23.5°, all to the nearest half degree). To the SSE, the ENE wall (az. 148.0°, alt. 0.5°, dec. −52.5°, all to the nearest half degree) is close to the rising point of Canopus (α Car, dec. −52.6° in AD 1800)—that is, the Hawaiian star Ke Aliʻi o Kona i Ka Lewa, "chief of the southern heavens" (Johnson, Mahelona, and Ruggles 2015, 180)—and the WSW wall (az. 153°, alt. 1°, dec. −56°) is close to the rising point of Gacrux (γ Cru, dec. −56.0° in AD 1800), the top star of Newe, or the Southern Cross (ibid., 200).

Additional remarks. Although there are various topographic and astronomical possibilities, none stands out as a prime candidate for motivating the orientation of this *heiau*.

8

Heiau of Kaupō *Moku*

Like Kahikinui, Kaupō was one of 12 districts, or *moku*, into which Maui Island was traditionally subdivided. There is some uncertainty as to whether the western boundary of Kaupō with Kahikinui was at Waiʻōpai Gulch or farther east at Pāhihi Gulch (Kirch 2014, 42). The eastern boundary of Kaupō with Kīpahulu was at Kālepa Gulch. The precise number and geographic location of *ahupuaʻa* and smaller *ʻili* land units within the *moku* of Kaupō are matters requiring further research. The *Indices of Awards* (Territory of Hawaiʻi 1929) list at least 18 land section names in which awards were made during the Mahele of 1846–1854. One of these was the large *ahupuaʻa* of Nuʻu on the western side of Kaupō, awarded to Kalaimoku (Land Commission Award 6239). Many other *ahupuaʻa* were apparently quite narrow. In addition, there seem to have been a significant number of *ʻili kūpono*, a special kind of *ʻili* reserved for the king or high-ranking *aliʻi* (chiefs).

Prior to 2000, relatively little archaeological work had been carried out in Kaupō District. Walker ([1930]) briefly described 30 *heiau* sites (his sites 140–169) based on his rapid survey in 1929 (see also Sterling 1998). He made detailed measurements and plan maps of only a few of these sites, such as Loʻaloʻa, Kou, and ʻOpihi (Walker sites 143, 155, and 158, respectively). Soehren (1963) compiled information on a number of Kaupō *heiau*, drawing primarily on Walker's manuscript. Kolb (1991) mapped and excavated at the two largest Kaupō *heiau*, Loʻaloʻa and Pōpōiwi, for his doctoral dissertation research at the University of California, Los Angeles. In June 2003, Kirch commenced an archaeological survey in the *ahupuaʻa* of Nuʻu, at the invitation of the late Charles P. Keau, who, along with his daughter Bernie and son-in-law Andy Graham, had purchased most of the *ahupuaʻa* and established Nuʻu Mauka Ranch. Kirch has continued fieldwork at Nuʻu intermittently since then, resulting in the intensive survey of much of the Nuʻu area below an elevation of 600 masl. More than 400 archaeological sites have now been documented. Sites within the *ahupuaʻa* of Nuʻu have been numbered with the prefix "NUU-."

In 2007 Alex Baer, a University of California, Berkeley, graduate student under Kirch's supervision, began an archaeological survey of portions of Kaupō District to the east of Nuʻu for his doctoral dissertation research (Baer 2015, 2016). Baer's surveys have covered approximately 3.5 km² and resulted in the discovery and recording of 580 archaeological sites. Sites recorded by Baer in areas other than Nuʻu are designated with the prefix "KAU-."

The *Ahupuaʻa* of Nuʻu

NUU-1, ʻOHEʻOHENUI (WALKER SITE 163)

Although we do not regard site NUU-1 as a *heiau*, we list it here because it is Walker's site 163 ([1930], 235), which he described as "a small heiau on top of a hill," adding the information that "one suggestion was that this was a heiau for tapa drying." Walker gives the name of the site as "Oheohenui." Kirch mapped NUU-1 with plane table and alidade, and two test pits were excavated in 2013. Although the architecture is quite impressive (consisting of a large terrace with high enclosing walls on two sides), it is more consistent with that of an elite residence site rather than a *heiau*. The test excavations also provided evidence for domestic occupation. Further details will be reported elsewhere.

NUU-78

NUU-78 is a probable *heiau* situated 40 m N of NUU-79. The site was noted during mapping of NUU-79 in October 2003, and a plane-table map was made of NUU-78.

Location and topographic setting. The site lies at an elevation of about 20 masl, 325 m inland of Nuʻu Bay. The level terrain consists of basanite of Puʻu Maile lava, about 10 m from the foot of the massive, young basanite of the Puʻu Nole *ʻaʻā* flow.

Architecture. NUU-78 was badly disturbed by prior bulldozing in the area by previous owners of Kaupō Ranch; the walls of the structure are largely collapsed, and only remnant facings remain (fig. 8.1). The structure consists of a double enclosure whose orientation deviates from cardinality by roughly 10° in a clockwise sense, with exterior dimensions of 30 m N–S by 20 m E–W. The walls, which were originally core-filled, are up to 2 m wide, but their original height can no longer be determined. It appears as though the E wall (running N–S) may originally have had a slight notch, but the area where the notch would have been situated has been bulldozed. It also appears that a low stone platform was situated in the smaller, N court of the enclosure, abutted to the inside of the N wall.

Archaeological context. The site is 40 m N of the large NUU-79 *heiau*. A complex of dryland agricultural field embankments lies immediately to the W of NUU-78.

Viewshed. No detailed observations were made; however, the prominent cinder cone of Puʻu Māneoneo is visible at azimuth 109°. The massive Puʻu Nole *ʻaʻā* flow, which rises steeply a short distance from NUU-78, would have blocked any views to the NE.

Orientation. The extant faces of the S enclosure wall, which was significantly skewed (see fig. 8.1), were observed by compass to be oriented with an eastward azimuth of 109°, the same bearing as for Puʻu Māneoneo. A remnant facing of the E wall has a northward azimuth of 7°.

NUU-79

NUU-79 is a substantial, double-court *heiau* that was partially modified during the post-contact period. Surprisingly, the site was not reported by Walker even though it is less than 200 m from Kaʻiliʻili Heiau (NUU-81), which he did record. Charles P. Keau first showed NUU-79 to Kirch in 2001; Kirch mapped the site with plane table and alidade in October 2003. Kirch and Ruggles made a brief compass-clinometer survey of NUU-79 on March 25, 2008, and returned to make total station and GPS surveys on November 6, 2011.

Location and topographic setting. NUU-79 sits on level ground about 10 masl, approxi-

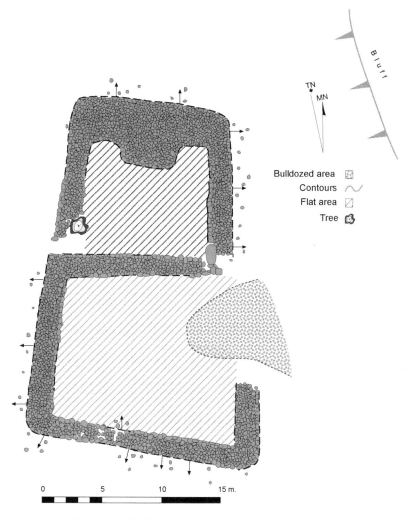

FIGURE 8.1. Plan of site NUU-78. (Based on plane-table survey by P. V. Kirch)

mately 265 m inland of Nuʻu Bay. Immediately E of the structure the terrain drops away steeply into a dry ravine or intermittent watercourse.

Architecture. The site originally consisted of a large rectangular enclosure, skewed from cardinality by about 8° in a clockwise sense, with exterior dimensions of 33 m E–W by 19 m N–S and subdivided into two courts (fig. 8.2). The N and E walls are largely intact (fig. 8.3), as is the eastern part of the S wall, but the W wall and the western part of the S walls were largely robbed of their stones and today can only be traced as low rubble stubs. The robbed stones were evidently used to construct a small rectangular enclosure immediately W of the *heiau*, almost certainly a postcontact house site, as well as a cattle wall that abuts the N *heiau* wall. Further postcontact modifications consist of a series of low, roughly defined terraces that abut the S wall on its exterior face; these probably contain nineteenth-century burials.

The N and E walls are massive, up to 3 m wide and 1.25 m high in the case of the N wall; they incorporate large ʻaʻā boulders in their base courses. The wall subdividing the E and W courts within the enclosure is low (~0.4 m high), abutting a small room or enclosure (interior dimensions

Kaupō • 287

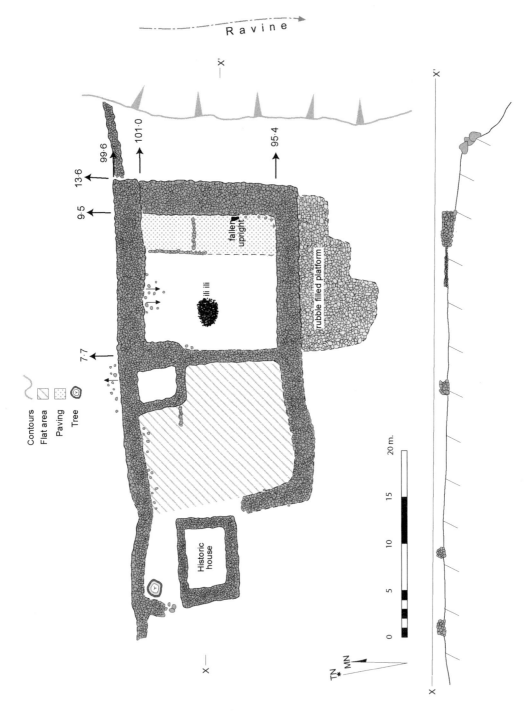

FIGURE 8.2. Plan of *heiau* NUU-79, showing the azimuths of intact straight segments of wall facing as determined by the total station survey. (Based on plane-table survey by P. V. Kirch)

FIGURE 8.3. View of the exterior face of the N wall of NUU-79. (Photo by P. V. Kirch)

4 by 3 m) attached to the inner face of the N wall. Along the inner face of the E wall within the E court are two discrete, paved areas; one of these has a fallen upright stone resting on it. Near the center of the E court is a concentration of 'ili'ili stones that might have been the interior paving for a small thatched structure. No interior features were noted within the W court.

Excavation. A test pit measuring 1 by 0.5 m was excavated against the interior face of the N wall in the E court in March 2011. Layer I consisted of 4–5 cm of eolian overburden. The cultural deposit (layer II) had an upper component (5–10 cm below surface) with a well-laid pavement of waterworn beach pebbles and smaller gravel ('ili'ili) ranging from 5 to 12 cm in diameter, mixed with 'a'ā clinkers in the same size range, all in a reddish brown compact clay matrix. Beneath the pavement the reddish brown cultural deposit contained charcoal flecks and chunks throughout. The deposit extended to a depth of 18–20 cm below surface, at which depth a yellowish red subsoil and angular 'a'ā rocks appeared. The large base cobble of the *heiau* wall rested on this subsoil, and the cultural deposit had accumulated against the base course.

The cultural deposit yielded 35 basalt flakes, a fractured basalt cobble, 1 flake of volcanic glass, and a small hammerstone (probably used for cracking open shellfish, rather than for lithic production). There were 18 pieces of coral, mostly small and weathered, including 1 piece of branch coral. Faunal material included a talus bone and a burned incisor of pig and 14 NISP of small weathered fragments of marine shell that could not be further identified. There were also 12 pieces of noncarbonized candlenut shell.

Dating and chronology. A sample of wood charcoal from the endemic shrub or small tree *Sophora chrysophylla* from level 4 of TP-1 (adjacent to the base of the N wall and thus postdating wall construction) was radiocarbon dated (Beta-308931) with a conventional age of 150 ± 30 BP, yielding 2σ calibrated age ranges of AD 1667–1709 (16.3%), 1717–1784 (31.4%), 1796–1890 (30.0%), and 1910–1950 (17.7%), the last of which can be rejected. This relatively late date suggests that NUU-79 was in use sometime from the late seventeenth through the eighteenth century. This period corresponds to the reigns of the late Maui *aliʻi nui* from Kaʻulahea II through Kahekili (Kirch 2014, table 2).

Archaeological context. A complex of dryland agricultural fields lies to the W of the *heiau*. Site NUU-78 is situated 40 m to the N.

Viewshed. The view to the N is dominated by the narrow ridges and valley rising steeply up to the Haleakalā Ridge. In other directions the view is currently obstructed by vegetation such as introduced *kiawe* (mesquite) trees, but formerly, in their absence, there would have been a clear view round to the W, down almost to the foot of the Haleakalā slopes. A number of small conspicuous features would have been visible on this skyline, including the twin peaks of the Lualaʻilua Hills at azimuths 263° and 266° and Manukani at 277°; there is also a prominent *puʻu* within a broad shallow dip up on the high ridgeline itself, at azimuth 344°. To the NE and E, the ground level rises steeply between 100 and 200 m from the site, restricting the view even in the absence of vegetation; but the cinder cone of Puʻu Māneoneo, its summit 0.9 km distant, would have been prominent between azimuths 100° and 115°, with the precipitous cliffs on its southern side forming a particularly conspicuous feature. Flat ground extending away toward the SE, S, and SW would probably have obscured the view of the sea in these directions, even without vegetation, except between azimuths 185° and 225°, where the sea horizon might have been visible.

Orientation. The longer E–W axis and the paved areas along the inner face of the E wall strongly suggest that this *heiau* faced eastward. The outer and inner faces of the N wall yield eastward azimuths of 99.6° and 101.0°, respectively, for a mean orientation of 100.3°, while the inner face of the S wall yields 95.4°. Thus, the *heiau* narrows somewhat toward the E, and our best estimate of the mean axial orientation is 97.8°. The inner and outer faces of the E wall yield northward azimuths of 9.5° and 13.6°, respectively, for a mean orientation of 11.6°, while the straighter, eastern face of the wall dividing the courts yields 7.7°.

To the E, the *heiau* faces just to the left of the conspicuous promontory of Puʻu Māneoneo (see fig. 4.13). The computer reconstruction of the closer horizon to the left of azimuth 92° may not be reliable, so the apparent rising point of the equinoctial sun at a small hill junction may be misleading, but it is clear that the December solstice sun would have been seen to rise directly at the foot of the cliffs on the southern side of the *puʻu*, certainly the most conspicuous point on this profile. This suggests that there could well have been an interest in tracking the position of sunrise from site NUU-79. The sun would have been seen to rise in line with the *heiau* orientation in early October and early March, the latter being the approximate time of the end of the Makahiki season. In late November, the time of the onset of the Makahiki season, when its declination would have been around −21°, the sun would have risen just above the tops of the cliffs, as can be seen by locating the declination −21° line in figure 4.13.

To the W, the principal axis (az. 277.8°) is aligned to within a degree of the peak of Manukani (az. 276.9°, alt. +5.0°, dec. +8.1°), which coincides with the setting point of Altair (α Aql), or Humu (Johnson, Mahelona, and Ruggles 2015, 161), with a declination of +8.1° in AD 1700, rising to +8.4° in AD 1800. To the N, the E wall is oriented upon a stretch of the Haleakalā ridgeline lacking distinctive features and centered upon a declination (~+77.5°) without obvious significance; to the S, it is aligned upon open sea, and again the declination (−67°) is without obvious significance.

Additional remarks. It is possible that observations from this *heiau* might have been used to mark the onset of the Makahiki season by reference to the position of sunrise rather than the timing of Pleiades rise, but because there is no conspicuous horizon feature directly at the relevant point, and the *heiau* does not directly face the direction in question, the evidence is equivocal.

NUU-81, KAʻILIʻILI (WALKER SITE 160)

Walker reported this site as "Heiau at Kailiili," describing it as "a large heiau measuring 50 × 124 feet…built of rough lava construction with pebbles and coral in its pavements" ([1930], 232). Walker's phrasing here, as well as his subsequent reference to Halekou Heiau (his site 161) as being "at Kailiili" ([1930], 233) make it clear that "Kaʻiliʻili" is a place-name (possibly an *ʻili* within Nuʻu *ahupuaʻa*) rather than the name of the *heiau* proper. He noted the presence of a "high platform at the south and one at the west" and provided a schematic sketch plan. Kirch was shown this site by Charles P. Keau in 2001 and returned to map it with plane table and alidade on October 22, 2003. Kirch and Ruggles made total station and GPS surveys of the structure on March 25, 2008.

Location and topographic setting. The site sits on a lobe of *ʻaʻā* lava of the Puʻu Maile basanite, elevated about 5–10 m and overlooking the terrain to the south and east of the *heiau*. At an elevation of 27 masl, the *heiau* would have had a view down the large swale seaward of the structure to Nuʻu Bay, some 450 m distant, as well as a longer view westward across much of Kahikinui District.

Architecture. The site consists of an elongated rectangular enclosure whose orientation deviates from cardinality by about 16° in a clockwise sense. The exterior dimensions are 38 m ESE–WNW by 20 m NNE–SSW; the NNE wall has a notched configuration (fig. 8.4). The core-filled walls range from 1.5 to 3 m thick and are up to 1 m high in places. As can be seen in the plan, there are several interior features. At the ESE end of the enclosure, a low wall partially sets off a level soil area. Midway along the SSW wall, there is an elevated stone platform (about 4 by 6 m) abutting the wall on the interior face. The WNW interior of the main enclosure consists of an elevated platform of *ʻaʻā* cobbles, with two rooms or space cells within it (these may have originally been covered over with thatch superstructures).

Excavation. Two test units, each 0.5 by 1 m, were excavated in NUU-81 in March 2011. TP-1 was situated against the inner face of the E wall (fig. 8.5), 1.5 m from the NE corner. Under 5 cm of eolian overburden, a single cultural deposit (layer II) extended to a depth of 26–28 cm below surface. The cultural deposit was a dark brown sediment with considerable charcoal throughout; several thin ash and charcoal lenses exposed in the western face of the unit may have derived from a hearth nearby. The base course of the E wall lay on top of layer III—the natural subsoil, which is dark yellowish brown—and the cultural deposit had accumulated against the wall.

TP-2 was located in the western part of the *heiau*, in an open, level soil area, and did not abut standing architecture. There was virtually no cultural deposit in this unit, with the subsoil appearing at 5–10 cm below surface.

Almost all of the cultural material came from TP-1. This included 91 basalt flakes and 1 broken basalt adze preform; some of the material is of high quality, and the assemblage is indicative of adze or other stone tool production. There was also 1 small flake of volcanic glass. There

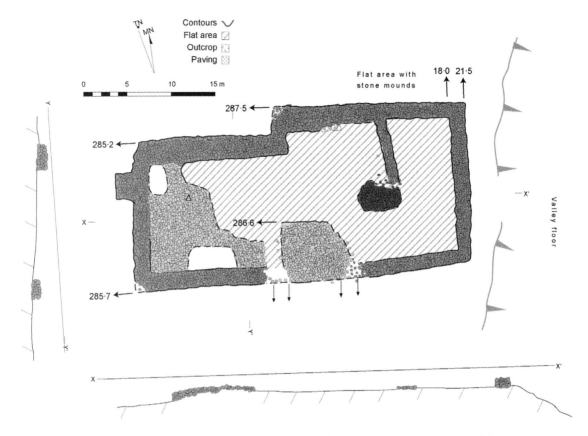

FIGURE 8.4. Plan of *heiau* NUU-81, showing the azimuths of intact straight segments of wall facing as determined by the total station survey. (Based on plane-table survey by P. V. Kirch)

were 18.9 g of charcoal, 3 pieces of noncarbonized candlenut endocarp, and 11 pieces of coral (mostly small and waterworn). Faunal material from TP-1 included 2 pieces of bone (1 shaft fragment from a juvenile medium mammal and 1 maxillary fragment from a pig) and 89 NISP of marine shell and sea urchin. The identifiable marine shells included ʻopihi (*Cellana exarata*), leho (*Cypraea caputserpentis*), pipipi (*Nerita picea*), and *Littorina* sp., and the sea urchin included wana (*Echinothrix diadema*). TP-2 yielded just 4 basalt flakes (of poor-quality material) and a small amount of charcoal.

Dating and chronology. A sample of wood charcoal from the short-lived endemic shrub *Chamaesyce* cf. *celastroides* from level 3 (near the base of the cultural deposit, layer II) of TP-1 (Beta-308932) was radiocarbon dated with a conventional age of 160 ± 30 BP, yielding 2σ calibrated age ranges of AD 1664–1707 (16.7%), 1719–1826 (47.4%), 1832–1884 (12.6%), and 1914–1950 (18.6%), the last two of which can be rejected because the *heiau* system was overthrown in 1819. As with NUU-79, the relatively late date for NUU-81 overlaps with the reigns of the late Maui *aliʻi nui* from Kaʻulahea II through Kekaulike.

Archaeological context. NUU-81 lies at the downslope end of a large series of dryland agricultural field embankments. There are also a number of probable residential features within a 100 m radius of the structure.

Viewshed. The Haleakalā Ridge is fully visible beyond higher ground from the NW

FIGURE 8.5. View of interior wall face and test excavation at site NUU-81. (Photo by P. V. Kirch)

round to the NE, and in the absence of the trees that now surround the site, its western slopes would be clearly visible down at least to the northernmost of the Luala'ilua Hills peaks at azimuth 265°, with the lowest part of the western slopes being obscured by rising ground some 250–300 m from the site. From the ENE round to the S (az. ~55° to ~175°), the view is restricted by higher ground, less than 200 m from the site, that obscures the view of Nu'u Bay beyond. The sea horizon would be visible from the S round to the SW (az. ~175° to ~240).

Orientation. The elevation and prominent architecture of the WNW end of the *heiau* suggest that it was oriented in this direction. In the western part of the structure, the outer faces of the NNE and SSW walls have westward orientations of 285.2° and 285.7° respectively, for a mean of 285.4°, while the outer faces of the NNE wall (east of the notch) and the NNE wall of the elevated stone platform yield 287.5° and 286.6° respectively (see fig. 8.4). The mean axial orientation, estimated as the average of the four azimuths, is 286.3°. It was not possible to obtain reliable orientation estimates for the WNW wall, but the ESE wall has intact straight inner and outer faces with northward orientations of 18.0° and 21.5° respectively, for a mean of 19.7°.

The *heiau* faces part of the western Haleakalā slopes, the azimuth range 285.0° to 287.5° corresponding to declination range +16.5° to +19.0° (all quoted to the nearest half degree). Although this is a featureless stretch of horizon, it does coincide with the setting place of the prominent V-shaped formation of stars known to Hawaiians as Ka Nuku o Ka Puahi, "the opening of the

Kaupō • 293

fireplace" (Johnson, Mahelona, and Ruggles 2015, 171) and to modern astronomers as the Hyades cluster together with Aldebaran, at the head of Taurus (dec. range around ~+15.0° to ~+18.5° in the eighteenth century; see table 4.2). It may also be significant that, a little to the right, the Pleiades (dec. range +23.3° to +23.8° around AD 1800) would have set directly above the cinder cone Puʻu Pane. There is no distant view in the opposite direction, to the ESE.

The ESE wall is oriented to the NNE upon an unremarkable stretch of the Haleakalā Ridge with declination range +68.0° to +71.5° (to the nearest half degree). There are no significant Hawaiian stars in this range, and it seems unlikely that this orientation had any astronomical connection. The range is some degrees to the left of a conspicuous notch where the ground drops away on the far side of Kaupō Gap. The notch itself has a declination of +65.4°, again without obvious significance. To the SSW, the wall is oriented upon open sea: the declination (~−62.5°) is close to the setting point of Acrux (α Cru, dec. around −61.5° in AD 1700 and −62.0° in AD 1800)—the bottom star of Newe, the Southern Cross (Johnson, Mahelona, and Ruggles 2015, 200).

Additional remarks. The unusual westerly orientation of this *heiau* may be related to the setting of the Hyades or, perhaps more likely, that of the Pleiades behind the cinder cone Puʻu Pane. However, this westerly orientation—which is also matched by NUU-100—may have to do with the fact that Nuʻu is on the western edge of the main Kaupō farming and habitation area, so these temples may have served as boundary *heiau*. In that case, their orientations toward the neighboring district of Kahikinui may have had something to do with warding off potential aggression from the chiefs and people of Kahikinui.

NUU-100, Halekou (Walker Site 161)

This substantial temple site was first recorded by Walker ([1930], 233), who specified its name as "Halekou" and stated that it was located "at Kailiili up the gulch from Nuu at an elevation of about 600 feet." He made a sketch plan of the structure, noting that "it belongs to the open platform type with a wall at the back and across one end" and that there is a "high platform" in the western end. This is one of several sites shown to Kirch by Charles P. Keau in 2001; Kirch mapped the structure with plane table and alidade on June 20, 2003. Kirch and Ruggles undertook total station and GPS surveys on March 25, 2008.

Location and topographic setting. At an elevation of 200 masl, NUU-100 sits on a substrate of Puʻu Maile basanite lava. Topographically, it occupies a low rise within a large swale that was probably well suited to intensive dryland cultivation.

Architecture. The ground plan of NUU-100 (fig. 8.6) is in many respects similar to that of NUU-81, such that one wonders whether the two *heiau* might have been constructed under the direction of the same *kahuna kuhi-kuhi-puʻu-one*, or priest-architect (see chapter 2). The structure consists of an elongated rectangular enclosure, its orientation deviating from cardinality by about 8° in a clockwise sense, with a distinct notch midway along the N wall. The maximum exterior dimensions are 47 m E–W by 24 m N–S. As with NUU-81, the stonework is distinctly more prominent on the W, an indication that the *heiau* was oriented to the W. The N, W, and S walls are buttressed by a secondary terrace in the W part of the structure, whereas the E end of the *heiau* is demarcated only by a relatively low wall. During the site survey, a unique flaked basalt artifact (fig. 8.7) was found lying on this W wall. A historic-period cattle wall adjoins the *heiau* on the NE and again on the SE.

The E portion of the enclosure interior, which is somewhat lower than the W part, was sub-

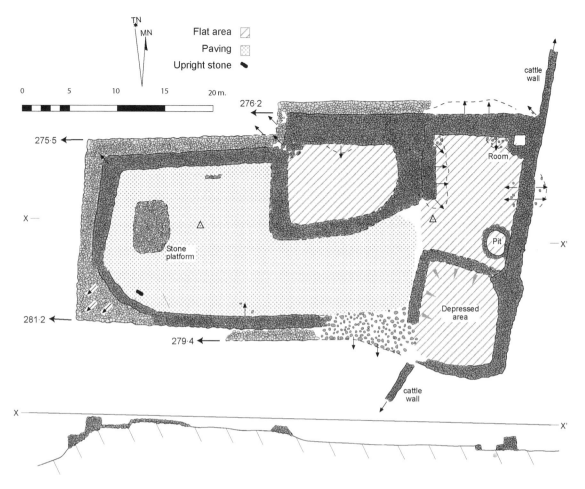

FIGURE 8.6. Plan of Halekou Heiau (NUU-100), showing the azimuths of intact straight segments of wall facing as determined by the total station survey. (Based on plane-table survey by P. V. Kirch)

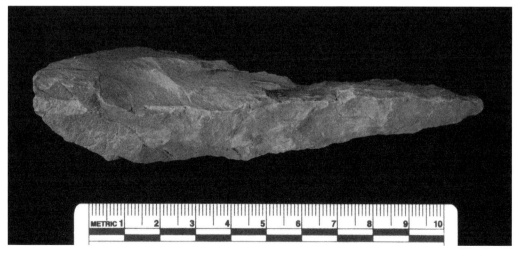

FIGURE 8.7. Basalt core tool, possibly an awl or graver, from Halekou Heiau (NUU-100). (Photo by P. V. Kirch)

divided into at least three discrete areas by low stone walls, which are now largely rubble heaps due to decades of cattle trampling. A small chamber, about 1 m square, is situated in the NE corner of the enclosure; there is also a stone-lined pit 0.4 m deep and 2 m in diameter near the center of the E wall, possibly for the disposal of offerings. The W part of the enclosure is elevated and probably originally consisted of a discrete platform of ʻaʻā cobbles and boulders, although no clear facings can now be discerned.

Excavation. In March 2011 a test pit, 0.5 by 1 m, was excavated under Kirch's direction; the unit was situated adjacent to the inner face of the E wall, 5.3 m from the NE corner. The stratigraphy here consisted of 4–6 cm of eolian overburden, followed by a cultural deposit (layer II) 25–30 cm thick. The cultural deposit consisted of a dark yellowish brown sediment but with more rock (estimated at 75% by volume) than sediment. This deposit abutted the base course of the enclosure wall, which rested on the original land surface.

The test excavation yielded a substantial amount of cultural material indicative of several kinds of male-associated activities having been carried out within the *heiau* enclosure. Formal artifacts included part of a quadrangular adze preform (broken due to end shock), the bend and point of a small rotating fishhook of bone, the shaft and partial bend of an unfinished bone fishhook, and a shaft of bone (medium mammal, possibly dog) that had been sawn and cut at one end. A basalt core and 143 basalt flakes were also found within the temple. The good quality of much of this lithic material suggests that it derives from adze manufacture. There were also 2 small flakes of volcanic glass. Twenty small waterworn pebbles (ʻiliʻili) and 59 pieces of rounded coral were also recovered from the deposit. Faunal material included 35 NISP of marine shell (mostly weathered and chalky), of which 1 fragment of a large species of *Cypraea* and 2 fragments of the smaller *Cypraea caputserpentis*, along with 1 *Cellana* sp. fragment and 3 *Littorina* sp. fragments, could be identified. Bone was limited to 9 NISP, which included 1 fish scale and 1 fish bone (unidentifiable to taxon), as well as 1 dog phalanx.

Dating and chronology. A sample of carbonized endocarp of the candlenut (*Aleurites moluccana*) from level 5 of TP-1, abutting the base course of the enclosure wall, was radiocarbon dated (Beta-308933) with a conventional age of 150 ± 30 BP, yielding 2σ calibrated age ranges of AD 1667–1709 (16.3%), 1717–1784 (31.4%), 1796–1890 (30.0%), and 1910–1950 (17.7%). This relatively late date is similar to those from NUU-79 and NUU-81, suggesting that all three temples were relatively late constructions, corresponding to the reigns of the last four Maui *aliʻi nui*.

Archaeological context. NUU-100 lies within an upland zone of dense habitation and agricultural features. The large elite residential site NUU-1 is about 230 m to the NE, and many other residential features occupy the lava ridge to the north and west of the site.

Viewshed. The Haleakalā Ridge dominates the skyline between the NW and the ENE, its western slopes—some 10 km distant—stretching away down to the sea beyond Alena Point to the WSW (az. ~245°). This western profile is almost, but not quite, obscured by a closer ridge ~150 m away, rising up between azimuths 276° and 281°. The distant peaks of the Lualaʻilua Hills (14.5 km away) are conspicuous at azimuths 258.6° and 262.0°, as is that of Kahua (11.4 km away) at 291.8°. To the NE the view is restricted by a hill just over 100 m from the site, and to the ENE, E, and ESE by successive ridges ~250 m and ~500 m distant that obscure the view of Nuʻu Bay to the E. On clear days, and in the absence of tall vegetation, there would be good views of Hawaiʻi Island to the SSE, between azimuths 130° and 170°, with open sea to the S and around to the WSW.

Orientation. As discussed earlier, this *heiau* is clearly oriented to the W. Reliable orientation estimates were obtained from segments of intact wall facing as follows: N wall W of the notch, outer face (az. 275.5°); N wall E of the notch, inner face (276.2°); S wall close to the W end, outer face (281.2°); and S wall, lower terrace further to the E, outer face (279.4°). The wall segments on the western side of the structure yield a mean axial orientation of 278.3°, while the more easterly wall segments, which converge less markedly toward the W, yield 277.8°. It was not possible to reliable obtain orientation estimates for the W wall. From the plane-table survey, it appears to be skewed significantly from the direction orthogonal to the main axis, oriented with a northerly azimuth of approximately 12°.

The *heiau* faces a part of the western horizon where a local ridge rises up just to the level of the distant sloping ridgeline, between the Luala'ilua and Kahua peaks. The declination, ~+9.5°, is close to the setting point of Altair (α Aql, dec. ~+8.5° around AD 1800), or Humu (Johnson, Mahelona, and Ruggles 2015, 161), and between the positions of equinoctial and solstitial sunset. It is likely that the changing position of the setting sun was noted on this horizon, especially given that its two limits coincide with conspicuous features: at the June solstice the sun set into the *pu'u* at Kahua, while at the December solstice it set directly above Alena Point (fig. 8.8). It is also noteworthy that around AD 1800 the Pleiades would have been seen to set behind Kahua.

On the eastern horizon, the main axial orientation coincides with a stretch where a ridge falls away steeply and the rising paths of celestial bodies would be more or less perpendicular to it. The declination, between −7° and −6°, does not correspond specifically to the rising point of any known bright Hawaiian star around AD 1800 but falls between that of all or part of Nā Kao (Johnson, Mahelona, and Ruggles 2015, 198)—that is, Orion's Belt (dec. in AD 1800 −2.0° to −0.5°, to the nearest half degree) —and Puanakau (ibid., 205), or Rigel (β Ori, dec. ~−8.5° in AD 1800). Although the exact orientation of the W wall is unclear, this wall was evidently oriented to the N upon a stretch of the Haleakalā Ridge without obvious topographic or astronomical significance, and to the S toward open sea, again without obvious astronomical significance.

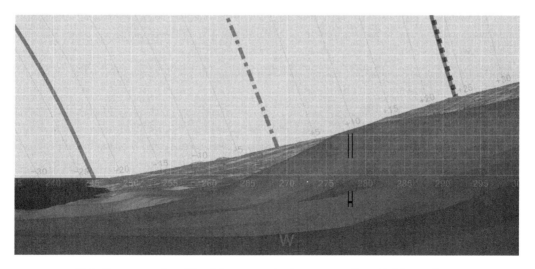

FIGURE 8.8. Digitally generated profile of the western horizon as seen from Halekou Heiau (NUU-100). The horizontal bar indicates the main axial orientation of the *heiau*. The solid shaded lines indicate the setting paths of the sun at the December solstice (*left*) and the June solstice (*right*); the dash-and-dot line indicates the setting path of the sun at the equinoxes; and the dotted line indicates the setting path of the Pleiades in AD 1800. (Graphic by Andrew Smith)

NUU-101

Surprisingly, given the prominent nature of its surviving architecture, NUU-101 was not recorded by Walker ([1930]), nor is it mentioned in the compendium of traditional places by Sterling (1998). The site—located at the old Nuʻu Landing on the E side of Nuʻu Bay, where the interisland steamships once loaded cattle and off-loaded cargo during the late nineteenth and early twentieth centuries—was discovered rather by accident. In 1999, Kirch and his students began to visit the Nuʻu Landing as a place to swim and relax at the end of a week of fieldwork. They soon recognized that the remnants of a major *heiau* were situated just inland of the bulldozed area that is today a favorite haunt of island fishermen, who cast their lines off the low cliffs. Kirch made a plane-table and alidade map of the remaining portions of the *heiau* on June 1, 2006. Ruggles and Kirch made a brief visit to the site in March 2008 and made prismatic compass measurements.

Location and topographic setting. The structure lies at an elevation of 8–10 masl, on the promontory of young Puʻu Nole ʻaʻā that forms the E side of Nuʻu Bay. Prior to the destruction of the W side of the *heiau* when Nuʻu Landing was constructed, the *heiau* probably extended to the edge of the rocky cliffs.

Architecture. As can be seen in the site plan (fig. 8.9), all that remains today of this once quite large *heiau* is the well-constructed, double-terraced E wall, as well as portions of the abutting S and N walls. It appears that the orientation deviated from cardinality by about 9° in a clockwise sense. The remaining E wall is 55 m long, and its double-faced exterior stands fully 2.45 m above the adjacent terrain. Assuming that the original *heiau* enclosure extended to the edge of the cliffs, as seems likely, the *heiau* would have had a basal area of about 1,870 m^2.

Nothing remains intact of the *heiau* interior, which probably was destroyed when the Nuʻu Landing was established. However, the area immediately W of the long N–S wall exhibits a high concentration of ʻiliʻili stones, suggesting that the *heiau* court was paved with these waterworn cobbles. A considerable number of branch coral pieces were also observed in and on the N–S wall.

Viewshed. This site has superb, unbroken views northward up to Haleakalā and Kaupō Gap, the horizon altitude rising to 16.5° directly to the N. The western slopes, 10–15 km distant, extend right down to the sea beyond Nuʻu Bay, to the WSW (az. 250°); and the eastern slopes extend down as far as an altitude of 5.5° (reached at az. 56°). Farther around toward the E, the horizon is formed by Puʻu Māneoneo, ~1.0 km away, its flattish top running almost level for some 25° before starting to fall away, and then dropping away sharply between azimuths 92° and 93°. The flat terrain to the ESE, SE, and SSE of the site would probably have been sufficient (even without tall vegetation) to obscure the sea horizon in those directions (although the peaks of Hawaiʻi Island could have been visible beyond in clear weather conditions), but the open sea was certainly visible from the S round to the WSW (az. ~180° to ~250°).

Orientation. Given the site's poor state of preservation, it is not possible to be confident about the direction of orientation. The E wall has a northward azimuth of 9° ± 1°, and the surviving segments of the S wall suggest an eastward orientation of 94° ± 2°.

To the E, the S wall faces the cliffs on the southern side of Puʻu Māneoneo cliffs, where the horizon drops away sharply. This is close to the position of equinoctial sunrise, but the declination range, –1° down to –5°, corresponds to that of Orion's Belt—part or all of Nā Kao (Johnson, Mahelona, and Ruggles 2015, 198)—with a declination of –2.0° to –0.5° (to the nearest half degree) between AD 1600 and 1800. At this time, the three stars of Nā Kao would have risen almost

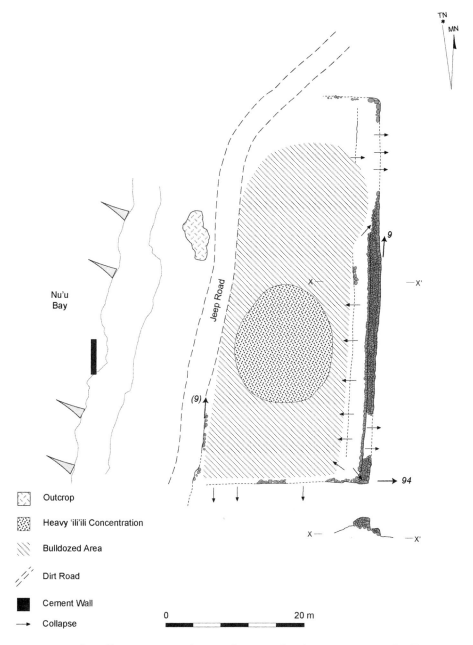

FIGURE 8.9. Plan of *heiau* NUU-101, showing the azimuths of intact segments of wall as determined by prismatic compass corrected for mean magnetic declination. (Based on plane-table survey by P. V. Kirch)

exactly out from the cliff, and the whole constellation of Orion would have been conspicuous just above this distinctive stretch of skyline (fig. 8.10). In the opposite direction, the wall aligns upon a featureless stretch of horizon between the Luala'ilua Hills (az. 264° and 268°) and Manukani (az. 279°). In the seventeenth and eighteenth centuries this stretch (dec. +3° to +7°) included the setting points of Bellatrix (γ Ori, dec. +6°); Procyon (α CMi, dec. +6°), or Ka 'Ōnohi Ali'i (ibid., 172); and Betelgeuse (α Ori, dec. ~+7.5°), or 'Aua (ibid., 146).

Kaupō • 299

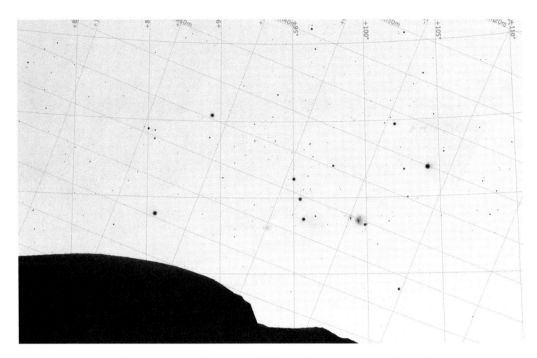

FIGURE 8.10. Eastern horizon as seen from *heiau* NUU-101. Photograph overlaid on a sky visualization for AD 1700 generated by Stellarium, showing the stars of Nā Kao (Orion's Belt) and the remainder of Orion, shortly after their rise above Puʻu Māneoneo. (Composite by Clive Ruggles)

To the N, the E wall faces an undistinctive stretch of the Haleakalā skyline between the summit and Kaupō Gap. The declination range, +79° to +81°, is of no obvious significance. In the opposite direction the wall faces the open sea. Again, the declination (around −68°) is without obvious significance.

NUU-151, Papakea (Walker Site 159)

This site was originally recorded by Walker [1930], 231. Ruggles briefly visited the site on April 8, 2002, and Kirch mapped the site with plane table and alidade on November 14, 2005.

Location and topographic setting. The site lies on a ridge of massive ʻaʻā lava overlooking the low promontory between Nuʻu Bay and Walu Bay, at an elevation of about 45 masl.

Architecture. Walker ([1930], 231) described this site as "a medium-sized heiau 73 × 64 feet...walled all the way round [with] a large paved court and two higher terraces" and noted that "the entrance is on the N.W. side." The main structure consists of a nearly square, leveled terrace (10 by 11 m) walled on the N and E with well-constructed core-filled walls. The NE corner has been widened to form a kind of platform 4 by 5 m in area, with a pit or depression in the center. A smaller terrace (4 by 7 m) is situated just to the SE of the main structure. To the N of the main structure, an irregular area in the rough ʻaʻā flow about 32 m long and 4–6 m wide has been leveled and roughly paved with ʻaʻā clinkers; this space is bounded on the N by a low wall of stacked ʻaʻā cobbles about 40–50 cm high. This leveled and roughly paved space may have been a seating area for persons assembled to witness rituals in the main structure.

Archaeological context. A small rectangular enclosure (possible habitation site) is located about 50 m to the E of NUU-151, while a cluster of probable burial mounds lies about 45 m to the S.

Viewshed. There are striking views in all directions, although these are somewhat obscured at present by surrounding trees. Northward, the broad sweep of the Haleakalā Ridge extends down to the sea beyond Alena Point in the WSW (az. ~250°) and down beyond Kaupō Gap as far as the NE (az. ~45°). From the NE round to the SE the horizon is dominated by Puʻu Māneoneo, about 200 m distant, its cliffs dropping away at azimuth 143° to reveal, in clear weather, the southern mountains of Hawaiʻi Island beyond, ~100 km distant, to the SSE. The sea forms the horizon from the S round to the WSW.

Orientation. From Walker's description it is likely that this *heiau* was oriented to either the N or the E, but given the poor state of preservation, we cannot be certain which.

The orientation of the extant wall, as determined by prismatic compass, is 358° (corrected for mean magnetic declination). This is close to the highest point on the visible horizon (alt. 17.0°, reached at az. ~359°).

The entire solar rising arc falls within the part of the horizon formed by the high ground around Puʻu Māneoneo, only about 200 m distant. The highest point (alt. 14.1°) is reached at about azimuth 70° and marks the rising point of the June solstitial sun (and the Pleiades) around the likely time of construction (see fig. 4.13).

NUU-153

NUU-153 was originally discovered during a reconnaissance survey by Sidsel Millerstrom and John Holson in November 2005. Situated at the inland head of a broad swale extensively terraced for dryland agriculture, NUU-153 appeared to be a small agricultural shrine or temple. The site was plane-table mapped by Kirch on March 23, 2010; at the same time Alex Baer excavated two test pits in the structure.

Location and topographic setting. NUU-153 lies at an elevation of 90 masl on the basanite of the Puʻu Maile *ʻaʻā* lava flow (substrate about 3,000–5,000 years old; Sherrod et al. 2007). The structure lies at the inland end of a swale, roughly 250 m long by 75 m wide, with good soil; the gently sloping floor of the swale is covered in E–W-trending embankments that divide the swale into agricultural terraces. Constructed on a 5 m elevation rise at the head of this swale, NUU-153 commanded a view over this productive agricultural landscape.

Architecture. The structure consists of a series of stone-faced terraces descending a slope approximately 5 m high (fig. 8.11). The uppermost terrace (7 m E–W by 4.5 m N–S) defines a level soil area. Below this is a main terrace 11 m long and 5 m wide, bounded on the E side by a low, L-shaped wall with a pronounced canoe-prow corner. This wall may have served as the foundation for a thatch superstructure on the E part of the main terrace. A number of waterworn cobbles and smaller *ʻiliʻili* pebbles were scattered over this terrace, along with 11 lithic artifacts, including 5 basalt adze blanks or preforms, several basalt cores, and a hammerstone. On the downslope (S) side, this main terrace was buttressed by a lower stone terrace, 2 m wide and running about 14 m E–W. Viewed from the floor of the swale, the steeply rising terraces present an impressive facade.

Excavation. Two test pits, each 1m², were excavated in NUU-153. TP-1 was situated on the upper terrace in an area of level soil. The cultural deposit here was relatively shallow, ending at about 25 cm below surface, and contained 43 basalt flakes and 1 waterworn *ʻiliʻili* pebble.

TP-2 was located on the eastern end of the main terrace, about 75 cm from the N–S-trending wall that defines the end of the terrace. Here layer I consisted of about 10 cm of eolian overburden, which partially obscured several large paving slabs. Removal of the pavement exposed a

FIGURE 8.11. Plan of site NUU-153, a small agricultural *heiau* or men's house. (Based on plane-table survey by P. V. Kirch)

deeper cultural deposit (layer II) extending to 45–50 cm below surface, at which depth large *ʻaʻā* boulder fill of the terrace foundation was encountered. The cultural deposit consisted of a fine, silty loam. At the base of the cultural deposit, directly overlying the rock fill, was a small basin-shaped combustion feature containing considerable charcoal and ash.

TP-2 yielded substantial evidence for adze production, including 1 hammerstone, 1 finished adze, 2 adze preforms ready for grinding (fig. 8.12), 76 basalt flakes, and 691 smaller pieces of angular basalt shatter. There were also 4 small flakes of volcanic glass. Vertebrate faunal material included 15 NISP of bone, with 2 medium mammal fragments, 1 possible bird bone fragment, and the remainder being unidentified fish bone. Invertebrate faunal material consisted of 207 NISP of marine shell, with the main identifiable taxa being *Cellana exarata*, *Conus* spp., *Cypraea caputserpentis*, and a large *Cypraea* sp. (possibly *C. mauritiana*), *Drupa* sp., *Littorina* sp., and *Nerita picea*, as well as 165 NISP of sea urchin (mostly test fragments but also a few spines of *Echinothrix diadema*).

Dating and chronology. A sample of wood charcoal from the short-lived endemic shrub

FIGURE 8.12. Basalt adze preforms and a finished adze from site NUU-153: *a*, adze preform (NUU-153-TP-2-2-1); *b*, adze preform (NUU-153-TP-2-2-3); *c*, finished adze (NUU-153-TP-2-3-1). (Photo by P. V. Kirch)

Chamaesyce cf. *celastroides*, from the small hearth at the base of the cultural deposit (layer II) in TP-2, was radiocarbon dated (Beta-308934), with a conventional age of 310 ± 30 BP. This corresponds to a 2σ calibrated age range of AD 1485–1650, putting the site in a relatively early period for the Kaupō region.

Archaeological context. As noted earlier, NUU-153 is associated with an extensive set of agricultural terraces. There are also a number of residential features within a distance of 20–80 m from the site, located to the N and W on the low ridges overlooking the swale.

Viewshed. NUU-153 commands an excellent view across the broad swale below it, which was originally covered in agricultural terraces. When the terrain was free of the exotic vegetation that now cloaks it, the site would also have commanded a view down to the sea at Nu'u Bay.

Orientation. No detailed observations regarding orientation were made at NUU-153. However, the plane-table map shows that the main terrace facings are oriented almost exactly upon cardinal E–W.

Additional remarks. NUU-153 most likely functioned as a small *heiau ho'oulu 'ai* (fertility temple) associated with the major agricultural field complex that it overlooks. The extensive amount of lithic production evidenced at the site also suggests that it may have had a simultaneous function as a men's house (*mua*) for the several residential complexes in the immediate vicinity. The ethnohistorical literature suggests that such dual functions need not have been mutually exclusive.

NUU-172 (WALKER SITE 165)

Despite being recorded as a *heiau* by Walker ([1930], 237), for several reasons we do not regard this site as such. The structure consists of a large terrace, 26 by 12 m, faced along the W, S,

Kaupō • 303

and E sides, and buttressed with a lower terrace on the E. The SE corner of the main terrace is marked by an acute angle, which seems to point directly to the northern tip of Hawai'i Island. There is a low, stacked wall along the N edge of the terrace. An L-shaped wall and adjacent pavement in the SW part of the terrace (about 4 by 4 m) and a smaller C-shaped shelter adjacent were presumably the foundations for two thatched houses. The site lacks branch coral or 'ili'ili offering stones. In its configuration it is similar to site KIP-394, and as with that structure, we believe that NUU-172 was most likely a residential platform for an elite household.

NUU-175

NUU-175, a classic notched *heiau*, was discovered during an archaeological survey by Kirch in 2005 and subsequently mapped by Kirch with plane table and alidade on June 18, 2005. Kirch and Ruggles visited the site on March 25, 2008, to make a total station survey. A good GPS survey could not be undertaken, owing to the site's position in a low swale with surrounding ridges that blocked satellite reception.

Location and topographic setting. NUU-175 is situated within a grove of *kukui* (candlenut) trees at the top of a long, narrow swale in the massive Pu'u Maile basanite 'a'ā lava flow, at an elevation of about 330 masl. A high ridge of the very young Pu'u Nole basanite rises steeply directly behind the structure to the north and also to the west.

Architecture. This notched *heiau*, whose orientation deviates from cardinality by about 10° in a counterclockwise sense, has its notch in the NW corner. It has a very well constructed E enclosure and a much less distinct W terrace, fading out into level ground on the far W end (fig. 8.13). The maximum dimensions are 28 m E–W by 16 m N–S, with the E enclosure measuring 12 by 12 m on the interior. The walls are of the core-filled type, with the interior face of the N wall reaching a height of 1.5 m (fig. 8.14). No features were observed within the E enclosure.

Excavation. A test pit, 1 m^2, was excavated within the E enclosure by Kirch in March 2011; the pit was positioned in a level soil area about 1 m in from the inner face of the S wall. Under an overburden consisting of about 10 cm of eolian sediment mixed with many candlenuts, a cultural deposit of dark reddish brown loam was encountered, extending to a depth of 40 cm below surface. The upper part of this deposit consisted of a rough pavement of fist-size 'a'ā clinkers and waterworn cobbles ('ili'ili), about 15–20 cm thick. Beneath this pavement, several large cobbles were encountered, four of which appeared to form an alignment running across the unit in a N–S direction; several pieces of branch coral were associated with this alignment, along with a pig phalanx bone. At 40 cm below surface the entire unit had exposed large 'a'ā boulders, which appeared to be basal fill for the structure.

The cultural deposit yielded 1 core and 67 flakes of basalt, some of it of good quality and all indicative of adze or other lithic tool production. Fifteen pieces of coral were recovered (all quite chalky and weathered), 3 of them of pieces of *Pocillopora* branch coral that were probably placed within the site as offerings. Faunal material was limited to 5 NISP of mostly small and unidentifiable fragments, except for the pig phalanx. There was no shell present, but this is probably due to the higher elevation of this site, where rainfall and soil acidity are not conducive to preservation of calcium carbonate materials.

Dating and chronology. A sample of wood charcoal from the endemic shrub or small tree *Psychotria* sp. (Beta-308935) from level 5 at the base of TP-1 was radiocarbon dated, with a conventional age of 300 ± 30 BP, yielding 2σ calibrated age ranges of AD 1489–1604 (69.6%) and 1611–1654

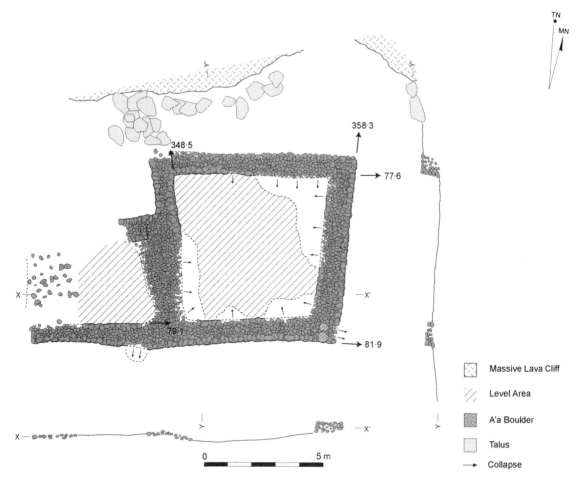

FIGURE 8.13. Plan of *heiau* NUU-175, showing the azimuths of intact straight segments of wall facing as determined by the total station survey. (Based on plane-table survey by P. V. Kirch)

FIGURE 8.14. View of the interior wall faces of site NUU-175. (Photo by P. V. Kirch)

(25.8%). This date indicates construction of the *heiau* sometime during the sixteenth or possibly early seventeenth century, corresponding to the reigns of the four Maui *aliʻi nui*, from Piʻilani to Kalanikaumakaōwakea (Kirch 2014, table 2). This *heiau*, with its classic notched form, is clearly considerably older than NUU-79, -81, or -100, all of which have elongated double-court plans.

Archaeological context. The temple sits at the head of a long swale that shows evidence of intensive dryland agriculture, in the form of field embankments.

Viewshed. The views from this site are highly restricted, owing to high lava ridges immediately to the west, north, and east, with altitudes as great as ~15°. There is a limited view to the WSW down the narrow valley, with its agricultural terracing, toward the sea.

Orientation. The longer W–E axis and the well-constructed E chamber show this *heiau* to have been oriented to the E. Well-preserved segments of inner facing on the N wall and outer facing on the S wall yield eastward azimuths of 77.6° and 81.9° respectively, for a mean axial orientation of 79.7°. The inner face of the W wall, with a northward azimuth of 348.5°, is close to perpendicular to this axis, while the E wall is skewed by some 10°, with a northward azimuth of 358.3°.

In the main direction of orientation, the approximate horizon altitude (measured by clinometer) is 15°, corresponding to a declination of about +15°. In the opposite direction the horizon altitude is approximately 8°, yielding a declination of about −7°. Given the close horizons, it is not possible to be more specific.

Only the southerly declination is better defined, but the value, ~−69.5°, is without obvious astronomical significance.

Additional remarks. There is no evidence to suggest that the axial orientation of this *heiau* was associated with specific astronomical phenomena. This structure probably functioned as a *heiau hoʻoulu ʻai*, or agricultural temple, given its position at the *mauka* head of a large swale with good soil, which had been terraced for cultivation.

NUU-188

NUU-188, a small notched *heiau*, was discovered during a survey in 2006 and was subsequently mapped by Kirch with plane table and alidade on June 2, 2006.

Location and topographic setting. At an elevation of about 250 masl, site NUU-188 lies in the base of a large natural swale at the interface of the massive Puʻu Nole ʻaʻā flow and the somewhat older Puʻu Maile flow (on which the *heiau* sits). The swale, which widens out downslope from the *heiau*, contains a large set of dryland agricultural terraces.

Architecture. As seen in the plan (fig. 8.15), NUU-188 consists of a notched enclosure whose orientation deviates from cardinality by about 30° in a clockwise sense, with exterior dimensions of 18 m ESE–WNW by 16 m NNE–SSW; the notch is in the SSE corner. The NNE and ESE walls join in the ENE corner at an acute angle, thus forming a point, or canoe-prow corner, toward the ENE. The NNW corner of the enclosure consists of a natural outcrop of angular boulders. Adjoining the enclosure on the WNW side is a level terrace, faced on the SSW and WNW. The ESE edge of this terrace, where it joins the WNW wall of the enclosure, has an alignment of five small boulders set upright; a flaked basalt cobble was found at the SSW end of this alignment.

Archaeological context. The *heiau* lies at the head of a major swale that contains a large complex of dryland agricultural terraces, suggesting that the site is likely to have functioned as an agricultural temple (*heiau hoʻoulu ʻai*).

FIGURE 8.15. Plan of *heiau* NUU-188. (Based on plane-table survey by P. V. Kirch)

Viewshed. Large *wiliwili* trees surrounding the structure block the viewshed today, but from what we could determine, the views to the ESE and NNE would be largely blocked by the steeping rising front of the massive Puʻu Nole basanite lava flow. Likewise, the view to the WNW is limited to the ridge of Puʻu Maile ʻaʻā lava a few meters away. Thus, the main view would have been to the SSW, down the swale and overlooking the complex of agricultural terraces.

Orientation. The prominent NNE wall has an eastward azimuth of about 117°, while the ESE wall has a northward azimuth of about 35°. As noted above, the NNE and ESE walls join at a prominent acute angle, pointing toward the ENE; the azimuth is approximately 65°.

NUU-424

This unusual site was reported to Kirch by the property owner, Andy Graham. Kirch made a brief reconnaissance and a GPS survey of the structure on June 7, 2013.

Location and topographic setting. The site is located at an elevation of 500 masl, on a tongue of very young Puʻu Nole basanite ʻaʻā lava.

Architecture. The structure consists of a large terraced platform skewed from cardinality by approximately 10° in a counterclockwise sense, with maximum dimensions of 32 m N–S by 13 m E–W. The lower terrace to the S is about 10 m square, faced on the downslope side. The main upper terrace, about 22 m long, is very level, paved with uniform-sized ʻaʻā cobbles. There is a large pit or sunken "room" at the N end of the main terrace, about 6 m in diameter and 1 m deep, faced around the perimeter. A possible small hearth or image posthole, marked by two upright slabs, was noted nearby.

To the N of the main terrace is a substantial enclosure with walls 1–1.5 m high and thick). Huge old *Cordyline* plants were growing inside the enclosure.

Archaeological context. No intensive survey has been carried out in the vicinity of the site.

Viewshed. Dense Christmas berry (*Schinus terebinthifolia*) shrubs and other exotic vegetation obscure the viewshed from the site today, but it would seem to have had an excellent view downward over much of the Nuʻu landscape, as well as up to the ridge line of Haleakalā.

Orientation. Our GPS survey indicates that the central axis of the main terrace has a northward azimuth of about 350°, which appears to be very close to the visible summit of Haleakalā from this viewpoint.

Heiau Elsewhere in Kaupō *Moku*

KAU-324 PŌPŌIWI (WALKER SITE 140, BISHOP MUSEUM SITE MA-A27-1)

This large and complex site, whose various components cover an area estimated at about 7,500 m² (thus making it the second-largest *heiau* on Maui, after Piʻilanihale in Hāna District), is variously known by the names Pōpōiwi (Basket of Bones) and Kanemalohemo (Man [or the god Kāne] with Loincloth Removed). Kamakau (1961, 66) writes that Maui *aliʻi nui* Kekaulike "built Kane-malo-hemo at Popoiwi," suggesting that the latter was a place-name and the former was the proper name of the *heiau*. Kamakau (1961, 188) also identifies this *heiau* as one of those rededicated during Liholiho's tour of Maui temples around 1802. Thrum, however, writes:

> Popoiwi was said to be a heiau, not the location of Kanemalohemo. This latter was simply a sacred place at Mokulau, makai of the road, famed as the spot where a

certain high priest of the Popoiwi temple stood and decried the overthrow of the kapu system and abandonment of the gods, which would result in the extinction of the order, and in his distress and despair he disrobed at this spot before all the people, hence the name, and foretold his own death, which occurred mysteriously the next day.

Popoiwi is referred to as a heiau on a land of the same name just above the road, though known to some of the old residents as Hanakalauai. This was found to be so irregular and dilapidated as rendered it difficult to approximate its shape, or size. (T. G. Thrum 1917, "Maui's Heiaus and Heiau Sites," quoted in Sterling 1998, 176)

Surprisingly—given the size and importance of Pōpōiwi—Walker does not mention the site by this name, although it certainly appears to be his site 140, to which he gives the name "Keakalauae Heiau" ([1930], 210; see also Soehren [1963, 86]). Walker gives the location of site 140 as "on west bank of Punahoʻa valley, overlooking the sea at Mokulau," which corresponds precisely with the location of Pōpōiwi. Walker provides only a cursory description: "One of the largest of the Kaupo heiaus and also credited to Kekaulike about 1730. Its greatest dimensions are 168 × 330 feet. The interior of the platform has been utilized for a pig pen and walls have been built around it. The heiau was also used as a Puuhonua, or Hill of Refuge, and two approaches have been built on the front."

Pōpōiwi was among the Maui *heiau* investigated by Kolb (1991) for his doctoral dissertation. Kolb produced the first map of the structure (1991, fig. 5.9), a difficult task given the size and complexity of the site and the fact that it is heavily overgrown. In addition to being rather schematic, Kolb's map omitted a number of key architectural elements, such as the main WSW wall of the *heiau* proper. Kolb seems to have mistaken the irregular, postcontact pigpen enclosure wall (his wall W3) for a late-stage *heiau* wall. Kolb excavated 14 test pits and obtained 5 radiocarbon dates.

Kirch, assisted by Alex Baer, mapped Pōpōiwi over a three-day period using plane table and alidade in July 2009 (fig. 8.16). Prior vegetation clearing by Baer and his assistant greatly facilitated identifying the many architectural features. Kirch and Ruggles visited the site on November 8, 2011, and made a partial total station survey of some of the main walls by means of a traverse across the site from the ESE corner (*I* in fig. 8.16). An accurate GPS survey was not possible because the dense tree canopy over the site repeatedly blocked satellite reception; however, we were able to establish the general outline of the site with GPS readings.

Location and topographic setting. Pōpōiwi sits astride a ridge that forms the western side of Punahoʻa Valley, at an elevation of 20 masl, about 150 m inland of Mokulau Bay and a short distance inland of the road. The geological substrate consists of deeply weathered, old Kula volcanic series.

Architecture. The architecture of Pōpōiwi is complex, with no fewer than 20 space cells or rooms (identified by capital letters in fig. 8.16). Broadly speaking, the orientations of the structures at Pōpōiwi deviate from cardinality by about 18° in a counterclockwise sense. The largest features are two large courts (features *G* and *H*), together forming a rectangular enclosure about 65 m NNW–SSE by 48 m ENE–WSW. Court *H* is about 2 m lower than court *G*; they are separated by a stone-faced terrace that has collapsed in many places. The courts are defined around the perimeter by a core-filled wall up to 4 m wide and ranging in height from 1 to 2 m.

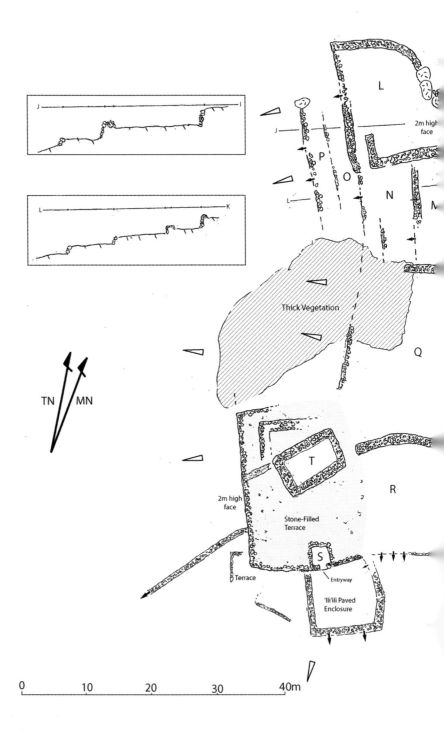

FIGURE 8.16. Plan of Pōpōiwi Heiau (KAU-324), showing the azimuths of intact straight segments of

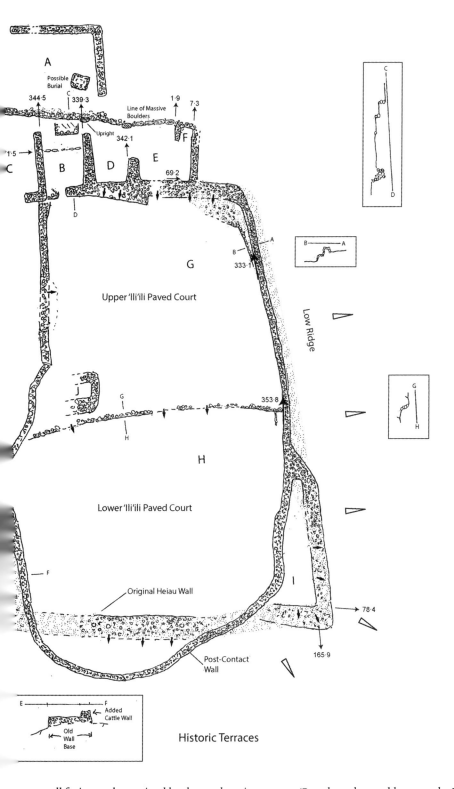

wall facing as determined by the total station survey. (Based on plane-table survey by P. V. Kirch)

This wall, however, has collapsed in places and was also robbed of stone in some areas to construct the postcontact pig enclosure wall, which follows the course of the original ENE *heiau* wall but then meanders irregularly. Kolb (1991, fig. 5.9) did not observe the massive, original *heiau* wall along the SSE and WSW sides, leading him to misinterpret key aspects of the site. The two courts are extensively paved with *'ili'ili* gravel, which was presumably carried up in baskets from the basalt gravel beach at Mokulau Bay. A further noteworthy aspect of the massive enclosing wall is that the ESE corner is not square but comes to a point formed by the acute angle of the SSE and ENE wall joins. This kind of canoe-prow configuration is seen at other southeastern Maui temples and chiefly residential features (e.g., at KIP-1010).

To the NNW of the upper court G is a broad terrace, originally well faced on the downslope side and about 1.5–2 m higher than the court; this terrace is divided into four space cells, designated C, B, D, and E on the plan. Feature B (8 by 6 m), which has an entryway from the lower court leading to it, is defined on the WSW and ENE by exceptionally well-constructed walls 1.2 m high; these walls have flat tops and appear to be designed to support the roof beams of a thatched superstructure. The interior of B is subdivided by a low terrace faced with a single-course alignment; a kind of terraced platform or "dais" is situated at the NNW end of B. The walls defining the ENE and WSW sides of B have gaps between them and the retaining wall of the terrace defining the NNW side of B; these gaps appear to have functioned as entryways into B from adjacent space cells C and D. Feature B has the appearance of a sanctum sanctorum, with the dais possibly being the resting place for a sacred object such as a feathered god image or possibly being a seating place for a person of extremely high rank.

The well-constructed retaining wall defining the NNW side of features D and E notably incorporates a line of five massive boulders (each 1.1–1.4 m in diameter) set adjacent to each other. These stones—which are essentially uprights set into the wall—end at a small wall jutting S and defining a small space cell, feature F. This small chamber, measuring 1.5 by 1.5 m on the inside, was also likely to have been roofed over and may have functioned either as a shrine or as a room in which sacred objects were kept. Notably, feature F occupies the NNE corner of the *heiau*.

Feature A is a level terrace upslope of features B, D, and E, defined on the ENE and NNW by a freestanding stacked wall. A platform on terrace A appears to be a postcontact burial. Several postcontact artifacts such as ceramic sherds and rusted iron were observed on this terrace, suggesting that it was used as a residence in the nineteenth century.

The remaining space cells or rooms are all situated to the W of what we believe to have been the original *heiau* proper, centered around the large courts G and H and the upper terrace with features C, B, D, and E. These additional features (K, L, M, N, O, P, Q, S, and T) consist of a series of terraces, some very well constructed with high facings (up to 2 m high in the case of L), that descend a fairly steep slope to the WSW. It is possible that they functioned as a residential area for priests or, indeed, for Kekaulike or other highly ranked *ali'i*.

Feature T is not a terrace but, rather, a rectangular enclosure (8 by 10 m) constructed atop terrace R. This enclosure has the appearance of a postcontact house site and may have been built over an earlier enclosure, traces of which can be seen outside of feature T on the NW.

Excavation. Kolb (1991, 510, table B.14) excavated 14 test units in various parts of Pōpōiwi Heiau. Unfortunately, other than depths below surface, no stratigraphic profiles or descriptions are provided.

Dating and chronology. Kolb (1991, 234) obtained five radiocarbon dates from unidentified wood charcoal samples; it is not indicated whether $\delta^{13}C$ values were determined. The two oldest dates (Beta-40369 and -40640) came from test unit PP5, excavated into the fill of the large court designated as feature H. The context was a "charcoal concentration" 70–84 cm in depth; no other stratigraphic details are given. However, Kolb (2006, table 1) categorizes these samples as "basal," so we presume that they came from beneath the terrace fill. Both samples yielded rather old ages of 790 ± 70 and 720 ± 90 BP respectively. The first date has calibrated age ranges of AD 1043–1104 (8.6%), 1118–1302 (85.2%), and 1367–1382 (1.5%); the second date has calibrated age ranges of AD 1051–1082 (2.3%), 1129–1133 (0.2%), and 1151–1417 (92.9%). (Kolb used a weighted average of the two dates, 761 ± 66 BP, as an estimate for the first construction phase of the *heiau*, "Building Episode Popoiwi I" [1991, 236, fig. 6.6], but this is not statistically justifiable.) Given that the species of dated wood were not determined (hence allowing the possibility of an inbuilt age effect) and that the context of the samples was "basal" (i.e., underlying the actual *heiau* architecture), we believe that this age should be taken as a *terminus post quem* rather than an estimate of initial temple construction.

A sample (Beta-40368) recovered from 95 cm depth in test unit PP1 in "the massive *makai* terrace" (Kolb 1991, 234)—that is, the S end of feature H—also reported in Kolb (2006, table 1) as "basal," yielded a conventional age of 580 ± 70 BP, with a calibrated age range of AD 1284–1438. This too must cautiously be interpreted as a *terminus post quem* rather than the date of actual construction.

Finally, two samples (Beta-40370 and -40641) were recovered from test unit PP12 (at depths of 136–162 cm below surface), which was excavated inside feature T and penetrated down through the fill of the large terrace designated as feature R. Kolb (2006, table 1) reports both of these samples as coming from "firepit" features. The samples returned conventional ages of 390 ± 70 (calibrated 2σ age range of AD 1421–1646) and 520 ± 70 BP (calibrated age ranges of 1286–1492 [94.6%] and 1602–1612 [0.8%]). Kolb combined both dates with a weighted average of 455 ± 61 BP, but this procedure is statistically unjustifiable. A more cautious approach would be to take the younger date as more likely representative (given that both samples supposedly come from the same fire pit) of the true age of the combustion feature. This would then provide an estimate for the use of an early stage of the feature R terrace, sometime in the fifteenth to the seventeenth century.

A careful reconsideration of the contexts from which Kolb's five radiocarbon samples were obtained (combined with the uncertainties regarding possible inbuilt age effects from botanically unidentified samples) raises questions regarding his proposed chronology for Pōpōiwi Heiau. Rather than a long sequence beginning as early as the twelfth or thirteenth century, the first acceptable date for site use is that from the feature R terrace, dating to the fifteenth to the mid-seventeenth century. This, of course, is a subsidiary terrace to the west of the main *heiau* proper. Until further excavations are conducted and additional short-lived taxa, stratigraphically controlled and botanically identified, are dated, the architectural chronology of Pōpōiwi will remain uncertain.

Archaeological context. No intensive archaeological survey has been conducted around Pōpōiwi Heiau, so the larger settlement context remains undefined.

Viewshed. From the ESE corner of the main *heiau*, currently the only part free of thick vegetation, there is a fine view northward and eastward, from the conspicuous Puʻu Ahulili, 5 km away on the skyline virtually due N (az. 357°), down the Mikimiki ridge to cliffs, ~0.9 km distant, that fall to the sea beyond Mokulau Bay, at an azimuth of 87°. Beyond the sea to the SE and SSE, between azimuths ~135° and 170°, Hawai'i Island could be seen on clear days, with the summit of Mauna Kea at azimuth 143°. In the absence of vegetation, the peak of Haleakalā (9.3 km) would be clearly visible to the NW from all parts of the *heiau*, rising sharply behind the closer ridgeline (~4.5 km), with the summit itself at azimuth 317°. Even in the absence of trees, the higher ground immediately to the west of the site would block the view to WNW, W, and SW at least as far around as azimuth ~200°.

Orientation. The higher terrace on the NNW side and particularly the dais in feature B clearly show that this *heiau* was NNW-facing, but the direction faced by the canoe-prow ESE corner may also have been significant.

We were able to obtain orientation measurements of various intact wall segments during our traverse from the ESE corner through to the northern terrace. The ENE and WSW walls of feature B have northward azimuths of 344.5° and 339.3° respectively, for a mean of 341.9°, which is close to that of the ENE wall of feature D (342.1°). The single-course alignment subdividing B has an eastward azimuth of 71.5°, which is orthogonal to this mean direction to within 1°, while the SSE wall of feature E (az. 69.2°) is orthogonal to the wall dividing B and D. In other words, the elements most closely related to the sacred focus of the site were all oriented with azimuths close to 342°. The remaining measured NNW–SSE orientations vary from 333.1° for a segment of the ENE perimeter wall toward the NNW end of court G to 7.3° for the E wall of the northern terrace. The mean of these values, 350.2°, is quite close to the orientation of central part of the ENE perimeter wall, around the division between courts G and H, which is 353.8°. The inner face of the SSE wall of acute corner I has an eastward azimuth of 78.4°, while the ENE wall has a southward azimuth of 165.9°, so that the bisecting line has an azimuth of 122.2°.

The northerly orientation of the upper terrace features coincides with the highest apparent point on the ridgeline to the NNW, a rounded summit between the Haleakalā peak to the left (which appears lower because of its greater distance) and the conspicuous Pu'u Ahulili on the right (fig. 8.17). The declination range, +69.5° to +74.5° (to the nearest half degree), does not include any bright stars known to have been of significance to the Hawaiians and it is more likely that the topographic association was of greater significance here. To the SSE, the dividing walls align upon the slopes of Hualālai in the south of Hawai'i Island, but this is unlikely to have had

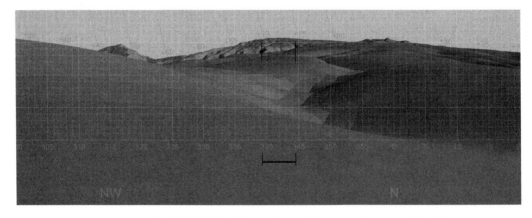

FIGURE 8.17. DTM-generated profile of the horizon to the north as seen from Pōpōiwi Heiau (KAU-324). The horizontal bar indicates the principal direction of orientation of the features in the northern terrace. (Graphic by Andrew Smith)

any significance both because this would involve sighting across the entire *heiau* and because this part of the Hawaiʻi Island, lower and farther away than the slopes of Mauna Kea and Mauna Loa, would rarely if ever have been visible in practice.

To the ENE, the northern terrace is aligned upon an unremarkable segment of the ridgeline about 1 km away, a few degrees to the south of the rising point of the June solstice sun and the Pleiades, which seems unlikely to have been significant.

It is likely that the ESE canoe prow corner of the *heiau* was intended to face Hawaiʻi Island, whose ruling chiefs were the target of Kahekili's attempted conquest (Kamakau 1961). But the line bisecting the corner angle has an azimuth of 122°, while the northern tip of Hawaiʻi Island (and the Hamakua coastline behind) bears between 130° and 140°, with Mauna Kea peaking at 143°, so the alignment, if intentional, was not accurate.

Additional remarks. Regarding the orientation, both the northerly and southeasterly directions encapsulated in the architecture are likely to have been determined topographically rather than astronomically.

KAU-333 Opihi (Walker Site 158)

Walker ([1930], 229–230) recorded this site as a "heiau at Opihi," describing it as "a large open platform of rough lava with several enclosures on it." Walker also provided a sketch map (reproduced in Sterling 1998, 182). Kirch briefly visited the site on March 31, 2013, making a GPS survey and general observations. Kirch and Alex Baer revisited the site on May 21, 2015, making a plane-table map as well as additional observations on wall orientations and viewsheds.

Walker ([1930], 229–230), referring to Thrum (1909), felt that this *heiau* had been built in two construction stages, the latter possibly associated with the tour of Maui temples undertaken by the young heir apparent Liholiho in 1801, when he was five years old, to rededicate the temples (Kamakau 1961, 187–188). This, however, appears to be just speculation on Walker's part, as Thrum refers only to Liholiho's rededications of the major *luakini heiau* (temples where human sacrifice was performed) of Loʻaloʻa, Pumakaʻa, and Kanemalohemo (1909, 47).

Location and topographic setting. KAU-333 sits at an elevation of about 160 masl on the lip of the very young Puʻu Nole ʻaʻā flow.

Architecture. The site is fairly complex, incorporating two discrete enclosures about 23 m apart, with the space between them consisting of an artificially leveled surface, roughly paved with ʻaʻā clinkers (fig. 8.18). The more inland (N) enclosure, designated KAU-333A, is a classic notched *heiau*, broadly cardinally oriented, with exterior dimensions of 17 m E–W by 14 m N–S. The notch is located in the NW corner. The well-built, core-filled walls are up to 2 m thick on the E and N sides. There is a formal entryway in the S wall, leading into the main court. An interior wall running N–S separates the interior of the structure into two discrete rooms. Branch coral was noted on the walls and inside the enclosure.

To the S of the notched *heiau* is a second enclosure (KAU-333B), skewed from cardinality by approximately 10° in the clockwise sense and measuring about 14 m E–W by 13 m N–S. The walls are core-filled, and the E and S walls have a second, buttressing terrace on the exterior. There is also a low terrace running along the interior face of the W wall. The interior has ʻaʻā clinker and ʻiliʻili paving, and branch coral was noted on the structure.

Excavation. Baer (2015, 173–182) excavated five test units within KAU-333, four of them in KAU-333B and one in the notched *heiau*, KAU-333A.

FIGURE 8.18. Plan of Opihi Heiau (KAU-333). (Based on plane-table survey by P. V. Kirch)

Dating and chronology. Four radiocarbon dates were obtained from Baer's test excavations, all on botanically identified materials (Baer 2015, table 5.7). The oldest date (AA-102212) is from the notched *heiau*, KAU-333A, with a conventional age of 380 ± 39 BP, which yields 2σ calibrated age ranges of AD 1441–1530 (55.7%) and 1531–1635 (39.7%). This date indicates construction of the notched *heiau* sometime between the mid-fifteenth and early seventeenth centuries. Of the dates from KAU-333B, the oldest (AA-102210), from the center of the enclosure, has a conventional age of 273 ± 38 BP, yielding calibrated age ranges of AD 1488–1604 (50.6%), 1610–1670 (37.4%), 1780–1799 (6.3%), and 1944–1950 (1.1%). A second date (AA-102209), from a test pit abutting the E wall, has an age of 122 ± 38, yielding calibrated age ranges of AD 1675–1778 (36.2%) and 1799–1942 (59.2%). The third date (AA-102211), from a test pit outside the S retaining wall, has an age of 81 ± 38 BP, yielding calibrated age ranges of AD 1682–1736 (26.4%) and 1805–1935 (69.0%). The multiple calibration intercepts make precise chronological assignment difficult, but the generally younger age ranges for the dates from KAU-333B suggest that it was constructed somewhat later than the notched *heiau*, KAU-333A.

Viewshed. Situated on the lip of a massive ʻaʻā flow, this *heiau* commanded a view southward out over a major portion of the Kaupō dryland agricultural field system. The view northward, however, may have been more significant, as this looks directly up to the rim of Haleakalā and to the major topographic feature of Kaupō Gap. The gap is framed, from this perspective, by the summit of Haleakalā (3,055 masl) on the left and by the peak of Kūiki (2,314 masl) on the right. Indeed, the main E wall of the notched *heiau*, KAU-333A, is oriented directly toward the low point between the two summits. The viewshed eastward is restricted by a low ridge in the foreground. Westward, however, there is a fine, broad view across most of Kahikinui District as far as the Lualaʻilua Hills.

Orientation. Based on the formal entryway on the S as well as the massive construction of the N and E walls, the notched *heiau* (KAU-333A) appears to be oriented to the N. The main E wall (running N–S) is oriented within 1° of true N–S, whereas the main S wall has an azimuth of 105° (thus actually running WNW–ESE). In contrast with the notched *heiau*, the rectangular enclosure (KAU-333B) may to be oriented to the E (given the elaborate buttressing of the E wall); its N and S walls have a mean axial orientation between 97° and 102°.

KAU-411

KAU-411 was discovered during an archaeological survey by Alex Baer and briefly visited by Kirch on October 5, 2013, when there was only sufficient time to make a few general observations and a rough sketch of the structure. Kirch and Baer revisited the site on May 20, 2015, when a plane-table map was made and detailed observations were taken of orientations and viewsheds.

Location and topographic setting. The site sits on a high ridge of the relatively old Kaupo mudflow formation, at an elevation of about 200 masl.

Architecture. The main structure consists of a large U-shaped enclosure open on the seaward side, whose orientation deviates from cardinality by roughly 20° in a clockwise sense (fig. 8.19). Its exterior dimensions are 24 m WNW–ESE by 24 m NNE–SSW. The core-filled walls are 2–4 m thick and average about 0.7–1 m high. There is a small room defined by three walls attached to the inside of the NNE wall. As seen in the plan (fig. 8.19), the main U-shaped enclosure abuts a smaller structure on the southwest. This smaller structure, whose walls are more collapsed, is actually a notched enclosure with two interior courts separated by two low, parallel

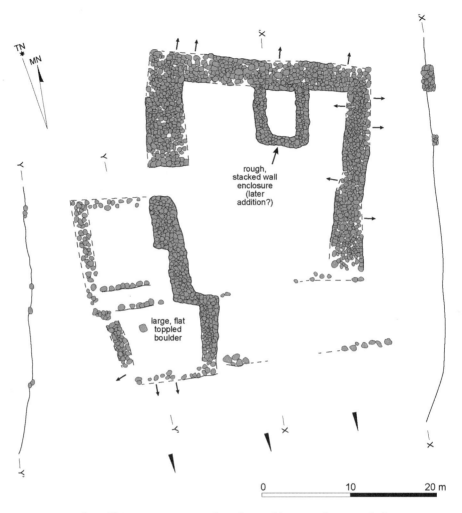

FIGURE 8.19. Plan of *heiau* KAU-411. (Based on plane-table survey by P. V. Kirch)

terrace facings. This smaller structure, whose condition suggests that it is older than the main U-shaped enclosure, appears to have been a notched *heiau* to which the large U-shaped enclosure was added.

Viewshed. This *heiau* commands one of the most superb viewsheds of any temple site in either Kahikinui or Kaupō. Perched on the lip of a prominent ridge, the view southward is a commanding sweep over the entire core of the Kaupō field system (the Pauku area), framed by Ka Lae o Ka ʻIlio to the southeast and by Puʻu Māneoneo to the southwest. (It is worth considering as well that when viewed from the perspective of someone laboring in the field system below the site, the *heiau* would have been highly visible up on this prominent ridge.) Northward, the view is equally stunning, as one looks directly up at Kaupō Gap, framed by the peaks of Haleakalā and Kūiki, with the slopes of the Kaupō fan's lava flows emanating out of Kaupō Gap in the middle distance. Westward, there is a clear view over most of Kahikinui as far as the Lualaʻilua Hills. The only direction in which there is not a view is eastward, because that view is blocked by the low ridge immediately adjacent to the *heiau*.

Orientation. Given the opening of the enclosure to the SSW, as well as the superb view up to Kaupō Gap and the twin mountain peaks, this *heiau* clearly is oriented to the NNE. The main ESE wall (running NNE–SSW) has a northward azimuth of 20°–21°, while the NNE wall (running WNW–ESE) has an eastward azimuth of about 109°. From the center of the NNE wall, the center of Kaupō Gap bears 16° while the high peak of Kūiki bears 19°, nearly the same as the azimuth of the ESE wall of the temple. This orientation to the mountain's prominent peak could well suggest that the temple was dedicated to Kū.

KAU-536 (Walker Site 147)

This site was first recorded by Walker ([1930], 218), who described it as "a walled heiau of good size," although he neglected to note its notched configuration. Kirch made a brief visit to the site on March 31, 2013.

Location and topographic setting. KAU-536 lies on a ridgetop about 150 m from the shore, at an elevation of about 50 masl. The substrate consists of the basanite of Loaloa lava flow.

Architecture. The structure consists of a notched enclosure, its orientation deviating from cardinality by about 14° in a counterclockwise sense. Its exterior dimensions are about 18 m NNW–SSE by 14 m ENE–WSW; the notch is in the ESE. The enclosing walls are quite massive, up to 2 m thick. The interior is sunken down nearly 2 m below the height of the walls. Thick vegetation obscured any internal features, although it is clear that there is a terrace or raised bench along the inside of the E wall.

Excavation. Baer (2015, 207–210) excavated three test units at KAU-536, two inside the enclosure and one in a flat area just outside the structure. Baer reports significant quantities of bone, shell, charcoal, coral, basalt flakes, and volcanic glass flakes from these test excavation units (ibid., table 5.19).

Dating and chronology. Three radiocarbon dates were obtained by Baer (2015, table 5.20) from his test excavations, all on candlenut endocarp (*Aleurites moluccana*). The first (AA-101581), from within the enclosure, has a conventional age of 379 ± 26 BP, yielding 2σ calibrated age ranges of AD 1446–1524 (64.8%), 1559–1563 (0.7%), and 1571–1631 (29.9%). The second sample (AA-101580), also from within the enclosure, has a date of 388 ± 26 BP, yielding calibrated age ranges of AD 1442–1522 (72.4%) and 1575–1625 (23.0%). The third sample (AA-102276) came from the exterior test pit, with an age of 350 ± 39 BP and a calibrated age range of AD 1456–1638. The calibrated age ranges associated with these dates indicate that KAU-536 was constructed relatively early in the sequence of Kaupō *heiau*, between the mid-fifteenth and early seventeenth centuries.

Viewshed. No detailed observations of the viewshed were made, but the most open views are to the east, toward Mokulau and the Kīpahulu coastline, and north to the ridgeline of Haleakalā.

Orientation. The structure is most likely oriented to the ENE, with an azimuth of about 76°.

KAU-994, Loʻaloʻa (Walker Site 143, Bishop Museum Site MA-A27-2)

Loʻaloʻa Heiau, the third-largest temple on Maui Island, is associated in Hawaiian oral traditions with *aliʻi nui* Kekaulike, who ruled the island from his royal center at Kaupō in the early part of the eighteenth century (Kirch 2010, 110–111). Kamakau (1961, 66) states that Kekaulike built Loʻaloʻa, which Kamakau calls a "war heiau (*mamala koa*)." Around 1802, Kamehameha I sent his son and heir, Liholiho (then only about five years old), on a tour "to be proclaimed in the heiaus of

Maui and Oahu" (ibid., 188; see also Thrum 1909; Sterling 1998, 176), which included Loʻaloʻa. It is likely that the last modifications to the temple were made during this visit by Liholiho.

Walker ([1930], 214; Sterling 1998, 173–174) surveyed Loʻaloʻa, producing for this structure one of his better site maps. He calls it "the great heiau of the district," characterizing the foundation as "a great open platform constructed of rough lava blocks built up in three step-terraces on the S.E. side and four on the N.E. side to a height of 30 feet." Soehren (1963, 87–88) also provides a description of the site. Kolb investigated Loʻaloʻa as part of his doctoral dissertation research (1991, 168–176), excavating a number of test pits and obtaining six radiocarbon dates. Kolb provides a schematic map of the structure (1991, fig. 5.8), but he significantly overestimated the size of the foundation, reporting it as 5,115 m² whereas the foundation as measured by GPS survey is approximately 4,200 m².

Ruggles visited Loʻaloʻa on April 8, 2002, undertaking a compass-clinometer survey, and returned three days later to attempt a theodolite survey, which had to be abandoned when deteriorating weather conditions prevented the acquisition of sun sightings (needed to determine absolute azimuths). Kirch made an initial GPS survey of Loʻaloʻa in 2003. Kirch and Ruggles visited the site together on March 24, 2008, undertaking both total station and GPS surveys, and Kirch returned to map the structure with plane table and alidade on July 27, 2008.

Location and topographic setting. Loʻaloʻa Heiau lies near the E edge of the Kaupō fan, about 70 m W of Manawainui Stream and overlooking the incised stream valley (fig. 8.20). At an elevation of 100 masl, it sits on a low rise of the massive basanite of Loaloa lava flow.

Architecture. Walker ([1930, 214]) gives the total length of the site as 510 feet (155 m),

FIGURE 8.20. Aerial view of Loʻaloʻa Heiau (KAU-994) from the north, showing the *heiau* in its landscape setting, situated immediately west of Manawainui Stream. (Photo by David Waynar)

whereas the extant structure measures about 105 m long, suggesting that a level area on the W, which Walker designated "Area E," was destroyed at some point after his 1929 survey. However, Walker referred to the walls in that area as being "modern," so it is likely that not much of significance has been lost.

In essence, Loʻaloʻa consists of a massive platform whose orientation deviates from cardinality by about 24° in a counterclockwise sense (fig. 8.21). It is elevated above the surrounding terrain on three sides, but most prominently on the ENE, where there is a succession of three stepped terraces rising to a height of about 7 m (fig. 8.22). The main platform is 105 m long by 45–52 m wide. The NNW side of the platform is also buttressed by a long, lower terrace. The platform seems to have been built over a natural rise in the ʻaʻā lava terrain but nonetheless incorporates an estimated 10,400 m^3 of fill.

There are a number of architectural features, which we describe here progressing from ENE to WSW (see fig. 8.21). Of particular note is the presence of 16 small rock cairns (*ahu*), arrayed along the SSE and ENE edges of the main platform; the cairns range from about 0.6 to 1.5 m in diameter and are up to 1 m high. When viewed from the surrounding terrain, the cairns give a crenellated appearance to the edge of the *heiau* platform. These cairns are quite likely associated with the final use of the *heiau* during Liholiho's tour in 1802.

The ENE portion of the platform, extending back about 25 m from the ENE edge, is level and was originally well paved with waterworn cobbles (Area A of Walker's [1930] map). A slightly raised rectangular area, faced with a single-course alignment on the WSW side, was probably the foundation for a thatched superstructure. Another probable superstructure foundation is situated closer to the NNW edge of the platform. There are as well four pits or depressions, ranging from 1 to 2 m in diameter; these are possibly postholes for large wooden images. Near the SSE edge of the platform there is a small circular enclosure, about 3 m in diameter. Finally, a large waterworn upright stone is located near the center of the leveled and paved area; the substantial lichen growth on this stone and the adjacent paving stones suggests that it has not been disturbed for a long period of time.

Approximately 40 m WSW of the ENE edge of the platform a well-built rock wall about 1 m high and 23 m long runs perpendicularly across the platform, separating the *heiau* platform into two sectors. Walker ([1930, 214]) opined that this wall "marks off the limits of the ceremonial part of the structure."

To the WSW of the dividing wall, the platform surface slopes down slightly. A small rectangular enclosure is partly attached to the WSW side of the dividing wall, while about 16 m WSW of the wall is a low terrace, faced on three sides, probably the foundation for a thatched structure. Two waterworn upright stones and a small pit or depression are also located in this area (referred to as Area B in Walker's map).

The most westerly part of the site still extant (Area C on Walker's map) is again somewhat lower in elevation than those areas to the ENE and contains two abutting enclosures and a smaller nearly square enclosure. There is also a sunken rectangular area measuring about 4 by 14 m (probably a house foundation) and several pits. Walker ([1930, 214]) reports that in 1929 he saw in this area "some rusty cooking pots and a sharpening stone and two stone pestles." He was told by Joseph Marciel of Kaupō Ranch that "32 years ago [i.e., ca. 1897]...there was a very old Hawaiian living here, possibly the last of the kahus or keepers of the heiau." This information suggests that the enclosures in this area were postcontact additions.

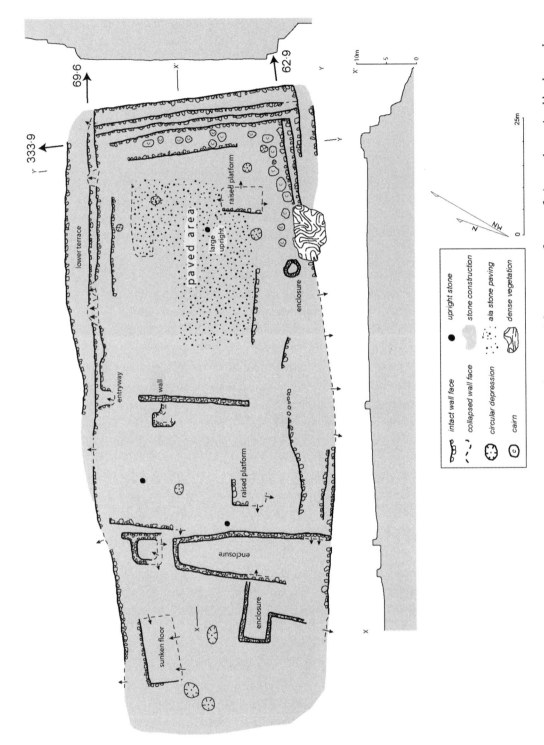

FIGURE 8.21. Plan of Loʻaloʻa Heiau (KAU-994), showing the azimuths of intact straight segments of terrace facing as determined by the total station survey. (Based on plane-table survey by P. V. Kirch)

FIGURE 8.22. View along the ENE side of the platform at Loʻaloʻa Heiau (KAU-994), toward the SSE, showing the stepped terraces. (Photo by Clive Ruggles)

Walker ([1930]) also mentions the pits in this area (which he calls Area D), along with *kukui* (candlenut), pride-of-India, and mango trees; the mango still remains. As noted earlier, a further leveled space to the WSW, called Area E by Walker, now lies outside the fenced area and has been disturbed by pasture bulldozing.

Excavation.   Kolb excavated 21 test pits at Loʻaloʻa (with a total area of 25.5 m²). Unfortunately, Kolb (1991, 505–508) offers only minimal details of these excavations; no stratigraphic sections are described, and faunal materials and other excavated objects are summarized only in the most general terms.

Dating and chronology.   Kolb obtained eight radiocarbon dates from wood charcoal samples recovered during his test excavations. None of the samples was botanically identified, and their stratigraphic contexts are difficult to discern precisely due to the lack of stratigraphic details or cross sections in Kolb (1991). The two oldest dates (Beta-40366 and -40639), with ages of 320 ± 70 and 350 ± 70 BP, are both said by Kolb to "represent a concentration of charcoal located at the base of LL-EL2, which was situated directly upon bedrock" (ibid., 232). Kolb used a weighted average of these two dates, 335 ± 61 BP, yielding a 2σ calibrated age range of AD 1445–1660, to estimate the date of "building element LL-EL2." Kolb reports that this building element is "situated directly atop the underlying bedrock of the west terrace" (175).

Another sample (Beta-40637), with a conventional age of 280 ± 80 BP (2σ calibrated ranges of AD 1443–1695 [74.0%], 1726–1814 [14.8%], 1838–1843 [0.3%], 1852–1868 [0.9%], and 1918–1950 [5.4%]), is reported as coming from "near the bottom of Test Unit LL9 [depth of 1.22 m], near the base of this subsurface occupation floor, slightly above bedrock," and associated with "Building Element LL-EL1" (Kolb 1991, 232). Kolb's map shows test pit LL9 as situated near the middle of the ENE part of the platform, and he describes building element LL-EL1 as "comprised of the original ʻaʻā and soil hillock which was paved with ʻiliʻili stones" (ibid.). Another sample (Beta-40364) from test pit LL9, nearer the surface (21–36 cm deep), yielded a much younger age of 110 ± 50 BP (calibrated age ranges of AD 1674–1778 [35.8%] and 1799–1942 [59.6%]). Kolb associates this date with his building element LL-EL3, which along with LL-EL4 "represent the largest modification phase of Loaloa," when the substantial ENE platform was constructed.

Two radiocarbon samples (Beta-40367 and Beta-40365) came from test unit LL10, also in the ENE part of the main platform. According to Kolb, these were associated with building element LL-EL4, "the large rock terraces which were added in order to expand the size of the east platform" (1991, 233). These samples yielded conventional ages of 190 ± 70 BP (from a depth of 102 cm) and 90 ± 70 BP (from a depth of 63 cm). Kolb combined both samples with a weighted average of 140 ± 61 BP, which yields calibrated age ranges of AD 1665–1785 (41.4%) and 1793–1950 (54.0%).

The final two samples (Beta-40638 and -43014) came from the "base of Test Unit LL19, excavated in the rubble fill of Building Element LL-EL5, a subsurface ʻaʻā pavement situated upon bedrock" (Kolb 1991, 233) underlying the rock terrace in the WSW part of the structure (Walker's Area D). Both samples are reported to come from a charcoal concentration at a depth of 95 cm, but they yielded quite disparate ages of 320 ± 80 BP (calibrated age ranges of AD 1431–1683 [87.1%], 1736–1805 [6.5%], and 1936–1950 [1.8%]) and 110 ± 60 BP (calibrated age ranges of AD 1669–1780 [37.6%] and 1798–1945 [57.8%]). Kolb combined these two dates for a weighted average of 195 ± 60 BP, a statistically unjustifiable procedure.

Kolb (1991, 239–240, figs. 6.6 and 6.7) interpreted his radiocarbon dates in terms of three "building episodes." Episode Loaloa I, between AD 1478 and 1651, marked initial construction at the site. This was followed, sometime after AD 1669, by episode Loaloa II, "a period of major modification," including the "addition of the massive rock terraces of the east platform" (239). Finally, Loaloa III, from AD 1778–1955, was that of "post-Contact modification" (240).

Because Kolb's radiocarbon dates were on unidentified wood charcoal rather than identified short-lived taxa, we cannot rule out inbuilt age in one or more of the samples. Nonetheless, the eight dates are mostly consistent with the reported construction phases (as best as can be determined in the absence of stratigraphic sections), with the exception of the two disparate dates from test unit LL19. In our view, the dates support an interpretation of initial construction at Loʻaloʻa sometime between the late fifteenth and the sixteenth centuries, followed by a major phase of expansion (the massive E platform and terraces), mostly between the mid-seventeenth and early eighteenth centuries. It seems likely that this major expansion phase was associated with the reign of Kekaulike, given the historical reference to his construction of war temples during his period of residence in Kaupō.

Archaeological context. Unfortunately, the terrain surrounding Loʻaloʻa Heiau appears to have been bulldozed for "pasture improvement" at some time in the past, obliterating any structures in the immediate vicinity of the site. There has also been no systematic archaeological survey of the area surrounding the *heiau*.

Viewshed. The northern half of the view from Loʻaloʻa is truly panoramic. To the NNW and N, one looks almost directly up the Manawainui Stream valley, with the ridgeline behind it (about 7 km distant) reaching an altitude of 15.8° at azimuth 347.6°. From the NNE to the E, a long, straight ridgeline 2–3 km distant extends down to meet the sea horizon at azimuth 86°. To the NW, the sharp peak of Haleakalā, 8.7 km away, is prominent at azimuth 321.7°, falling away gradually toward the W until it disappears behind a closer ridge (~500 m away) around azimuth 277°. The view is also restricted by rising local ground to the WSW and SW: this would be true between azimuths ~220° and ~265° even in the absence of vegetation. Hawai'i Island extends from azimuth 130° to 172°, with the peaks of Mauna Kea, Mauna Loa, and Hualālai at visible at 143°, 157°, and 166° respectively.

Orientation. With its superbly constructed triple-terraced facade, 7 m high, at the eastern end, there can be no doubt that Lo'alo'a Heiau faced ENE.

The facing of the long NNW side of the platform is intact at several points along almost its entire length, enabling a very good estimate to be obtained of the intended orientation. EDM readings were taken at six locations roughly equally spaced along a 36 m stretch, yielding a best-fit eastward azimuth of 69.6°. Individual segments yielded orientations varying from 68.1° to 71.2°, reflecting some sinuosity (whether present from the outset or arising because of slumping of the facade over time) that is evident on the plan; but the facing is not bent or curved overall. The SSE face exists intact for some 15 m westward from the ESE corner, and four points along the base of the first (highest) facing on the top of the second terrace yielded a best-fit orientation of 62.9°. This gives a mean orientation for the ENE–WSW axis of 66.3°.

The ENE face is well preserved, although the lower part is overgrown. The lowest facing is straight, but the intermediate faces bow outward in the center. Readings were taken at seven points roughly equally spaced along the base of the first (highest) facing—that is, along the back of the second terrace down—yielding a best-fit northward orientation of 333.9°. Thus, the orientation of the ENE face is not perpendicular to the main axis but deviates from it by some 2.4°. Possibly, the ENE face was intended to be perpendicular to the SSW face—the deviation here is only 1.0°. However, it also seems possible—given the size and evident care taken in the construction of this immense monument—that the orientation had significance in its own right.

Lo'alo'a Heiau faces a topographically unremarkable stretch of the eastern ridgeline. However, it also faces precisely the direction of June solstice sunrise (fig. 8.23). The horizon altitude in the axial direction (az. 66.3°) is 3.8°, which yields a declination of +23.4°; this is just below the center of the rising sun at the June solstice. Furthermore, the declination span of the Pleiades rising in AD 1600 is +22.6° to +23.1° (as shown in fig. 8.23), rising to +23.0° to +23.5° in AD 1700 and +23.3° to +23.8° in AD 1800. Thus, this *heiau* also precisely marked the rising position of the Pleiades from the later part of the seventeenth century through to the earlier part of the nineteenth. While it is not possible to distinguish between these two possibilities purely on the basis of the orientation data (and there is no reason why both should not have been significant), there can be absolutely no doubt that the principal orientation of this great *heiau* was astronomically determined.

In the northward direction, the NNW–SSE axis is aligned upon a relatively low and flat part of the ridge to the right of the Haleakalā peak (az. 334.0°, alt. 13.5°, dec. +64.0°, all to the nearest half degree). The star Hiku Kahi, or Dubhe (α UMa)—the first of the seven stars in Nā Hiku, or Ursa Major (Johnson, Mahelona, and Ruggles 2015, 150)—set close to this spot (dec. +63.8° in AD 1600, +62.8° in AD 1800). In the opposite direction the alignment is about halfway down the left-hand slope of Mauna Loa. Yet the direction orthogonal to the main axis, 156.3°, would have been almost exactly upon the peak of that mountain, thus undermining any sugges-

FIGURE 8.23. DTM-generated profile of the horizon to the ENE and E as seen from Loʻaloʻa Heiau (KAU-994). The horizontal bar indicates the *heiau*'s principal direction of orientation. The solid shaded line indicates the rising path of the sun at the June solstice; the dash-and-dot line indicates the rising path of the sun at the equinoxes; and the dotted line indicates the rising path of the Pleiades in AD 1600. (Graphic by Andrew Smith)

tion that the orientation of the ENE facade might have been deliberately skewed from the orthogonal so as to face a conspicuous feature on Hawaiʻi Island. It therefore seems more likely that this facade was constructed roughly orthogonally to the SSE facade, although it remains possible that the broad orientation toward Hawaiʻi Island was held to be significant. Another possibility is that the orientation of the facade (az. 154°, alt. 1°, dec. −57°) was determined by the rising of the Southern Cross, or Newe (ibid., 200), with whose rising point it coincides (dec. −61° to −55° in AD 1600, −62° to −56° in AD 1800).

Additional remarks. In addition to the huge importance of Loʻaloʻa Heiau in other respects, the main orientation of this immense *heiau* seems to have been carefully determined astronomically, in relation to the rising of the Pleiades, the rising of the June solstice sun, or possibly both.

KAU-995, Kou (Walker Site 155)

This large, impressive *heiau* was reported by Walker ([1930], 226), who provided a reasonably accurate plan (reproduced in Sterling 1998, 180). Kirch first visited the site on October 18, 2007, making a GPS survey, and later mapped the site with plane table and alidade on April 28, 2008. Ruggles and Kirch made a total station survey on March 24, 2008.

Location and topographic setting. Kou Heiau lies about 75 m from the shoreline at an elevation of about 10 masl, on the east side of a small bay at Waiū. The geological substrate here is the very young Puʻu Nole ʻaʻā lava flow.

Architecture. The architecture of Kou is quite different from that of other southeastern Maui *heiau*, consisting of a large three-sided rectangular enclosure, with a high altar on the eastern side, that is completely open in the opposite direction (fig. 8.24). The orientation of this main enclosure deviates from cardinality by about 18° in a clockwise sense. It has exterior dimensions of approximately 70 m WNW–ESE by 26 m NNE–SSW. The walls are massive, core-filled in construction technique, up to 4 m wide, and as much as 3 m high (fig. 8.25). In addition, the NNE wall makes an abrupt jog at the WNW end and then runs inland to the N for 55 m; it has a formal entryway gap midway along this N extension. Outside the *heiau*, about 6 m ENE from the jog in the NNE wall, is a large, distinctive ‘a‘ā boulder 3–4 m in diameter. The manner in which the *heiau* wall jogs around the boulder suggests that it may have had some sacred significance.

The ESE part of the main enclosure, which is clearly the altar (*ahu*), has a stepped upper terrace rising to a height of 3.6 m above the adjacent terrain. Immediately WNW of the high ESE

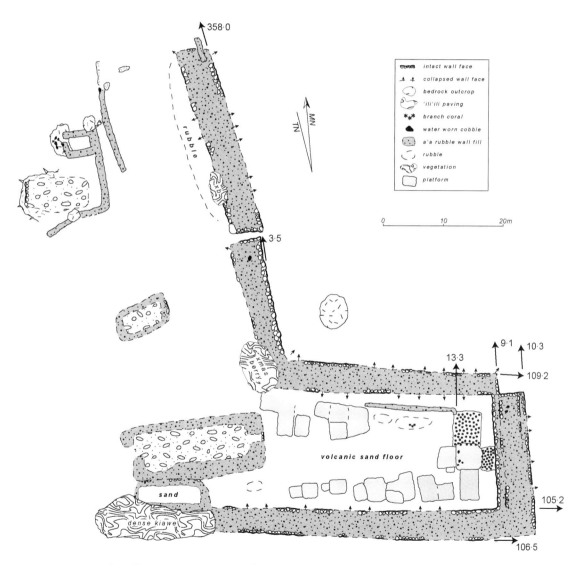

FIGURE 8.24. Plan of Kou Heiau (KAU-995), showing the azimuths of intact straight segments of wall facing and the altar platform edge as determined by the total station survey. (Based on plane-table survey by P. V. Kirch)

Kaupō • 327

FIGURE 8.25. View of the exterior of the S wall at Kou Heiau (KAU-995), with Kaupō Gap in the distance. (Photo by P. V. Kirch)

wall are 5 low, well-faced platforms in a sort of cruciform arrangement. Two of these are covered in large quantities of branch coral fragments. Not all of these platforms appear to be of the same age, and at least 2 may be historic-period graves attached to earlier platforms. Another 16 rock-filled platforms are arrayed along the inside of the main enclosure, mostly on the SSW but also the NNE side; these are almost certainly postcontact-period burials (as also noted on Walker's [1930] plan map).

At the WNW end of the main enclosure there is a rectangular enclosure whose low walls of ʻaʻā rubble have largely collapsed; this measures 20 by 5 m and is skewed so as to be closer to true cardinal orientation. The enclosure, which has an entryway on the E and an interior paved with ʻiliʻili gravel, was presumably the foundation for a thatched superstructure. Another, somewhat smaller enclosure (10 by 3 m) with walls up to 1.75 m high and lacking an entrance is attached to the inside of the SSW wall of the main enclosure. This was also likely roofed over when the heiau was in use. Finally, there are traces of a rectangular paved area of ʻiliʻili gravel to the west of the long NNE wall extension, again probably marking the foundations of a thatched structure. Several walls of stacked ʻaʻā cobbles in the NNW part of the site are most likely postcontact constructions.

Dating and chronology. Kirch, Mertz-Kraus, and Sharp (2015, table 1) reported ^{230}Th dates on five coral samples from Kou Heiau, three from surface contexts and two from architecturally integral contexts. Two early dates came from corals obtained from collapsed segments of the temple's NNE wall, where they were part of wall fill. Sample KOU-CS-5A yielded a date of AD 1099 ± 8, while KOU-CS-4 had a date of AD 1396 ± 7. Although both samples came from the same architectural context, we believe that it is unlikely that either sample dates the time of construction of the massive NNE wall. Kou has presumably had been a locus of ritual activity from a very early time (see, e.g., the oral tradition recounted by Kamakau [1991, 112] regarding the arrival of Kāne and Kanaloa from Kahiki [Tahiti] and opening a freshwater source at Kou), during which coral offerings were likely made at this location. A large quantity of lava rubble was collected and heaped up to construct the NNE wall, presumably incorporating these and other pieces of coral that were already present on the site.

In contrast to the two wall-fill samples, the three surface context samples all suggest a relatively late age for the main temple architecture. Sample KOU-CS-3, with a date of AD 1634 ± 9, comes from the main altar platform. This date would be coincident with the reign of either Kamalālāwalu or Kauhi-a-Kama, two Maui rulers preceding Kekaulike. Samples KOU-CS-1 and KOU-CS-2 have dates of AD 1779 ± 5 and 1794 ± 4 respectively. These indicate very late continued use of the temple, into the period following initial European contact (AD 1778), a period notable for interisland warfare between the Maui and Hawai'i kingdoms, with several major battles occurring at Kaupō (Kamakau 1961). Continued offerings at a major *luakini* temple such as Kou during this period are consistent with the historical record.

Archaeological context. A thorough archaeological survey of the area around Kou Heiau has yet to be conducted. However, about 50 m SW of the *heiau* there is a double-chambered enclosure with a notched plan, which may be a small notched *heiau* or fishing shrine. Unfortunately, the structure has had considerable modification owing to its use as a campsite by modern fishermen.

Viewshed. The broad sweep of the Haleakalā ridgeline is especially impressive from this site, extending in a wide dome all the way from Alena Point (15 km away) in the WSW (az. ~253°), round to the NE (az. 50°). The Luala'ilua Hills stand out in the distance on the lower western slopes, peaking at azimuths 266° and 269°. The Kaupō Gap is a spectacular feature, creating a large dip in the high horizon around azimuth 13°. A series of local ridges, between 300 and 800 m from the site, break the distant skyline to the ENE, between azimuths ~50° and ~70°, and the lower slopes, descending down to Ka Lae o Ka 'Ilio in the E (az. 100°) are formed by a ridgeline 2–2.5 km distant. Hawai'i Island extends low above the sea horizon from azimuth 130° to azimuth 167°, its highest peak, Mauna Kea, reaching an altitude of 1.7°.

It should be noted, however, that from the interior of the main enclosure, the horizon beyond the high ESE wall is not visible, with the wall itself forming the horizon); likewise, the sea horizon to the south and part of the horizon to the north are hidden by the side walls, depending on the position of the observer.

Orientation. Kou's orientation is unusual among southeastern Maui *heiau* in being to the south of east. Best-fit orientations were obtained from several measurements of intact outer facings on the NNE and SSW walls, yielding eastward azimuths of 109.2° and 106.5° respectively and a mean axial orientation of 107.8°. A shorter segment of the inner face of the SSW wall yielded 105.2°. The main outer facing of the ESE wall has a northward orientation of 10.3°, while an intact segment of the upper terrace yielded 9.1°, for a mean of 9.7°, more than 8° away from the direction orthogonal to the main axis (17.8°). The western edge of the main altar platform has an orientation of 13.3°, still more than 4° off orthogonality. Short facing segments on the outer (E) side of the NNE wall extension, S and N of the entryway gap, yield northward azimuths of 3.5° and 358.0° respectively.

To the ESE, the *heiau* faces open sea. The segment of horizon bracketed by the NNE and

SSW wall orientations corresponds to a declination range of −15.7° to −18.1°, with the main axial orientation upon −16.9°. This is close to the rising position of Sirius (α CMa, dec. −16.2° in AD 1600, −16.4° in AD 1800), known to Hawaiians as Hōkūhoʻokelewaʻa, the canoe-guiding star (Johnson, Mahelona, and Ruggles 2015, 154, 211). The horizon in this direction would have been visible to a priest standing on the stepped upper terrace of the altar, atop the ESE wall, but (as already noted) it was not visible from the interior. From a position to the WNW of the altar, about 10 m from the ESE wall, this wall would have had an altitude of approximately 10°, and Sirius would eventually have appeared above this several degrees to the right of its center, at around azimuth 112° (fig. 8.26).

FIGURE 8.26. Altar and high ESE wall of the main enclosure at Kou viewed from the interior along the axis. Photograph overlaid on a sky visualization for AD 1600 generated by Stellarium, showing the canoe-guiding star (Hōkūhoʻokelewaʻa, or Sirius) rising with a declination of −16° and the stars of Nā Kao (Orion's Belt) and the remainder of Orion high in the sky. Note that from a position close to the altar the altitude of this wall would be much higher. (Composite of photo by P. V. Kirch and graphic by Clive Ruggles)

Although Sirius is known as the canoe-guiding star, the orientation of Kou to the ESE is not a direction that correlates with Polynesian voyaging to "Kahiki." However, Noyes (2013, table 1) points out that other Hawaiian names for Sirius (such as Loaʻa Ke Kāne, meaning "procure the man") refer to human sacrifice. Given that Kou is a large temple likely to have functioned as a *luakini*, this association with sacrifice may be significant.

To the WNW, the equinox sun sets directly into the northern, upslope Lualaʻilua summit, but the principal axis of the *heiau* (az. 288.0°, alt. 6.5°, dec. +19.0°, to the nearest half degree) aligns upon an unremarkable stretch of the sloping ridgeline between ~2° and ~4° to the left of the setting position of the Pleiades and the June solstice sun, close to the setting position of Ka Nuku o Ka Puahi (Hyades-Aldebaran, dec. +14.5° to +18.0° in AD 1600 and +15.0° to +18.5° in AD 1800, to the nearest half degree) and Hōkūleʻa (Arcturus, α Boo, dec. +21.5° in AD 1600 and +20.0° in AD 1800, to the nearest half degree) (Johnson, Mahelona, and Ruggles 2015, 156).

The ESE wall is oriented northward to the left of the large dip at azimuth 13° and southward upon open sea. The declinations concerned, about +77° to the N and −68° to the south, are of no obvious significance. The NNE wall extension is oriented northward toward a uniformly sloping part of the Haleakalā skyline, devoid of any particularly conspicuous features. The southern and northern parts of the wall align upon points with declinations +82.2° and +83.7° respectively; again, there are no obvious astronomical correlates.

KAU-996 Paukela (Walker Site 141)

This *heiau* was originally recorded by Walker ([1930], 211) as "heiau at Paukela." Walker states that "the whole hill-top has been used as a heiau," and his description suggests a combination of both residential and ritual features. Kirch briefly visited the site on March 31, 2013.

Location and topographic setting. The site lies on gently sloping terrain about 80 m from the shoreline, at an elevation of about 30 masl. The substrate consists of the relatively young basanite of Loaloa lava flow.

Architecture. The main structure evident was a substantial terraced rectangular platform, measuring about 21 by 10 m, with its longer axis bearing ENE–WSW. It is well faced (1–2 m high) on the ENE side. Some large, flat paving stones were noted on top of the platform.

Viewshed. There are an excellent views to the NE, ENE, and E, down to the point and bay at Mokulau and beyond to the Kīpahulu coastline.

Orientation. The structure appears to have been oriented to the ENE, the azimuth of the central axis being about 57°.

KAU-999 Keanawai (Walker Site 157)

This structure was first recorded by Walker ([1930], 228) as a *"heiau* at Keanawai." As Walker observed, the structure has "an unusual plan of two enclosures, one at each end." Kirch visited the site on March 31, 2013, making a quick GPS survey and observations. Kirch and Alex Baer revisited the site on May 20, 2015, when a plane-table map was made along with observations on viewsheds and orientation.

Location and topographic setting. Site KAU-999 lies on a tongue of the young Puʻu Nole ʻaʻā flow at an elevation of about 160 masl.

Architecture. The structure consists of two enclosures, each roughly 10 m square and skewed from cardinality by approximately 10° in a clockwise sense, joined by a core-filled wall

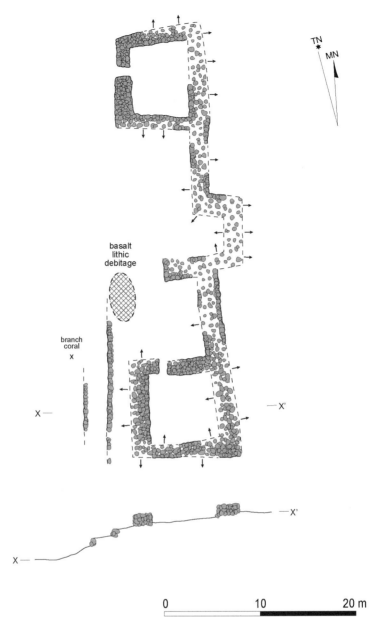

FIGURE 8.27. Plan of Keanawai Heiau (KAU-999). (Based on plane-table survey by P. V. Kirch)

(up to 2 m wide and about 26 m long) that connects the SE corner of the N enclosure with the NE corner of the S enclosure (fig. 8.27). The N enclosure has a formal entryway in its W wall, while the S enclosure has a formal entryway in its N wall. A large terrace lies between the two enclosures. The center of the connecting wall is interrupted by a formal, U-shaped niche about 5 by 4 m. Two secondary terraces buttress the S enclosure on the steep W face descending down into a swale.

Excavation.   Baer (2015, 239–247) excavated four test units in KAU-999, two in each of the walled enclosures.

Dating and chronology. Baer (2015, table 5.30) obtained three radiocarbon dates from his test excavations, all on botanically identified, short-lived taxa. Two samples were recovered from unit 2, in the northern enclosure. A basal sample (AA-102213) yielded a conventional age of 514 ± 60 BP and a calibrated 2σ age range of AD 1296–1477. A second sample (AA-102214) from higher up within the stratigraphic section (and more likely to be associated with the construction of the enclosure itself) had a date of 211 ± 38 BP, yielding calibrated age ranges of AD 1530–1538 (0.4%), 1634–1695 (29.4%), 1726–1814 (46.9%), 1837–1844 (0.4%), 1852–1868 (1.0%), and 1917–1950 (17.2%). A sample from unit 4 in the southern enclosure (AA-102215) yielded a conventional age of 203 ± 39 BP and calibrated age ranges of AD 1641–1697 (25.5%), 1725–1815 (47.8%), 1835–1878 (3.8%), and 1917–1950 (18.2%). The consistency between the latter two samples suggests that the main structure was most likely constructed sometime after the mid-seventeenth century.

Viewshed. As with KAU-333, site KAU-999 overlooks an area of the Kaupō dryland agricultural field system downslope from the *heiau*. However, the view to the S is largely obscured by a low ridge about 100 m distant. The most prominent view is undoubtedly to the N, where two peaks of Haleakalā and Kūiki frame Kaupō Gap. To the W there is a moderately good view of Kahikinui, as far as the Luala'ilua Hills.

Orientation. The orientation of this unusual structure is not clear. The most developed architecture is on the E side of the site, and our original impression was that it is E-oriented. It may be significant, however, that the main N–S-trending axis of the site is oriented toward Kaupō Gap and that the buttressing terraces below the S enclosure point directly to the peak of Kūiki. In this case, the orientation could be to the N.

Additional remarks. Although Walker ([1930]) called this site a *heiau*, we are not entirely convinced of its ritual function. It certainly does not correspond in form with most other southeastern Maui *heiau*. The alternative interpretation that it was an elite residence should not be discounted.

KAU-1001

This site, which was originally likely to have been quite extensive and prominent, now consists only of some remnant foundations; we record it here for the sake of completeness. The site was not recorded by Walker ([1930]) or Sterling (1998). Kirch has visited it numerous times and made a few notes and GPS observations on June 8, 2003.

Location and topographic setting. The site is situated on the southern headland of Mokulau Bay, about 100 m E of Huialoha Church. The remnant foundation alignments are perched at the edge of the lava cliff, 10 m high, that drops down to a boulder beach.

Architecture. All that remains of the structure today are some basal foundation alignments constructed of large waterworn boulders (~50 cm in diameter), which must have been hauled up from the boulder beach below the site. Two nearly parallel alignments (northward azimuth about 0°, corrected for magnetic declination—i.e., due N–S are situated just back of the bluff edge; these appear to be the lower courses of what was originally a higher terrace facing. To the south, near the point, are two set slabs next to a third boulder, which has an *X* incised in it (a probable nineteenth-century land survey marker). Just to the E of these stones (which may mark a corner of the original *heiau*) are two parallel rows of stones about 60 cm apart that appear to be the base of a core-filled wall now robbed of its upper courses; this may have been a south wall of the original *heiau*. There is a great deal of *'ili'ili* gravel scattered over the area west of the main alignments; a number of basalt flakes were also noted. It is impossible to determine the area of the original structure, but it is likely to have had N–S dimensions of at least 60 m.

The most likely explanation for the removal of most of the stones making up the original *heiau* would have been construction of the nearby Huialoha Church and the substantial stone wall that encloses the church and its graveyard.

Viewshed. From this prominent headland there is a superb view to the east, with the Kīpahulu coastline and headland of Mokuʻia Point visible. There is also an uninterrupted view of the Haleakalā ridgeline and up into Kaupō Gap.

Orientation. It is impossible with any certainty to determine the orientation of the original structure, although the approximately 0° bearings of the two remnant alignments suggest that it was close to cardinal orientation.

Epilogue

It is hard to believe now that, as late as the early 1970s, considerable doubts lingered among anthropologists and other scholars that the Pacific islands had been discovered and settled by people who voyaged purposefully out into uncharted seas, putting their lives in the hands of highly skilled and knowledgeable navigators. The devastating effects of two centuries of European colonization and acculturation had resulted in the loss of much indigenous knowledge of astronomy and navigation. Few recalled that when Captain James Cook had visited Tahiti in 1769, the Ra'iatean priest and navigator Tupaia had astounded the British captain with his knowledge of the star paths and voyaging routes to more than seventy islands. By the mid-twentieth century, only a handful of indigenous Pacific navigators, such as Mau Piailug of Satawal and Tevake of Taumako, still possessed such traditional knowledge. Academics argued whether the Pacific islands might simply have been discovered by a gradual accumulation of accidental, "drift" voyages.

In 1972, David Lewis published his pathbreaking book, *We, the Navigators*, documenting the noninstrumental methods of wayfinding once used by Polynesian and Micronesian canoe navigators. At the same time in Hawai'i, renewed interest in ancient Polynesian voyaging led to the formation in 1973 of the Polynesian Voyaging Society and to the construction of a replicated double-hulled canoe named *Hōkūle'a*. The year 1975 saw the publication of *Nā Inoa Hōkū*, a compilation of Hawaiian and Pacific star names and sky knowledge by Hawaiian scholars Rubellite Kawena Johnson and John Kaipo Mahelona, a work that attempted "to document what remains of a once-flourishing mastery of celestial navigation by accumulating the star lore which has managed to survive centuries of meagre regard" (Johnson, Mahelona, and Ruggles 2015, 3). Just a year later, *Hōkūle'a* made its epic maiden voyage from Hawai'i to Tahiti and back, navigated by Mau Piailug, providing an irrefutable demonstration of effective long-distance ocean wayfinding using the stars, ocean swells, and other natural indicators. Subsequent voyages of *Hōkūle'a* between most of the archipelagoes of Polynesia have continued to demonstrate the voyaging capabilities of ancient Polynesian watercraft, as well as the accuracy with which a wayfinder well trained in noninstrumental navigation could take a canoe from island to island (Finney 2003). Four decades on, the reality of Polynesian and Micronesian navigation skills and how these were used to colonize the Pacific is now supported by an abundance of ethnohistorical, archaeological, and linguistic evidence (e.g., Kirch 2000, 2012; Kirch and Green 2001).

Just as scholars once doubted the abilities of Pacific navigators to sail effectively between islands, archaeologists in Hawai'i and elsewhere in Polynesia were for decades reluctant to consider the possibility that *heiau* and *marae* had any associations with astronomical knowledge or were constructed and oriented to take account of anything other than mundane topography. The idea that *heiau* might have been positioned so as to take account of the appearance of the sky or, indeed, that *heiau* shared any common orientation patterns had been flatly rejected, yet without ever really being empirically tested. As our surveys in southeastern Maui have shown, these old assumptions must likewise now be rejected.

Our work in Kahikinui and Kaupō underlines the prominent place occupied by the sky in Hawaiian history, quite apart from star knowledge used in navigation and voyaging. It is in some respects ironic, then, that tensions have arisen in recent years regarding modern astronomers' desire to build ever more powerful telescopes (such as the proposed Thirty-Meter Telescope on Mauna Kea) on the highest mountains of Hawai'i and Maui. Both Polynesian navigators and twentieth-century astronomers have, in their different ways, hugely expanded the boundaries of human knowledge and endeavor, and there are common concerns—for example, to preserve dark starry skies free from light pollution. Perhaps the roots of a consensus may yet emerge from the unique relationship that exists in the Hawaiian Islands between cutting-edge science and technology and indigenous culture through the medium of the sky and astronomy.

What the revival of long-distance voyaging showed to be possible, archaeology has now confirmed as a part of the Polynesian historical legacy. The revival of Pacific wayfinding by the Polynesian Voyaging Society sparked a complete rethinking of the history of Polynesian exploration and settlement, spurring archaeologists to employ new methods to empirically trace the numerous contacts and exchanges between distant islands and archipelagoes, such as through the geochemical analysis and sourcing of stone adzes. Such symbioses between living tradition and historical research should be encouraged whenever possible. It is our hope that the research we have been privileged to carry out in Kahikinui and Kaupō will similarly inspire and encourage new avenues of inquiry into the rich connections between Hawaiian culture and the heavens.

As contemporary practice develops, it makes constant reference to the past. At the same time, knowledge and awareness of past practices help to generate respect for, and hence help to protect and preserve, the living tradition. Looking backward provides a solid basis for looking forward, while looking forward provides the motive for looking backward. An old Hawaiian saying sums it all up: *I ka wā ma mua, ka wā ma hope.* From the way of the past comes the way forward.

GLOSSARY OF HAWAIIAN WORDS

'a'ā	A kind of lava characterized by broken, jagged cobbles and boulders.
'aha	A sennit or cord of braided coconut fiber; also the name of a temple rite involving the recitation of a long prayer, during which the priest held a cord.
'ahi	Yellowfin tuna (*Thunnus albacares*).
ahu	A cairn, altar, or shrine.
ahupua'a	A territorial land unit, under the control of a subchief (the *ali'i 'ai ahupua'a*), typically extending from the mountains to the sea.
'ai kapu	A pre-Christian eating taboo, under which men and women ate separately.
'āina	Land.
'āina malo'o	Arid or dry lands, especially on the leeward sides of the Hawaiian Islands.
aku	Bonito or skipjack tuna (*Katsuwonus pelamis*).
akua	A god, spirit, or deity.
akua loa	The "long god," an image of Lono carried on a clockwise circuit of the *moku* during the annual Makahiki period.
akua poko	The "short god," similar to the *akua loa* but taken only to the lands of the chief (*poko*).
ali'i	An elite; a member of the chiefly class.
ali'i 'ai moku	The chief in charge of a district (*moku*).
ali'i akua	Literally, god king; an expression reserved for the highest-ranking *ali'i*, especially those of *pi'o* rank.
ali'i nui	Literally, great chief; the highest-ranked chief within a polity; a king.
'anā'anā	Sorcery, practiced by means of prayer and incantation.
'anu'u	On certain *heiau*, a tower constructed of wooden poles and covered in white '*ōloa* barkcloth, into which the priest entered to receive the god's oracle.
'aumākua	Ancestral deities; literally, collective parents.
'awa	Kava (*Piper methysticum*), the root of which was used to prepare a psychoactive (but nonalcoholic) beverage consumed in substantial quantity by the elites.
hai	An offering or sacrifice; to sacrifice.
haiau	Alternate spelling of *heiau*.
hale	A house or thatched structure.

hale noa	Literally, free house; a house without taboo, in which both sexes could mingle freely; the main dwelling house of a *kauhale* complex.
hale o Lono	A temple dedicated to the god Lono; literally, house of Lono.
hale pahu	A house where the *pahu* drum was kept within a *heiau*.
hale umu	An oven house within a *heiau*.
hale waiea	A small house within a *heiau* where the *'aha* ritual was performed.
haole	A white person; a Caucasian.
hā'uku'uke	An edible sea urchin (*Colobocentrotus atratus*) common on the rocky shores of Kahikinui.
heiau	The general term for a Hawaiian ritual place, where sacrifices of any kind were offered to the gods or to ancestral spirits.
heiau ho'oulu 'ai	An agricultural temple; literally, temple to increase food.
hōkū	A star or asterism.
hōlua	A sled with two wooden runners used on grassy slopes or on a specially prepared stone sled course; also refers to the sled course.
ho'oilo	Winter or the rainy season.
ho'okupu	Tribute (literally, to cause to grow or increase); in particular, tribute provided by an *ahupua'a* community during the annual Makahiki collection.
hula	Various forms of traditional dance.
hula pahu	Sacred dances performed to the accompaniment of a drum (*pahu*).
'ie'ie	An indigenous climbing shrub of the pandanus family (*Freycinetia arborea*). Its tough fibers were used to weave the foundations for feathered garments.
'ili	A land unit smaller than an *ahupua'a*.
'ili'ili	Pebbles or small stones; water-rolled gravel used to pave house floors and also placed as offerings in certain contexts such as on *heiau* altars.
'ili kūpono	A land segment within an *ahupua'a* that paid tribute directly to the ruling chief (*ali'i nui*) and not to the chief of the *ahupua'a*.
ipu	Bottle gourd (*Lagenaria siceraria*).
kahua	A platform or stage used for *hula* performance, among other purposes.
kahuna (plural *kāhuna*)	A priest.
kahuna 'anā'anā	A sorcerer; a priest who specialized in praying persons to death.
kahuna kuhi-kuhi-pu'u-one	A priest who specialized in the design and construction of *heiau*.
kahuna lapa'au	A medicinal priest or curing expert.
kahuna nui	A high priest, usually of the Kū cult, who officiated at ceremonies in the king's temple.
kāhuna pule	Priests who officiated in formal temple ceremonies, especially of the Kū, Lono, and Kāne cults.
kāli'i	A ritual in which spears were hurled at the ruling chief as he landed from a canoe, dodging them and thus displaying his fearlessness and dexterity.

kalo	Taro (*Colocasia esculenta*).
kamaʻāina	Someone born to a particular locale; literally, child of the land.
Kanaloa	One of the four major gods of the Hawaiian pantheon.
Kāne	A principal god of the Hawaiian pantheon; the creator god; also the deity of flowing waters and irrigation, to whom taro (*kalo*) was sacred.
kapa	Barkcloth.
kapu	Sacred, prohibited, or forbidden.
kapu moe	A prostrating taboo, obligatory for the highest-ranked chiefs.
kapu pule	Four periods during the lunar month, of two or three nights duration each, when men were required to attend rituals within the *heiau* precincts.
kau	A period of time, but also with special reference to the drier season or summer.
kauhale	A cluster of functionally differentiated houses making up a household in ancient Hawaiʻi; literally, plural houses.
kaukau aliʻi	Lesser grades of chiefs.
kāula	A prophet or seer; one who has the ability to transmit oracles.
kī	The ti plant (*Cordyline fruticosa*), possessing an edible root and large, glabrous leaves used for a variety of purposes.
kīhei	A shawl or cape; traditionally a rectangular barkcloth garment worn over one shoulder and tied in a knot.
kiʻi	An image or idol; refers in particular to carved wooden images set up in temples.
kilo hōkū	An astronomer; one who specializes in observing the stars.
kino lau	The many forms taken by a supernatural body; literally, numerous bodies.
kipapa	A pavement.
kīpuka	A patch of older lava surrounded by younger lava flows.
koa	An endemic hardwood tree (*Acacia koa*).
koʻa	A fishing shrine, dedicated to the god Kūʻula; also a fishing spot in the ocean.
kona	The leeward sides of the Hawaiian Islands; the direction from which winter storms derive.
konohiki	The land manager for an *ahupuaʻa* land unit, representing the *aliʻi ʻai ahupuaʻa* chief.
Kū	One of the four principal gods of the Hawaiian pantheon; the god of war.
kuaʻāina	Literally, back land; refers to the countryside or rural lands and equally to the people who resided on those lands; backcountry people; a rustic backwoodsman.
kuahu	An altar.
kuapola	A *kapu* night during the annual Makahiki period.
kukui	The candlenut tree (*Aleurites moluccana*), a Polynesian introduction found in some of the wetter gulches in Kahikinui.

kula	Dryland cultivation areas. In Mahele land claims, *kula* lands are typically distinguished from irrigated taro lands (*loʻi*).
kuleana	A right, privilege, responsibility, property, or estate. In the Mahele division of lands, *kuleana* refers to a land claim made by a commoner, usually a house lot and associated gardens lands.
kūlolo	A sweet pudding made of grated taro and coconut cream.
kupuna (plural *kūpuna*)	An elder or ancestor.
lana-nuʻu-mamao	A wooden tower or framework on a *heiau*; also known as *ʻanuʻu*.
lauhala	Leaves of the *hala* tree (*Pandanus tectorius*), used in plaiting mats and other objects.
lele	The sacrificial altar in a temple.
lua	A pit, such as the pit for disposing of sacrificial offerings in a temple.
luakini	A state temple dedicated to the god Kū, at which human sacrifice was performed.
Lono	One of the four principal gods of the Hawaiian pantheon; the god of thunder, rain, and dryland agriculture, to whom the sweet potato (*ʻuala*, or *Ipomoea batatas*) was sacred.
mā	The group of kinsmen, friends, or retainers of a certain person whose name precedes the particle.
makaʻāinana	A commoner.
Makahiki	A four-month period that commenced with the acronychal rising of the Pleiades in November. This was the period of tribute collection, sacred to Lono, when war was forbidden.
Makaliʻi	The Pleiades.
makai	In the direction of the sea or ocean; the reciprocal of *mauka*.
makua	Parent.
mana	Supernatural or divine power; efficacy; also the term for a thatched house situated within a *heiau* enclosure and holding images and sacred objects.
māpele	A kind of fertility *heiau* dedicated to Lono.
mauka	Toward the mountain, inland; the reciprocal of *makai*.
mele	A song or chant.
moku	A political district. Most islands were divided into 6 *moku*, although Maui had 12.
moʻo	A land unit smaller than an *ʻili*; also the word for a gecko or lizard.
moʻolelo	Oral tradition; oral history.
mua	A men's eating house. The *mua* of a king was where he would hold court with other high-ranking males and advisers.
nīʻaupiʻo	The offspring of a royal incestuous marriage between a brother and a sister or between half-siblings; literally, recurved coconut midrib.
niho	Tooth; also refers to stones set into the ground or interlocking.
niho palaoa	A whale tooth pendant that was a symbol of the *aliʻi*.
noa	Free, without *kapu*; the opposite of *kapu*.
ʻōhiʻa	An endemic tree (*Metrosideros polymorpha*) common in the upland

	forest zone of Kahikinui; sometimes specifically referred to as *'ōhi'a lehua* to distinguish it from the Polynesian-introduced mountain apple (*'ōhi'a 'ai*).
oli	A form of chant that was not danced to.
'ōpelu	Mackerel (*Decapterus* spp.).
'opihi	Limpets (*Cellana exarata* and other species), which were a prized food.
paehumu la'au	A wooden fence or enclosure surrounding a *heiau*.
pā hale	A walled enclosure surrounding a house site.
pāhoehoe	A smooth kind of lava with a ropy surface.
pahu	A particular kind of cylindrical drum, with a tympanum of sharkskin, used during *luakini* temple ceremonies.
pānānā	A compass; literally, sighting wall.
pili	A native grass (*Heteropogon contortus*) formerly common in the lowland regions of Kahikinui, used to thatch roofs.
pi'o	A royal marriage between a brother and a sister or between half-siblings; the offspring of such a union. *Pi'o* unions were the exclusive privilege of the highest-ranked *ali'i*.
po'e kahiko	Literally, ancient people; the traditional population of Hawai'i.
pōhaku o Kāne	Literally, stone of Kāne; shrines with an upright stone or boulder, dedicated the creator god Kāne.
po'okanaka	A *heiau* of human sacrifice; an alternative name for a *luakini* temple (literally, human head).
pule	Prayer; to pray.
pu'u	A hill or cinder cone.
'uala	Sweet potato (*Ipomoea batatas*).
unu a Lono	A kind of agricultural temple dedicated to Lono.
waihau ipu-o-Lono	A kind of *heiau* dedicated to Lono.
waiwai	Wealth, goods, or property; literally, water (reduplicated).
wiliwili	A native tree (*Erythrina sandwicensis*) in the legume family, common in the lower to mid-elevations of Kahikinui.

REFERENCES

Abbott, I. A. 1992. *Lā'au Hawai'i: Traditional Hawaiian Uses of Plants.* Honolulu: Bishop Museum Press.

Andrews, L. 1865. *A Dictionary of the Hawaiian Language.* Honolulu: Henry M. Whitney.

Ashmore, W., and A. B. Knapp, eds. 1999. *Archaeologies of Landscape: Contemporary Perspectives.* Oxford: Blackwell.

Athens, J. S., T. M. Rieth, and T. S. Dye. 2014. A paleoenvironmental and archaeological model-based age estimate for the colonization of Hawai'i. *American Antiquity* 79:144–155.

Baer, A. U. 2015. On the cloak of kings: Agriculture, power, and community in Kaupō, Maui. PhD diss., University of California, Berkeley.

———. 2016. Ceremonial architecture and the spatial proscription of community: Location versus form and function in Kaupō, Maui, Hawaiian Islands. *Journal of the Polynesian Society* 125:289–305.

Beaglehole, J. C., ed. 1967. *The Journals of Captain James Cook on His Voyages of Discovery.* Vol. 3, *The Voyage of the* Resolution *and* Discovery, *1776–1780.* 2 parts. Cambridge: Cambridge University Press for the Hakluyt Society.

Beckwith, M. W., ed. 1932. *Kepelino's Traditions of Hawaii.* Bernice P. Bishop Museum Bulletin 95. Honolulu: Bishop Museum Press.

———. 1940. *Hawaiian Mythology.* Honolulu: University Press of Hawai'i.

———. 1951. *The Kumulipo: A Hawaiian Creation Chant.* Chicago: University of Chicago Press.

Bennett, W. C. 1931. *Archaeology of Kauai.* Bernice P. Bishop Museum Bulletin 80. Honolulu: Bishop Museum Press.

Bronk Ramsey, C., and S. Lee. 2013. Recent and planned developments of the program OxCal. *Radiocarbon* 55(2–3):720–730.

Buck, P. H. 1957. *Arts and Crafts of Hawaii.* Bernice P. Bishop Museum Special Publication 45. Honolulu: Bishop Museum Press.

Chapman, P. S. and P. V. Kirch. 1979. *Archaeological Excavations at Seven Sites, Southeast Maui, Hawaiian Islands.* Department of Anthropology Report 79-1. Honolulu: Bishop Museum.

Chun, M. N. 1993. *Na Kukui Pio 'Ole: The Inextinguishable Torches.* Honolulu: First People's Productions.

———. 2014. *Ka Na'i Aupuni: Kamehameha and His Feathered Gods, Tahitian Colonies and Sorcery.* Honolulu: First People's Productions.

Coil, J. H. 2004. "The beauty that was": Archaeological investigations of ancient Hawaiian agriculture and environmental change in Kahikinui, Maui. PhD diss., University of California, Berkeley.

Coil, J. H., and P. V. Kirch. 2005. An Ipomoean landscape: Archaeology and the sweet potato in Kahikinui, Maui, Hawaiian Islands. In C. Ballard, P. Brown, R. M. Bourke, and T. Harwood, eds., *The Sweet Potato in Oceania: A Reappraisal,* pp. 71–84. Oceania Monograph no. 56. Sydney: University of Sydney.

Cordy, R. 1970. *Piilanihale Heiau Project: Phase I Site Report.* Department of Anthropology Report 70-9. Honolulu: Bishop Museum.

Cox, J. H., and W. H. Davenport. 1974. *Hawaiian Sculpture.* Honolulu: University of Hawai'i Press.

Dean, J. 1978. Independent dating in archaeological analysis. In M. Schiffer, ed., *Advances in Archaeological Method and Theory,* vol. 1, pp. 223–265. New York: Academic Press.

Dixon, B., P. J. Conte, V. Nagahara, and W. K. Hodgins. 2000. *Kahikinui Mauka: Archaeological Research in the Lowland Dry Forest of Leeward East Maui.* Report prepared for the Department of Hawaiian Home

Lands. Honolulu: Historic Preservation Division, Department of Land and Natural Resources, State of Hawai'i.

Dunnell, R. C. 1971. *Systematics in Prehistory.* New York: Free Press.

Dye, T. S. 1989. Tales of two cultures: traditional historical and archaeological interpretations of Hawaiian prehistory. *Bishop Museum Occasional Papers* 29:3–22.

Earle, T. 1997. *How Chiefs Come to Power: The Political Economy in Prehistory.* Stanford, CA: Stanford University Press.

Ellis, W. (1827) 1963. *Journal of William Ellis: Narrative of a Tour of Hawaii, or Owhyhee; with Remarks on the History, Traditions, Manners, Customs and Language of the Inhabitants of the Sandwich Islands.* Reprint. Honolulu: Advertiser Publishing Co.

Emerson, N. B. 1909. *Unwritten Literature of Hawaii: The Sacred Songs of the Hula.* Bureau of American Ethnology Bulletin 38. Washington, DC: Smithsonian Institution.

———. 1915. *Pele and Hiiaka: A Myth from Hawaii.* Honolulu. Honolulu Star-Bulletin.

Emory, K. P. 1921. An archaeological survey of Haleakala. *Bishop Museum Occasional Papers* 7(11): 237–259.

———. 1924. *The Island of Lanai: A Survey of Native Culture.* Bernice P. Bishop Museum Bulletin 12. Honolulu: Bishop Museum Press.

———. 1928. *Archaeology of Nihoa and Necker Islands.* Bernice P. Bishop Museum Bulletin 53. Honolulu: Bishop Museum Press.

———. 1933. *Stone Remains in the Society Islands.* Bernice P. Bishop Museum Bulletin 116. Honolulu: Bishop Museum Press.

———. 1934. *Tuamotuan Stone Structures.* Bernice P. Bishop Museum Bulletin 118. Honolulu: Bishop Museum Press.

———. 1943. Polynesian stone remains. In C. S. Coon and J. M. Andrews IV, eds., *Studies in the Anthropology of Oceania and Asia,* pp. 9–21. Papers of the Peabody Museum of American Archaeology and Ethnology, Harvard University, vol. 20. Cambridge, MA.

———. 1947. *Tuamotuan Religious Structures and Ceremonies.* Bernice P. Bishop Museum Bulletin 191. Honolulu: Bishop Museum Press.

Emory, K. P., W. J. Bonk, and Y. H. Sinoto. 1959. *Hawaiian Archaeology: Fishhooks.* Honolulu: Bishop Museum Press.

Finney, B. 2003. *Sailing in the Wake of the Ancestors: Reviving Polynesian Voyaging.* Honolulu: Bishop Museum Press.

Firth, R. 1967. *The Work of the Gods in Tikopia.* New York: Humanities Press.

Fleming, A. 2006. Post-processual landscape archaeology: A critique. *Cambridge Archaeological Journal* 16:267–280.

Flexner, J., and P. V. Kirch. 2016. Field mapping and Polynesian prehistory: A methodological history and thoughts for the future. In F. Valentin and G. Molle, eds., *La pratique de l'espace en Océanie: Découverte, appropriation et émergence des systèmes sociaux traditionnels* [Spatial dynamics in Oceania: Discovery, appropriation and the emergence of traditional societies], pp. 15–30. Paris: Société préhistorique française.

Flexner, J., M. Mulrooney, M. McCoy, and P. V. Kirch. 2017. Visualizing Hawaiian sacred sites: The archives and J. F. G. Stokes's pioneering archaeological surveys, 1906–1913. *Journal of Pacific Archaeology* 8:63–76.

Fornander, A. (1878–1885) 1996. *Ancient History of the Hawaiian People to the Times of Kamehameha I.* Honolulu: Mutual Publishing. Reprint of *An Account of the Polynesian Race,* vol. 2.

———. 1916–1920. *Fornander Collection of Hawaiian Antiquities and Folk-Lore.* Edited by T. G. Thrum. Bernice P. Bishop Museum Memoirs, vols. 4–6. Honolulu: Bishop Museum Press.

Gesch, D. B. 2007. The National Elevation Dataset. In D. Maune, ed., *Digital Elevation Model Technologies and Applications: The DEM Users Manual,* 2nd ed., pp. 99–118. Bethesda, MD: American Society for Photogrammetry and Remote Sensing.

Gesch, D. B., M. Oimoen, S. Greenlee, C. Nelson, M. Steuck, and D. Tyler. 2002. The National Elevation Dataset. *Photogrammetric Engineering and Remote Sensing* 68:5–11.

Giambelluca, T. W., Q. Chen, A. G. Frazier, J. P. Price, Y.-L. Chen, P.-S. Chu, J. K. Eischeid, and D. M. Delparte. 2013. Online Rainfall Atlas of Hawai'i. *Bulletin of the American Meteorological Society* 94:313–316. doi:10.1175/BAMS-D-11-00228.1.

González-García, A. C. 2014. Lunar alignments—identification and analysis. In C. L. N. Ruggles, ed., *Handbook of Archaeoastronomy and Ethnoastronomy*, vol. 1, pp. 493–506. New York: Springer.

Hammatt, H. H., and W. Folk. 1994. *Archaeological Reconnaissance of an 8,300-Acre Project Area, Kahikinui, Maui, TMK 1-9-01:3*. Report prepared for R. M. Towill Corporation. Honolulu: Cultural Surveys Hawai'i.

Handy, E. S. C. 1927. *Polynesian Religion*. Bernice P. Bishop Museum Bulletin 34. Honolulu: Bishop Museum Press.

———. 1940. *The Hawaiian Planter*. Vol. 1. Bernice P. Bishop Museum Bulletin 161. Honolulu: Bishop Museum Press.

Handy, E. S. C., and E. G. Handy. 1972. *Native Planters in Old Hawai'i: Their Life, Lore, and Environment*. Bernice P. Bishop Museum Bulletin 233. Honolulu: Bishop Museum Press.

Hartshorn, A. S., O. A. Chadwick, P. M. Vitousek, and P. V. Kirch. 2006. Prehistoric agricultural depletion of soil nutrients in Hawai'i. *Proceedings of the National Academy of Sciences USA* 103:11092–11097.

Hiroa, T. R. 1945. *An Introduction to Polynesian Anthropology*. Bernice P. Bishop Museum Bulletin 187. Honolulu: Bernice P. Bishop Museum.

Holm, L. A. 2006. The archaeology and the 'aina of Mahamenui and Manawainui, Kahikinui, Maui Island. PhD diss., University of California, Berkeley.

Holm, L. A., and P. V. Kirch. 2007. Up in smoke: Assumptions of survey visibility and site recovery. *Hawaiian Archaeology* 11:83–100.

Hommon, R. 2013. *The Ancient Hawaiian State: Origins of a Political Society*. Oxford: Oxford University Press.

Hoskin, M. A. 2001. *Tombs, Temples and Their Orientations*. Bognor Regis, UK: Ocarina Books.

Hudson, A. E. [1931]. Archaeology of Hawaii. Typescript in Library and Archives, Bernice P. Bishop Museum, Honolulu.

'I'ī, J. P. 1959. *Fragments of Hawaiian History*. Translated by M. K. Pukui. Honolulu: Bishop Museum Press.

Ingold, T. 2000. *The Perception of the Environment: Essays in Livelihood, Dwelling and Skill*. New York: Routledge.

Johnson, R. K., J. K. Mahelona, and C. L. N. Ruggles. 2015. *Nā Inoa Hōkū: Hawaiian and Pacific Star Names*. Rev. ed. Bognor Regis, UK: Ocarina Books.

Joppien, R., and B. Smith. 1988. *The Art of Captain Cook's Voyages*. Vols. 3 and 4, *The Voyage of the* Resolution *and* Discovery *1776–1780*. New Haven, CT: Yale University Press.

Kahn, J., and P. V. Kirch. 2014. *Monumentality and Ritual Materialization in the Society Islands: The Archaeology of a Major Ceremonial Complex in the 'Opunohu Valley, Mo'orea*. Bishop Museum Bulletin in Anthropology 13. Honolulu: Bishop Museum Press.

Kamakau, S. M. 1961. *Ruling Chiefs of Hawaii*. Honolulu: Kamehameha Schools Press.

———. 1964. *Ka Po'e Kahiko: The People of Old*. Translated by M. K. Pukui. Bernice P. Bishop Museum Special Publication 51. Honolulu: Bishop Museum Press.

———. 1976. *The Works of the People of Old: Na Hana a ka Po'e Kahiko*. Bernice P. Bishop Museum Special Publication 61. Honolulu: Bishop Museum Press.

———. 1991. *Tales and Traditions of the People of Old: Nā Mo'olelo a ka Po'e Kahiko*. Translated by M. K. Pukui; edited by D. B. Barrère. Honolulu: Bishop Museum Press.

Kelly, M. 1980. *Hālau hula* and adjacent sites at Kē'ē, Kaua'i. In D. B. Barrère, M. K. Pukui, and M. Kelly, *Hula: Historical Perspectives*, pp. 95–121. Pacific Anthropological Records 30. Honolulu: Bernice P. Bishop Museum.

Kikiloi, K. S. T. 2012. Kūkulu Manamana: Ritual power and religious expansion in Hawai'i; The ethnohistorical and archaeological study of Mokumanamana and Nihoa Islands. PhD diss., University of Hawai'i at Mānoa.

Kirch, P. V. 1984. *The Evolution of the Polynesian Chiefdoms*. Cambridge: Cambridge University Press.

———. 1985. *Feathered Gods and Fishhooks: An Introduction to Hawaiian Archaeology and Prehistory*. Honolulu: University of Hawai'i Press.

———. 1990. Monumental architecture and power in Polynesian chiefdoms: A comparison of Tonga and Hawaii. *World Archaeology* 22:206–222.

———. 1994. The pre-Christian ritual cycle of Futuna, western Polynesia. *Journal of the Polynesian Society* 103:255–298.

———. 1997a. Kahikinui: An introduction. In P. V. Kirch, ed., *Nā Mea Kahiko o Kahikinui: Studies in the Archaeology of Kahikinui, Maui*, pp. 1–11. Berkeley: Oceanic Archaeology Laboratory, University of California.

———, ed. 1997b. *Nā Mea Kahiko O Kahikinui: Studies in the Archaeology of Kahikinui, Maui*. Oceanic Archaeology Laboratory, Special Publication no. 1. Berkeley: Archaeological Research Facility, University of California.

———. 2000. *On the Road of the Winds: An Archaeological History of the Pacific Islands before European Contact*. Berkeley: University of California Press.

———. 2004a. Solstice observation in Mangareva, French Polynesia: New perspectives from archaeology. *Archaeoastronomy: The Journal of Astronomy in Culture*, 17:1–19.

———. 2004b. Temple sites in Kahikinui, Maui, Hawaiian Islands: Their orientations decoded. *Antiquity* 78:102–114.

———. 2007. Paleodemography in Kahikinui, Maui: An archaeological approach. In P. V. Kirch and J.-L. Rallu, eds., *The Growth and Collapse of Pacific Island Societies: Archaeological and Demographic Perspectives*, pp. 90–107. Honolulu: University of Hawai'i Press.

———. 2010. *How Chiefs Became Kings: Divine Kingship and the Rise of Archaic States in Ancient Hawai'i*. Berkeley: University of California Press.

———. 2011a. The archaeology of dryland farming systems in southeastern Maui. In P. V. Kirch, ed., *Roots of Conflict: Soils, Agriculture, and Sociopolitical Complexity in Ancient Hawai'i*, pp. 65–88. Santa Fe, NM: School for Advanced Research Press.

———. 2011b. When did the Polynesians settle Hawaii? A review of 150 years of scholarly inquiry and a tentative answer. *Hawaiian Archaeology* 12: 3–26.

———. 2012. *A Shark Going Inland Is My Chief: The Island Civilization of Ancient Hawai'i*. Berkeley: University of California Press.

———. 2014. *Kua'āina Kahiko: Life and Land in Ancient Kahikinui, Maui*. Honolulu: University of Hawai'i Press.

Kirch, P. V., J. Coil, A. S. Hartshorn, M. Jeraj, P. M. Vitousek, and O. A. Chadwick. 2005. Intensive dryland farming on the leeward slopes of Haleakala, Maui, Hawaiian Islands: Archaeological, archaeobotanical, and geochemical perspectives. *World Archaeology* 37:240–258.

Kirch, P. V., and R. C. Green. 1987. History, phylogeny, and evolution in Polynesia. *Current Anthropology* 28:431–443, 452–456.

———. 2001. *Hawaiki, Ancestral Polynesia: An Essay in Historical Anthropology*. Cambridge: Cambridge University Press.

Kirch, P. V., A. S. Hartshorn, O. A. Chadwick, P. M. Vitousek, D. R. Sherrod, J. Coil, L. Holm, and W. D. Sharp. 2004. Environment, agriculture, and settlement patterns in a marginal Polynesian landscape. *Proceedings of the National Academy of Sciences USA* 101:9936–9941.

Kirch, P. V., J. Holson, and A. Baer. 2009. Intensive dryland agriculture in Kaupō, Maui, Hawaiian Islands. *Asian Perspectives* 48:265–290.

Kirch, P. V., S. Millerstrom, S. Jones, and M. McCoy. 2010. Dwelling among the gods: A late pre-contact priest's house in Kahikinui, Maui, Hawaiian Islands. *Journal of Pacific Archaeology* 1:144–160.

Kirch, P. V., R. Mertz-Kraus, and W. D. Sharp. 2015. Precise chronology of Polynesian temple construction and use for southeastern Maui, Hawaiian Islands determined by ^{230}Th dating of corals. *Journal of Archaeological Science* 53:166–177.

Kirch, P. V., P. Mills, J. Kahn, and S. Lundblad. 2012. Interpolity exchange of basalt tools facilitated via elite control in Hawaiian archaic states. *Proceedings of the National Academy of Sciences USA* 109:1056–1061.

Kirch, P. V., and S. J. O'Day. 2002. New archaeological insights into food and status: A case study from pre-contact Hawaii. *World Archaeology* 34:484–497.

Kirch, P. V., C. L. N. Ruggles, and W. D. Sharp. 2013. The *pānānā* or "sighting wall" at Hanamauloa, Kahikinui, Maui: Archaeological investigation of a possible navigational monument. *Journal of the Polynesian Society* 122:45–68.

Kirch, P. V., and M. Sahlins. 1992. *Anahulu: The Anthropology of History in the Kingdom of Hawaii.* 2 vols. Chicago: University of Chicago Press.

Kirch, P. V., and W. D. Sharp. 2005. Coral ^{230}Th dating of the imposition of a ritual control hierarchy in precontact Hawaii. *Science* 307:102–104.

Kolb, M. J. 1991. Social power, chiefly authority, and ceremonial architecture in an island polity, Maui, Hawaii. PhD diss., University of California, Los Angeles.

———. 1992. Diachronic design changes in *heiau* temple architecture on the island of Maui, Hawai'i. *Asian Perspectives* 31:9–38.

———. 1994. Monumentality and the rise of religious authority in precontact Hawai'i. *Current Anthropology* 35:521–548.

———. 1999. Monumental grandeur and political florescence in pre-contact Hawai'i: Excavations at Pi'ilanihale Heiau, Maui. *Archaeology in Oceania* 34(2):73–82.

———. 2006. The origins of monumental architecture in ancient Hawai'i. *Current Anthropology* 47:657–665.

Kolb, M. J., and E. C. Radewagen. 1997. Nā heiau o Kahikinui: The temples of Kahikinui. In P. V. Kirch, ed., *Nā Mea Kahiko o Kahikinui: Studies in the Archaeology of Kahikinui, Maui*, pp. 61–77. Oceanic Archaeology Laboratory, Special Publication no. 1. Berkeley: Archaeological Research Facility, University of California.

Kuykendall, R. S. 1938. *The Hawaiian Kingdom*. Vol. 1, 1778–1854: *Foundation and Transformation*. Honolulu: University of Hawai'i Press.

Ladd, E. J. 1969. 'Alealea temple site, Honaunau: Salvage report. In R. Pearson, ed., *Archaeology on the Island of Hawaii*, pp. 95–132. Asian and Pacific Archaeology Series no. 3. Honolulu: Social Science Research Institute, University of Hawai'i.

———. 1973. Kaneaki temple site—an excavation report. In E. J. Ladd, ed., *Makaha Valley Historical Project: Interim Report No. 4*, pp. 1–30. Pacific Anthropological Records 19. Honolulu: Bernice P. Bishop Museum.

———. 1985. *Hale-o-Keawe Archaeological Report*. Western Archaeological and Conservation Center, Publications in Anthropology no. 33. Honolulu: National Park Service.

Lewis, D. 1972. *We, the Navigators: The Ancient Art of Landfinding in the Pacific*. Honolulu: University of Hawai'i Press.

Liller, W. 1989. The megalithic astronomy of Easter Island: Orientations of *ahu* and *moai*. *Archaeoastronomy*, no. 13 (supplement to *Journal for the History of Astronomy* 20): S21–S48.

Macdonald, G. A., and A. T. Abbott. 1970. *Volcanoes in the Sea: The Geology of Hawaii*. Honolulu: University of Hawai'i Press.

Makemson, M. W. 1941. *The Morning Star Rises: An Account of Polynesian Astronomy*. New Haven, CT: Yale University Press.

Malo, D. 1951. *Hawaiian Antiquities (Moolelo Hawaii)*. Bernice P. Bishop Museum Special Publication 2. Honolulu: Bishop Museum Press.

Maunupau, T. K. 1998. *Huakai Makaikai a Kaupo, Maui: A Visit to Kaupō, Maui*. Edited by N. N. C. Losch. Honolulu: Bishop Museum Press.

McAllister, J. G. 1933a. *Archaeology of Kahoolawe*. Bernice P. Bishop Museum Bulletin 115. Honolulu: Bishop Museum Press.

———. 1933b. *Archaeology of Oahu*. Bernice P. Bishop Museum Bulletin 104. Honolulu: Bishop Museum Press.

McCoy, M. D. 2008. Life outside the temple: Reconstructing traditional Hawaiian ritual and religion through new studies of ritualized practices. In L. Fogelin, ed., *Religion, Archaeology, and the Material World*, pp. 261–278. Southern Illinois University Center for Archaeological Investigations, Occasional Paper no. 36. Carbondale, IL.

———. 2014. The significance of religious ritual in ancient Hawai'i. *Journal of Pacific Archaeology* 5:72–80.

McCoy, M. D., T. N. Ladefoged, M. W. Graves, and J. W. Stephen. 2011. Strategies for constructing religious authority in ancient Hawai'i. *Antiquity* 85:927–941.

McCoy, P. C., M. I. Weisler, J. Zhao, and Y-X. Feng. 2009. ^{230}Th dates for dedicatory corals from a remote alpine desert adze quarry on Mauna Kea, Hawai'i. *Antiquity* 83:445–457.

McKern, W. C. 1929. *Archaeology of Tonga*. Bernice P. Bishop Museum Bulletin 60. Honolulu: Bishop Museum Press.

Mulrooney, M. A., and T. Ladefoged. 2005. Hawaiian *heiau* and agricultural production in the Kohala dryland field system. *Journal of the Polynesian Society* 114:45–67.

Noyes, M. H. 2013. From Kūkaniloko: Sirius in the Hawaiian sky. *Time and Mind: The Journal of Archaeology, Consciousness and Culture* 6:157–174.

Oliveira, K.-A. R. K. N. 2014. *Ancestral Places: Understanding Kanaka Geographies*. Corvallis: Oregon State University Press.

Oliver, D. L. 1974. *Ancient Tahitian Society*. 3 vols. Honolulu: University of Hawai'i Press.

Palmer, H. S. 1927. *Geology of Kaula, Nihoa, Necker, and Gardner Islands, and French Frigates Shoal*. Bernice P. Bishop Museum Bulletin 35. Honolulu: Bishop Museum Press.

Phelps, S. [1937]. A regional study of Molokai. Typescript in Library and Archives, Bernice P. Bishop Museum, Honolulu.

Prendergast, F. 2014. Techniques of field survey. In C. L. N. Ruggles, ed., *Handbook of Archaeoastronomy and Ethnoastronomy*, vol. 1, pp. 389–409. New York: Springer.

Pukui, M. K. 1983. *'Olelo No'eau: Hawaiian Proverbs and Poetical Sayings*. Bernice P. Bishop Museum Special Publication 71. Honolulu: Bishop Museum Press.

Pukui, M. K., and S. H. Elbert. 1986. *Hawaiian Dictionary*. Rev. and enlarged ed. Honolulu: University of Hawai'i Press.

Pukui, M. K., S. H. Elbert, and E. T. Mookini. 1974. *Place Names of Hawaii*. Rev. ed. Honolulu: University of Hawai'i Press.

Reimer, P. J., E. Bard, A. Bayliss, J. W. Beck, P. G. Blackwell, C. Bronk Ramsey, C. E. Buck, et al. 2013. IntCal13 and Marine13 radiocarbon age calibration curves 0–50,000 years cal BP. *Radiocarbon* 55(4): 1869–1887.

Ridpath, I., and A. P. Norton, eds. 2004. *Norton's Star Atlas and Reference Handbook, Epoch 2000.0*. New York: Pearson.

Rowland, B. 2007. Growth of the coral, growth of the heavens: An investigation into the significance of ritual coral use in Hawai'i and Polynesia. Honours thesis, University of Queensland.

Ruggles, C. L. N. 1997. Whose equinox? *Archaeoastronomy*, no. 22 (supplement to *Journal for the History of Astronomy* 28): S45–S50.

———. 1999a. *Astronomy in Prehistoric Britain and Ireland*. New Haven, CT: Yale University Press.

———. 1999b. Astronomy, oral literature, and landscape in ancient Hawai'i. *Archaeoastronomy: The Journal of Astronomy in Culture* 14:33–86.

———. 2001. *Heiau* orientations and alignments in Kaua'i. *Archaeoastronomy: The Journal of Astronomy in Culture* 16:46–82.

———. 2007. Cosmology, calendar, and temple orientations in ancient Hawai'i. In C. L. N. Ruggles and G. Urton, eds., *Skywatching in the Ancient World: New Perspectives in Cultural Astronomy*, pp. 287–329. Boulder: University Press of Colorado.

———. 2011. Pushing back the frontiers or still running around the same circles? "Interpretative archaeoastronomy" thirty years on. In C. L. N. Ruggles, ed., *Archaeoastronomy and Ethnoastronomy: Building Bridges between Cultures*, pp. 1–18. Cambridge: Cambridge University Press.

———. 2014a. Analyzing orientations. In C. L. N. Ruggles, ed., *Handbook of Archaeoastronomy and Ethnoastronomy*, vol. 1, pp. 411–425. New York: Springer.

———. 2014b. Basic concepts of positional astronomy. In C. L. N. Ruggles, ed., *Handbook of Archaeoastronomy and Ethnoastronomy*, vol. 1, pp. 459–472. New York: Springer.

———. 2014c. Best practice for evaluating the astronomical significance of archaeological sites. In

C. L. N. Ruggles, ed., *Handbook of Archaeoastronomy and Ethnoastronomy*, vol. 1, pp. 373–388. New York: Springer.

———. 2014d. Long-term changes in the appearance of the sky. In C. L. N. Ruggles, ed., *Handbook of Archaeoastronomy and Ethnoastronomy*, vol. 1, pp. 473–482. New York: Springer.

———. 2014e. Nature and analysis of material evidence relevant to archaeoastronomy. In C. L. N. Ruggles, ed., *Handbook of Archaeoastronomy and Ethnoastronomy*, vol. 1, pp. 353–372. New York: Springer.

———. 2014f. Stellar alignments—identification and analysis. In C. L. N. Ruggles, ed., *Handbook of Archaeoastronomy and Ethnoastronomy*, vol. 1, 517–530. New York: Springer.

———. 2017. Postscript: Still our equinox? *Journal of Skyscape Archaeology* 3(1): 132–135.

Ruggles, C. L. N., and N. J. Saunders. 1993. The study of cultural astronomy. In C. L. N. Ruggles and N. J. Saunders, eds., *Astronomies and Cultures*, 1–31. Niwot: University Press of Colorado.

Sahlins, M. 1958. *Social Stratification in Polynesia*. Seattle: American Ethnological Society.

———. 1972. *Stone Age Economics*. Chicago: Aldine-Atherton.

———. 1981. *Historical Metaphors and Mythical Realities: Structure in the Early History of the Sandwich Islands Kingdom*. Ann Arbor: University of Michigan Press.

———. 1985a. Hierarchy and humanity in Polynesia. In A. Hooper and J. Huntsman, eds., *Transformations of Polynesian Culture*, pp. 195–217. Auckland: Polynesian Society.

———. 1985b. *Islands of History*. Chicago: University of Chicago Press.

———. 1989. Captain Cook at Hawaii. *Journal of the Polynesian Society* 98:371–423.

———. 1990. The political economy of grandeur in Hawaii from 1810 to 1830. In E. Ohnuki-Tierney, ed., *Culture through Time*, pp. 26–56. Stanford, CA: Stanford University Press.

———. 1992. *Historical Ethnography*. Vol. 1 of P. V. Kirch and M. Sahlins, *Anahulu: The Anthropology of History in the Kingdom of Hawaii*. Chicago: University of Chicago Press.

———. 1995. *How "Natives" Think: About Captain Cook, for Example*. Chicago: University of Chicago Press.

Scarre, C. 2005. *Monuments and Landscape in Atlantic Europe: Perception and Society during the Neolithic and Early Bronze Age*. New York: Routledge.

Schmitt, R. C. 1973. *The Missionary Censuses of Hawaii*. Pacific Anthropological Records 20. Honolulu: Bishop Museum.

Shapiro, W. A., P. L. Cleghorn, P. V. Kirch, L. Holm, E. L. Kahahane, and J. D. McIntosh. 2011. *Archaeological Inventory Survey for the Proposed Auwahi Wind Farm Ahupuaʻa of Auwahi, District of Kahikinui, Island of Maui, Hawaiʻi [TMK (2) 1-9-001:006]*. Report prepared for Sempra Generation Inc. Kailua, HI: Pacific Legacy.

Sharp, W. D., J. G. Kahn, C. M. Polito, and P. V. Kirch. 2010. Rapid evolution of ritual architecture in central Polynesia indicated by precise ^{230}Th/U coral dating. *Proceedings of the National Academy of Sciences USA* 107:13234–13239.

Sherrod, D. R., J. M. Sinton, S. E. Watkins, and K. M. Brunt. 2007. *Geologic Map of the State of Hawaiʻi*. U.S. Geological Survey Open-File Report 2007-1089. Washington, DC: U.S. Geological Survey. http://pubs.usgs.gov/of/2007/1089/.

Silva, F., and N. Campion, eds. 2015. *Skyscapes: The Role and Importance of the Sky in Archaeology*. Oxford: Oxbow.

Soehren, L. J. 1963. An archaeological survey of portions of East Maui, Hawaii. Mimeographed report prepared for the U.S. National Park Service. Honolulu: Bernice P. Bishop Museum.

Šprajc, I. 2014. Alignments upon Venus (and other planets)—identification and analysis. In C. L. N. Ruggles, ed., *Handbook of Archaeoastronomy and Ethnoastronomy*, vol. 1, pp. 507–516. New York: Springer.

Sterling, E. P. 1998. *Sites of Maui*. Honolulu: Bishop Museum Press.

Stock, J., J. Coil, and P. V. Kirch. 2003. Paleohydrology of arid southeastern Maui, Hawaiian Islands, and its implications for prehistoric human settlement. *Quaternary Research* 59:12–24.

Stokes, J. F. G. [1909]. Heiau of Molokai. Typescript in Library, Bernice P. Bishop Museum, Honolulu.

———. 1991. *Heiau of the Island of Hawai'i: A Historic Survey of Native Hawaiian Temple Sites*. Edited by T. Dye. Bishop Museum Bulletin in Anthropology 2. Honolulu: Bishop Museum Press.

Stuiver, M., and H. A. Polach. 1977. Discussion: Reporting of ^{14}C data. *Radiocarbon* 19(3):355–363.

Territory of Hawaii, Office of the Commissioner of Public Lands. 1929. *Indices of Awards Made by the Board of Commissioners to Quiet Land Titles in the Hawaiian Islands*. Honolulu: Territorial Government.

Thrum, T. G. 1906. Heiau and heiau sites throughout the Hawaiian Islands. In *Hawaiian Annual for 1907*, pp. 36–48.

———. 1909. Tales from the temples. In *Hawaiian Annual for 1909*, pp. 44–54.

Thurman, R. M. R. 2015. The archaeology of Maunawila Heiau, Hau'ula Ahupua'a, Ko'olauloa District, O'ahu. *Hawaiian Archaeology* 14:17–32.

Tilley, C. 1994. *A Phenomenology of Landscape*. Oxford: Berg.

Valeri, V. 1985. *Kingship and Sacrifice: Ritual and Society in Ancient Hawaii*. Chicago: University of Chicago Press.

Van Gilder, C. 2005. Families on the Land: Archaeology and Identity in Kahikinui, Maui. PhD diss., University of California, Berkeley.

Van Gilder, C., V. Nagahara, and K. Hodgins. 1999. *Preliminary Report on Coastal Survey of Kahikinui, Maui*. Prepared for the State Historic Preservation Division, Department of Land and Natural Resources. Honolulu: State of Hawai'i.

Walker, W. M. [1929]. Field Note Book, "Maui, South Shore, February 1929." In Library and Archives, Bernice P. Bishop Museum, Honolulu.

———. [1930]. Archaeology of Maui. Typescript in Library and Archives, Bernice P. Bishop Museum, Honolulu.

Weisler, M. I., K. D. Collerson, Y.-X. Feng, J.-X. Zhao, and K.-F. Yu. 2006. Thorium-230 coral chronology of a late prehistoric Hawaiian chiefdom. *Journal of Archaeological Science* 33:273–282.

INDEX

NOTE. This is a full index to Part I but only a skeleton index to the catalog presented in Part II (Chapters 6–8). Explanatory terms in square brackets are used to differentiate index terms with more than one meaning. Page numbers in **boldface** refer to illustrations.

^{14}C dating. *See* radiocarbon dating
^{230}Th dates, presentation of, 141–142
^{230}Th dating, 38, 39, 58, 62, 68–71
    of Kahikinui *heiau*, 68–70, 129, 131

'A'ā [star name]. *See* Sirius
'a'ā rocks, 40–41
accelerator mass-spectrometry (AMS), 66
access to sites, 142
accuracy:
    defined (*cf.* precision), 76
    of altitude readings, 88
    of compass measurements, 87
acronical rise. *See* acronychal rise
acronychal rise:
    explained, 78, 137
    of Pleiades, 27, 28, 132, 135, 137
    true vs. apparent, 28, 137
acronychal set, 78
adzes, 32, 35, 72
    production of, 28, 73, 132, 135
    sourcing of, 336
agricultural temples, 22–24, 57, 71, 133
    as focus for Makahiki rituals, 30
    associated with Lono, 23, 30
    vertical divisions within, 23
Ahu'ena Heiau, Hawai'i Island, 21, **22**

akua loa ("long god"), circuit of the, 31–32, 123
akua poko ("short god"), circuit of the, 31–32
'alā (waterworn) stones, 42, 48, 49
    as representations of deities, 42
ALE-1, 42, 50, 55, 90, 136, 186–189, **187, 188**
ALE-4 (Wailapa Heiau), 50, 58, 90, 114, 189–193, **191, 192**
ALE-121, 90, 110, 114, 193–194, **194**
ALE-140, 194–198, **195, 196**
    branch coral at, **51**
    'ili'ili paving at, 42
    orientation of, 91, 107, 108, 114, 116, 135
    Pleiades sighting at, 117, 122, 136, 137
    wall construction at, **41**
    waterworn cobbles at, 54
ALE-211, 55, 110, 198
'Āle'ale'a Heiau, Hōnaunau, Hawai'i Island, 35
Alena, 184–186. *See also* ALE-1 etc.
ali'i (chiefs), 17
ali'i nui. *See* kings
alignments:
    astronomical, 125–128
    astronomical potential of, 85
    expressed as declination ranges, 104
    identifying possibly significant, 85
    Pleiades, 136

    solstitial, 75, 82, 124, 137
    topographic, 110–114, 125–128
    within the solar arc, 109, 110, 122, **123**
altars, 17, 25, 49
    raised, indicating direction of orientation, 90
altitude, 86
    accuracy of measurements of, 88
    defined (*cf.* elevation), 76
ancestors ('aumākua), 19
Ancestral Polynesia, 15, 26, 30, 49
Andrews, Lorrin, 14
annual ritual cycle. *See* ritual cycle
anomalies, magnetic, 87
'anu'u (towers), 11, 21, **22**, 23, 25, 49, 54
archaeoastronomical surveys, 11. *See also* compass-clinometer surveys; total station surveys
archaeoastronomy, 10, 84, 138
    data-driven approaches in, 85, 138
    methodology, 85, 138
    survey procedure in, 85–87
archaic state formation, 15, 38, 39, 71
architects (*kāhuna kuhi-kuhi-pu'u-one*), 21
artifacts:
    from *heiau* excavations, 72–73, **73**
    sourcing using X-ray fluorescence, 73

astronomical knowledge:
    Hawaiian, 3–4, 335, 336
    modern, 336
    Polynesian, 3, 335, 336
astronomical observations, motivations for, 76
AUW-6, 143–144
AUW-9, 58, 72, 104, 136, 144–148, **145**
AUW-11, 148–150
    alignment upon Kanahau, 110
    date of, 100
    orientation of, 90, 91
    upright ʻalā stone at, 42, 49, **50**, 57
AUW-20, 44
Auwahi, 143. See also AUW-6 etc.; WF-AUW-100 etc.
azimuth:
    defined, 76, 141
    distribution for *heiau* orientations (four directions), **101**, 103, **104–105**
    distribution for sacred (*kapu*) direction, 98–100, **101**, **102**, 103, **104–105**
    measurements of, 86, 87–88
    ranges, 103, 125

Baer, Alex, xviii, 62, 63, 66
basal contexts, 63
basalt flakes, 72, 73
benches, raised, 49
Bennett, Wendell C., 35, 36–37
bird bones, 74
branch coral, **51**
    concentrations of, indicating direction of orientation, 91
    elsewhere in Polynesia, 52
    offerings, 48, 49, 50, 51–52, 62
    use of, for dating, 68, 129
Brigham, William T., 34
buttressing, 42–43
    indicating direction of orientation, 90

calendar, 4, 26–32, 83, 132–133, 137
    Proto-Polynesian, 26
    Polynesian, 138
    See also month
canoe building, 18, 134, 135
canoe-prow corners, 43–44, **59**, 116
    in relation to sacred (*kapu*) direction, 96–97
    orientation of, 100, 102, **102**
    orientation of in relation to prominent landmarks, 103
    tendency to face Hawaiʻi Island, 97
Cape Kumukahi, Hawaiʻi Island, 75
cardinality of *heiau* orientations, 100, 103, **104–105**
Castor (Nānāmua), 109
Chapman, Peter S., 8, 9
chiefs (*aliʻi*), 17
chronology, 61
    in relation to elevation, 131
    in relation to morphology, 129, 131
    in relation to orientation, 100, 130, 131
    of Kahikinui and Kaupō *heiau*, 66, 70–71
    of Kahikinui *heiau*, 38–39, 68, 129–130, 131
cinder cones, 77, 114–115, 135
    alignments upon, 111–113
    See also Kanahau; Puʻu Hōkūkano; Puʻu Māneoneo; Puʻu Pane; Puʻu Pīmoe
cists, 50
class distinctions, 19, 31
coastal promontories, 77
    alignments upon, 111–113
    See also Ka Lae o Ka ʻIlio
commoners (*makaʻāinana*):
    ancestor cult among, 19
    rituals, 34
compass directions, 141
compass-clinometer surveys, 11, 87
Cook, Captain James, 31, 335
coordinates, reason for withholding, 142
coral. *See* branch coral
core-filled walls, 42

cosmical set, 78
courts, 48
craft activities within *heiau*, 73, 132
cultural landscape studies, 9
cumulative probability histogram, 103, **104–105**, 107
curvigram. *See* cumulative probability histogram

dark sky preservation, 336
dating:
    of *heiau* on Lehua, 39
    of *heiau* on Mokumanamana, 39
    of *heiau* on Nihoa, 39
    of Kahikinui *heiau*, 39
    of Maui *heiau*, 38
    See also [230]Th dating; chronology; radiocarbon dating
dawn rites, 135–136
days of the moon, 27
December solstice:
    alignments upon sunset at, 124
    declination of sun at, 82
    explained, 82
declination [astronomical], 78, 86
    distinction from magnetic declination, 76
    distribution for *heiau* orientations (four directions), 104, **108**
    distribution for sacred (*kapu*) direction, 104, 105–106, 107–108, **107**
    of moon, 83, 84
    of principal Hawaiian stars, 79–81
    of sun, over the year, 83
    ranges, 104, 125
declination [magnetic], 87
Dibble, Sheldon, 14
digital terrain model (DTM) data, 88
districts (*moku*), 6
dog bones, 74
double *heiau*. *See* KIP-1010
DTM (digital terrain model) data, 88

earliest dates for Kahikinui *heiau*, 130
Easter Island (Rapa Nui), 16, 138
easterly orientation group, 100
    interpreted as Kāne temples, 134
east-northeasterly orientation group, 100, 104
    interpreted as Lono temples, 134–135
elevation:
    defined (*cf.* altitude), 76
Ellis, Rev. William, 13
elongated double-court enclosures, 55–56
    earliest dates for, 130
    other features at, 55–56
Emerson, Nathaniel, 14
Emory, Kenneth P., 34–35, 37–38
equinoxes:
    lack of evidence of interest in, 109, 134
    uncertain cultural significance of, 83
    variations in definition of, 83
errors:
    gross, 86
    random, 86
    systematic, 86–87
excavations, 62, 72–74
    activities within *heiau*, as revealed by, 71–74
    artifacts recovered from, 72–73, **73**
    basal contexts, 63

faunal remains, 73–74
fences, 25
fertility temples. *See* agricultural temples; Lono
fish bones, 74
fishhooks, 35, 72, **73**
    manufacture of, 132
fishing shrines (*koʻa*), 24, 43, 57
    orientation of, 97
Fornander, Abraham, 34, 129
functional typology of *heiau*, 20–24, 133
    relation to orientation, 134
    See also agricultural temples; fishing shrines; *pōhaku o Kāne*; war temples

gods, principal, 17–19
    directional associations of, 18, 133–134
    See also Kanaloa; Kāne; Kū; Lono
gods, representations of (*ʻaumakua*), 49

hale mua, 24
Hale o Keawe Heiau, Hōnaunau, Hawaiʻi Island, 35
hale o Lono. *See* agricultural temples; Lono
Haleakalā, 5, 76, 110, 135
    Crater, 77
    Ridge, 78, 110, 114–115, 135
Halekou Heiau (NUU-100), 56, 294–297, **295**
    adze and fishhook production at, 72
    basalt core tool from, **295**
    date of, 100, 130
    Pleiades sighting at, 136
    sunset viewed from, 115, **297**
    westerly orientation of, 135
Hawaiʻi Island, 34, 35, 103, 135. *See also* Mauna Kea; Mauna Loa
Hawaiian Annual, 34
heiau:
    activities within, 71–74
    architectural development of, 17
    architectural features, 48–51
    as gathering places, 73, 132
    as places for observing the heavens, 132
    as places of sacrifice, 3, 15, 17, 20, 23
    associated gods, 133–135
    association with astronomical knowledge, 336
    components of, 17
    construction techniques, 44
    courts, 48
    craft activities within, 73, 132
    dates of, 44–47
    definition of, 14–15
    development from Eastern Polynesian *marae*, 16
    direction faced, *see* sacred (*kapu*) direction
    divisions, internal, 48
    early accounts of, 13
    early surveys of, xvii, 34–35
    elevations of, 44–47, 60–61
    etymology, 3, 15
    evidence for butchering at, 72
    excavated artifacts, 72–72, **73**
    excavations within, 35, 62, 63, 71–74
    "four directionality" of, 89–90
    functional typology, 20–24, 133, *see also* agricultural temples, fishing shrines, *pōhaku o Kāne*, war temples
    geographical distribution, 60–61, **60**
    hierarchical system of, 71
    houses within, 25, 50
    identification by field survey, 10
    landscape setting, 76–77
    marking land boundaries, 135
    modified during postcontact times, 59–60
    morphology, 36–37, 44–47, 52–57, 129–131, *see also* elongated double-court enclosures, notched enclosures, platform *heiau*, square enclosures, terraced *heiau*, U-shaped enclosures
    notched, *see* notched enclosures
    on Mokumanamana (Necker Island), 37
    orientation of, *see* orientation
    pathways within, 50–51
    pioneering archaeological studies of, 4
    Proto-Polynesian origins, 15
    recent primary research, 35–36
    relation to Polynesian *marae*, 15

secondary platforms, 48–49
sizes of, 44–48
stonework, 40–44, see also a'ā
  rocks, 'ili'ili, pāhoehoe
  slabs, upright stones,
  waterworn ('alā) stones
superstructures, 24–25, 40,
  see also towers (anu'u)
terraces, 42–43
topographic settings, 61
twentieth-century studies,
  34–35
types of, see associated gods,
  functional typology,
  morphology
wall construction techniques,
  42
with multiple construction
  phases, 58–60
within the cultural land-
  scape, 9
heiau ho'oulu 'ai. See agricultural
  temples
Heiau Ridge complex, 134,
  203–205, **204**, **205**. See also
  KIP-75; KIP-76; KIP-77;
  KIP-115; KIP-405; KIP-410
heliacal rise:
  explained, 78
  of Pleiades, 137
heliacal set, 78
hōkū. See stars
Hōkūho'okelewa'a. See Sirius
Hōkūle'a voyaging canoe, 335
Hōkūpa'a (Polaris), 82, 107
Honouliuli, O'ahu Island, 75
Horizon [visualization
  program], 88–89
    limitations for close hori-
      zons, 89
horizon, visibility of, 88
Hyades plus Aldebaran (Ka Nuku
  o Ka Puahi), 109, 110, 136
    possible sighting device for,
      116–117, 118

'Ī'ī, John Papa, 14, 21, 23, 28
Ikiiki, 137
'ili'ili:
  flakes or pebbles, as offer-
    ings, 42, 49

gravel, 42
paving, 42, 50
images (ki'i), 11, 16, 21, 23, 25
  of Kū, 18
isotopic fractionation, 63

Johnson, Rubellite Kawena, 335
June solstice:
  alignments upon sunset at,
    75
  declination of sun at, 82
  explained, 82
  link to Pleiades heliacal rise,
    137
  possibly observed at Lono
    temples, 135

Kahikinui:
  archaeological surveys in,
    8–9
  cultural landscape, 9
  etymology, 6
  land units (ahupua'a) in, see
    Alena, Auwahi, Kīpapa-
    Nakaohu, Luala'ilua,
    Mahamenui, Manawainui,
    Nakaaha
  map of, 6
  physical landscape, 5, 7
  traditions relating to, 6–7
Kaho'olawe Island, 35, 77
kāhuna. See priests
Kailio Point. See Ka Lae o Ka 'Ilio
Kaiwiloa, Helio, 60
Ka Lae o Ka 'Ilio, 123, 124
  alignments upon, 109, 114
Kamakau, Kēlou, 13, 14, 31
Kamakau, Samuel, 6, 14, 17, 19,
  21, 23
Kamalālāwalu, 68, 70, 71
Kamehameha I, 20, 21
Kamoamoa Heiau (MAH-363),
  281–283, **282**
Kanahau (prominent red cinder
  cone), 77
  alignments upon, 110, 111–113,
    114
Kanaloa, 18, 19, 134, 135
  black shining road of, 19, 120,
    134
  linked with daily and

seasonal movements of
  sun, 19
much traveled red road of, 134
paired with Kāne, 19
Kāne [god], 18, 134
  black shining road of, 19, 120,
    134
  flaming path of, 134
  linked with the rising sun, 19
  linked with daily and
    seasonal movements of
    sun, 19
  paired with Kanaloa, 19
  temples, 120, 134, 136
  See also pōhaku o Kāne
Kāne [kapu period], 28
Kāne'ākī Heiau, O'ahu Island, 35
Ka Nuku o Ka Puahi. See Hyades
  plus Aldebaran
Kaoao, Simeon, 60
Kapoukahi, 20
kapu:
  direction, see sacred (kapu)
    direction
  periods during the month
    27–28, 73, 132
KAU-324. See Pōpōiwi Heiau
KAU-333, 315–317, **316**. See also
  KAU-333A; KAU-333B
KAU-333A, 48, 54, 315–317
KAU-333B, 90, 315–317
KAU-411, 54, 59, 317–319, **318**
KAU-536, 319
KAU-994. See Lo'alo'a Heiau
KAU-995. See Kou Heiau
KAU-996, 331
KAU-999, 66, 331–333, **332**
KAU-1001, 42, 333–334
Kaua'i Island, 35
  heiau typology in, 36–37
Kauhi-a-Kama (Maui king), 70
kāula (prophets), 20
Ka-Ulu-a-Pā'oa Heiau, Kaua'i
  Island, 75
Kaupō, 285
  archaeological surveys in, 9
  cultural landscape, 9
  etymology, 6
  Gap, 77
  historical traditions relating
    to, 7

land units (*ahupua'a*) in, 285, *see also* Nu'u
map of, **6**
physical landscape, 7
See also KAU-324 etc.
kava, 15, 19
Kekaulike, 7, 49, 66, 70, 130
Kepelino, 14, 19, 20, 23
ki'i. See images
Kiha-a-Pi'ilani, 68, 71
kings (*ali'i nui*), 130
   use of *luakini* temples by, 20, 32–33
   See also Kamalālāwalu; Kauhi-a-Kama; Kekaulike; Kiha-a-Pi'ilani; Pi'ilani
KIP-1, 200–203, **200**, **201**, **203**
   axis of orientation of, 11–12, 90
   date of, 70, 130
   'ili'ili paving at, 42
   internal features at, 48–49, 56, **56**
   non-orthogonality of walls at, 136
   sunrise viewed from, 122
KIP-75, 48, 54, 55, 74, 90, 108, 210–213, **211**
KIP-76, 213
KIP-77, **53**, 215–219, **216**
   alignments at, 108–109, 114
   axis of orientation of, 90
   bone working at, 72
   cist at, 50, 54
   faunal remains at, 74
   niche at, 51
   possible sighting device at, 119
   upright stones at, 49, 54
KIP-80, 57, 219–222, **220**
KIP-115, 57, 213–215, **214**
KIP-188, 91, 222–223
KIP-273, 42, 223–226. *See also* KIP-273A, KIP-273B
KIP-273A, 54, 90, 110, **224**, **225**
KIP-273B, 54, 90, 109, 118, 119, 122, 136, **224**
KIP-275, 226–228, **227**
   alignment upon Kanahau, 77, 110
   area of, 56

axis of orientation of, 90
Pleiades or Hyades sighting device at, 116–117, 122, 136, **229**
KIP-306, 136, 228–229
KIP-307, 60, 77, 91, 110, 114, 229–232, **231**, **233**
KIP-330, 77, 90, 91, 109, 110, 232–235, **234**, **235**
KIP-366, 90, 91, 109, 110, 114, 124, 235–237
KIP-394, 44, 237–238
KIP-405, 207–210, **208**
   alignment upon Ka Lae o Ka 'Ilio, 109, 110–114, **210**
   alignment upon Kanahau, 77, 110
   altar at, 49, 57, 90
   branch coral at, **209**
KIP-410, 90, 91, 110, 205–206
KIP-414, 238–240, **239**
   alignment upon Ka Lae o Ka 'Ilio, 109, 110–114
   axis of orientation of, 91
   canoe-prow corner at, 100–103
   stacked walls at, 43, 57–58
KIP-424, 110, 240–241
KIP-567, 49, 90, 91, 104, 107, 114, 241–243, **242**
KIP-728, 48, 59–60, 136, 243–246, **244**, **245**
KIP-1010, 58–59, **59**, 246–250, **247**, 251
   area of, 44, 54
   axis of orientation of, 90
   canoe-prow corner at, **59**, 100–103
   duplication of alignments in the two *heiau*, 115–116
   exterior terraces at, 42–43
   'ili'ili paving at, 42
   pathway at, 50–51
   secondary platform at, 48
   topographic alignments at, 110, 114
KIP-1025, 57, 90, 109, 110, 114, 124, 250–253, **252**
KIP-1146, 58, 90, 114, 123, **124**, 253–256, **255**, **256**
KIP-1151, 256–260, **258**

alignment upon Kanahau, 110
axis of orientation of, 90, 91
pit with stone facings at, 49
sighting navigation stars at, 119–120
sunrise viewed from, 122, **259**
KIP-1156, 260–263, **261**
   alignment upon Kanahau, 103, 110
   internal divisions at, 54, **55**
   June solstice sunset viewed from, 115
   orientation of canoe-prow corner at, 96–97, 103
   Pleiades sighting at, 107–108, 136, 138
   secondary platform at, 48
KIP-1306, 57, 263–264
KIP-1307, 58, 72, 264–266, **265**
KIP-1317, 44, 57, 66, 266–267, **267**
KIP-1398, 57, 267–268, **269**
Kīpapa. *See* Kīpapa-Nakaohu
Kīpapa-Nakaohu, 199–200. *See also* KIP-1 etc.
ko'a (fishing shrines), 24, 43, 57
   orientation of, 97
Kolb, Michael, 8, 35, 37, 38, 58, 62, 71, 129
   radiocarbon dates obtained by, 63–65, 66–67
Kou Heiau (KAU-995), 326–331, **327**, **328**
   alignment upon Sirius, 109, **330**
   area of, 44, 54
   associated with Kāne and Kanaloa, 19
   axis of orientation of, 90
   branch coral at, 52
   chronology of, 69–70, 100
   'ili'ili paving at, 42, 50
   large house within, 50
   secondary platforms at, 49
Kū [god], 3, 18, 130, 134
   rites after end of Makahiki period, 32–33
   worshipped at *luakini* temples, 20, 135, 136
Kū [*kapu* period], 28
Kū'ula, 24, 49, 57, 134

La'amaikahiki, 7
Ladd, Ed, 35
Lahainaluna seminary group, 14
Lāna'i Island, 34
landscape:
   physical, 5
   setting of Kahikinui *heiau*, 76–77
   setting of Kaupō *heiau*, 77
Lanikāula, 20
latitude, accuracy needed for determining declination, 88
lava rocks. See *a'ā* rocks; *pāhoehoe* slabs
Laval, Honoré, 75
lay of the land, in relation to *heiau* orientations, 75, 100, 104
Lewis, David, 335
lithic assemblages, 72, 73
Lo'alo'a Heiau (KAU-994), 57, 319–326, **320**, **322**
   alignment upon the Pleiades, 107–108, 136, **326**
   area of, 44, 56
   orientation of, 90, 91, 104, 108
   chronology of, 66, 67
   exterior buttressing at, 43, **323**
   history, 7
   multiple construction phases at, 58
   upright *'alā* stones at, 49
"long god" (*akua loa*), circuit of the, 31–32, 123
Lono, 18, 134
   associated with agricultural temples, 23, 30
   link to the Pleiades, 18
   rites at end of Makahiki period, 32
   role in reinforcing power relations, 33
   temples, 135
LUA-1, 90, 91, 114, 122, 168–172, **169**, **172**
LUA-3, 172–175, **174**, **175**
   altar at, 49
   axis of orientation of, 9
   internal divisions at, 54
   pāhoehoe walls at, 41–42
   pits or image holes at, 550

LUA-4, 175–178, **177**
   alignment on Pleiades, 89, 107–108, 136
   altar at, 49
   axis of orientation of, 90
   branch coral at, 54
   June solstice sunrise viewed from, **89**
   upright stones at, 43
LUA-29, 58, 178–183, **179**
   axis of orientation of, 91, 97
   branch coral at, 54
   furthest primary axis from cardinal orientation, 100
   Pleiades observation at, 136, 138
   sunset viewed from, **182**
   topographic features visible from, 115, **181**
LUA-36, 43, 57, 90, 109, 110, 114, 183
LUA-39, 57, 110, 184, **185**, **186**
luakini. See war temples
Luala'ilua, 168
   associated with Kāne and Kanaloa, 19
   Hills, 114, 115, 120, 121, 122, 138
   See also LUA-1 etc.
lunar:
   node cycle, 83
   phase cycle, 27
   standstill limits, 83, 84

magnetite, 87
MAH-231, 91, 278–281, **279**, **281**
MAH-363 (Kamoamoa Heiau), 281–283, **282**
Mahamenui, 278. See also MAH-231 etc.
Mahelona, John Kaipo, 335
maka'āinana (commoners):
   ancestor cult among, 19
   rituals, 34
Makahiki period, 26, 28, 123, 132–133, 137
   correlation with rainy season, 29, 30, 132–133
   rites during, 30–32, 71
   signal for the onset of, 28
   tribute collection during, 30–32, see also "long god," "short god"

Makali'i [asterism]. See Pleiades
Makali'i [month name], 26
malae:
   development of, in Western Polynesia, 15–16
   See also *marae*
Malo, Davida, 14, 20–21, 22–23, 26, 52, 129, 137–138
Manawainui, 283. See also MAW-2
Mangareva, 75–76, 138
marae, 15
   association with astronomical knowledge, 336
   development of, in Eastern Polynesia, 16
   orientation of, 138
   Proto-Polynesian origins, 15
Mau Piailug (navigator), 335
Maui Island, 35
   heiau typology, 37
Mauna Kea, 39, 88, 114, 336
   adze quarry, 73
   as possible orientation target, 103, 116
Mauna Loa, 88
   alignments upon, 114
Maunawila Heiau, O'ahu Island, 36, 39
MAW-2, 91, 114, 136, 283–284
McAllister, J. Gilbert, 35
Mo'okini Heiau, Hawai'i Island, 17
moai, 16
moku (districts), 6
Mokumanamana (Necker Island), 16, 35, 37
Moler, Mo, xix
mollusks, marine, 74
Moloka'i Island, 34, 35
month:
   days of the, 27
   kapu periods during, 27–28, 73, 132
   names, 26, 27
   See also Ikiiki
moon:
   nights of the, 27
   variations in declination of, 83

morphology of *heiau*, 36–37,
    44–47, 52–57, 129
  in relation to function, 130
  temporal patterns in, 17, 71,
      130–131
  See also elongated double-
      court enclosures; notched
      enclosures; platform
      heiau; square enclosures;
      terraced *heiau*; U-shaped
      enclosures
Muʻa, 15

Nā Hiku (Ursa Major), 110
NAK-27, 270–272
  axis of orientation of, 91
  furthest secondary axis from
      cardinal orientation, 100
  Pleiades sighting at, 118, 136,
      271
NAK-29, 272–275, **273**, **274**
  axis of orientation of, 91
  cist at, 54
  date of, 130
  pit at, 54
  Pleiades observation at, 136
  seating slab at, 51
  wall alignments, 109, 110,
      115
NAK-30, 275–277, **276**
  alignment on Pleiades,
      107–108, 136
  axis of orientation of, 90, 91
  Pleiades sighting at, 118, 136
  upright stones at, 49, **119**
NAK-34, 114, 277–278
Nakaaha, 268–270. *See also*
    NAK-27 etc.
Nakaohu. *See* Kīpapa-Nakaohu
Nānāhope. *See* Pollux
Nānāmua (Castor), 109
Native Hawaiian sources. *See* ʻĪʻī,
    John Papa; Kamakau, Kēlou;
    Kamakau, Samuel; Malo,
    Davida
navigation:
  Pacific, 335–336
  Polynesian, 336
  stars, 120
Necker Island (Mokumana-
    mana), 16, 35, 37

niches, 51
  indicating direction of
      orientation, 90
nights of the moon, 27
Nihoa Island, 16, 34
north celestial pole, 82
northerly orientation group, 100
  interpreted as Kū temples, 135
notched enclosures, 48, 53–54,
    136
  branch coral at, 54
  earliest dates for, 130
  function of, 130
  other features at, 54
  position of notch, 53
  position of notch in relation
      to sacred (*kapu*) direction,
      91, 96–97
Nuʻu, 285. *See also* NUU-1 etc.
NUU-1, 286
NUU-78, 91, 286, **287**
NUU-79, 286–291, **288**, **289**
  axis of orientation of, 91
  date of, 130
  hammerstone found at, 72
  ʻiliʻili paving at, 42
  postcontact house at, 60
  sunrise viewed from, 115, 122
NUU-81, 55, 291–294, **292**, **293**
  axis of orientation of, 90
  date of, 100, 130
  Pleiades viewed from, 136
  possible alignment upon
      Hyades plus Aldebaran
      (Ka Nuku o Ka Puahi), 109
  westerly orientation of, 135
NUU-100. *See* Halekou Heiau
NUU-101, 43, 44, 55, 56, 298–300,
    299, **300**
NUU-151, 122, 300–301
NUU-153, 72, 301–303, **302**, **303**
NUU-172, 44, 303–304
NUU-175, 90, 91, 136, 304–306,
    305
NUU-188, 306–308, **307**
NUU-424, 308

Oʻahu Island, 35
obliquity of the ecliptic, 82
offerings, 14, 19, 21, 22, 23, 24, 25
ʻOpunohu Valley, Moʻorea, 138

oracle towers. *See* towers (*ʻanuʻu*)
orientation, 26, 44–47, 336
  and the lay of the land, 75,
      100, 104
  axis of, 11–12, 75, 85, 133, *see
      also* primary axis,
      secondary axis
  azimuth distribution, **101**
  best estimates of, 90, 103
  broad cardinality of, 100, 103,
      104–105
  clustering of, 100, 133–135, *see
      also* easterly orientation
      group, east-northeasterly
      orientation group,
      northerly orientation
      group
  correlation with associated
      gods, xvii, 133–135
  data selection methodology,
      84–85
  determination of, 11–12, 85
  factors influencing, 75–76
  in relation to chronology, 100,
      130, 131
  of fishing shrines (*koʻa*), 97
  of temples elsewhere in
      Polynesia, 138
  principal direction of, *see*
      sacred (*kapu*) direction
  principles of, according to
      native Hawaiian writers,
      26
  summary of features
      influencing, 125–128
  systematic analysis of, 85, 89,
      97–104
  toward enemy territory, 26
  westerly, 135
  relation to functional
      typology, 134
  uncertainties in, 103

Pāʻao, 17, 34, 129
pāhoehoe slabs, 41–42, 50–51, 121
Pānānā (sighting wall), 7, **181**
Papaʻenaʻena Heiau, 21
pathways, 50–51
paving, 42
Phelps, Southwick, 35
Piʻilani, 38, 68, 71

Pi'ilanihale Heiau, 38
pig bones, 73
pits, 49–50
plane-table surveys, 10–11
planets, 84
platform *heiau*, 56–57
　other features at, 57
platforms, secondary, 48–49
Pleiades (Makali'i), 110, 118, 132, 135, 136
　acronychal rise of, 27, 28, 30, 132, 135, 137
　alignments upon, 108, 136
　cosmical set of, 137, 138
　heliacal rise of, 137
　heliacal set of, 138
　influence of upon *heiau* location and orientation, 137
　link to Lono, 18
　names in Hawaiian, 27
　observation of attested in ethnohistory, 75
　observation of at upland Auwahi sites, 120–121, 136
　observation of in ancestral Polynesia, 26
　observation of setting of, 137–138
　rising between Luala'ilua Hills, 120, 121
　rising or setting over distinctive horizon feature, 136
　sighting devices for, 116–118, **117**, 119, 136
　used to determine start of Makahiki period, 18, 28, 135, 137
　used to divide proto-Polynesian calendar, 137
po'okanaka. *See* war temples (*luakini*)
pōhaku o Kāne, 24, 49, 57, 66
Polaris (Hōkūpa'a), 82, 107
Pollux (Nānāhope), 110, 119
possible alignments upon, 108–109
Polynesian Voyaging Society, 335, 336

Pōpōiwi Heiau (KAU-324), 58, 308–315, **310–311**
　altar at, 49
　area of, 44
　axis of orientation of, 90
　canoe-prow corner at, 43–44, 103
　chronology of, 66–67
　history, 7
　horizon to north, **314**
　'ili'ili paving at, 42
　multiple construction phases at, 58
　postcontact house at, 60
population:
　historical, 7
precession (of the equinoxes), 78, 82, 84
precision, 89
　defined (*cf.* accuracy), 76
　in total station surveys, 87
priests (*kāhuna*), 19–20
　stones used for astronomical observations by, 76, *see also* seating slabs
　types of, 19
primary axis, 89–90
　longer, indicating direction of orientation, 91
principal gods, 17–19:
　directional associations of, 18, 133–134
　See also Kanaloa; Kāne; Kū; Lono
prophets (*kāula*), 20
Proto-Central Eastern Polynesian terms, 28, 30
Proto-Polynesian:
　calendar, 26, 137
　terms, 19
　terms for lunar months, 26
　terms for ritual places, 15, 16
Pukui, Mary Kawena, 14
Pu'u Hōkūkano, 51, 115, 121, 167
Pu'u Māneoneo, 115, 122
Pu'u Pane, 136
Pu'u Pīmoe, 115, 120, 138
　alignments upon, 114
Pu'ukoholā Heiau, Hawai'i Island, 20

radiocarbon dates:
　presentation of, 141
radiocarbon dating, 62, 70–71
　"inbuilt age" effect, 63
　accelerator mass-spectrometry (AMS), 66
　isotopic fractionation correction, 63
　of Kahikinui and Kaupō *heiau*, 63–67
Rapa Nui (Easter Island), 16, 138
rat:
　bones, 74
　consumption of, in Eastern Polynesia, 74
Red Hill (Haleakalā), 76, 77
refuse pits (*lua*), 25
Richards, William, 14
ritual cycle, 26
　role in reinforcing power relations, 33
　See also calendar; Makahiki period

sacred (*kapu*) direction:
　azimuth distribution, 98–100, **101**, **102**
　determination of, 90–91, 97
　factors influencing, 75–76
　in relation to canoe-prow corners, 96–97
　in relation to notched corners, 91, 96–97
　primary indicators of, 90, 92–95
　secondary indicators of, 90–91, 92–95
　See also orientation
sacrifice, 3, 15, 17, 24, 33
　human, 3, 17, 18, 20, 22, 23, 33
　pig, 73
seasons, 28–29, 137
seating slabs, 51, 120, 121
secondary axis, 89–90
Sharp, Warren, xvii
"short god" (*akua poko*), circuit of the, 31–32
sighting devices, 116–118, **117**, 119, 122, 135, 136, 137
Sirius ('A'ā; Hōkūho'okelewa'a), 110

orientation of Kou Heiau (KAU-995) upon, 109
possible sighting device for, 119
site:
  access, 142
  mapping techniques, 10–11, *see also* compass-clinometer surveys, plane-table surveys, total station surveys
  numbering systems, 10
size variations in *heiau*, 44–48
skyscape, 9–10
Smith, Andrew, 88
solar rising arc, 82, 122
  alignments within, 122
  viewed from site WF-AUW-403, 121–122
solar setting arc, 82
solstices:
  observations of in Mangareva, 75–76, 138
  See also June solstice; December solstice
square enclosures, 54
  function of, 130
  other features at, 54–55
stacked walls, 43
stars (*hōkū*):
  changing appearance over the year, 82
  circumpolar, 82
  names of, 4, 335
  principal Hawaiian, 78, 79–81
  seasonal events, 82
  See also acronychal rise; cosmical set; heliacal rise; heliacal set
state formation, 15, 38, 39, 71
Stellarium (sky visualization program), 89
stonework, 40–44. See also *a'ā* rocks; *'ili'ili*, *pāhoehoe* slabs; upright stones; waterworn (*'alā*) stones
Stokes, John F.G., 34, 129
sun:
  changing position of, over the year, 28, 75, 82, 83, 122, 132, 134

changing declination of, over the year, 83
first light of, 118
tracked against natural landmarks, 122, 124
tracked using alignments, 123
sun-azimuth determinations, 86
survey:
  fieldwork dates, xvii
  techniques, *see* compass-clinometer surveys, plane-table surveys, total station surveys

telescopes, 336
temple sites. See *heiau*; *malae*; *marae*
terraced *heiau*, 56–57
  other features at, 57
terraces, raised:
  indicating direction of orientation, 90
test pits. See excavations
Tevake, 335
Thrum, Thomas, 34, 35
topographic features. *See* visual landmarks
total station surveys, 11, 86
  determination of north point, 86
  time determination, 86–87
towers (*'anu'u*), 11, 21, **22**, 23, 25, 49, 54
tribute collection, 30–32. See also "long god," "short god"
Tupaia, 335
types of *heiau*. See functional typology of *heiau*; morphology of *heiau*

U/Th dating. See ^{230}Th dating
unu temples, 23–24
upland Auwahi sites, 120–121, 136
  Pleiades observation at, 120, 121
  See also WF-AUW-338; WF-AUW-343; WF-AUW-359; WF-AUW-391; WF-AUW-403; WF-AUW-493; WF-AUW-574

upright stones, 43, 49, 114, 123, **124**
  as representations of gods and ancestors, 49
  as sighting device, 118, **119**
  indicating direction of orientation, 90
  used as artificial foresights elsewhere in Polynesia, 76
Ursa Major (Nā Hiku), 110
U-series dating. *See* ^{230}Th dating
U-shaped enclosures, 54–55
  function of, 130
  other features at, 54–55

Valeri, Valerio, 3, 133
visibility:
  of horizons, 88
  of visual landmarks around dawn, 135–136
visual landmarks, 76–77, 125–128
  correlated with astronomy, 115, 120, 132
  visibility of, around dawn, 135–136
visualization of horizon profiles, 88–89
voyaging, Polynesian, 335–336

Waha'ula Heiau, Hawai'i Island, 17
Wailapa Heiau (ALE-4), 50, 58, 90, 114, 189–193, **191**, **192**
Waipi'o, Hawai'i Island, 103
Walker, Winslow, 8, 35
walls:
  core-filled, 42
  more substantial, indicating direction of orientation, 90
  orthogonality of, 43
  stacked, 43
Waolani, Nu'uano, O'ahu Island, as location of first *heiau*, 17
war temples (*luakini*), 20–22, 32–33, 130, 135
  features of, 21
  houses within, 25
  orientation upon enemy territory, 75
  rituals 32–34

Index • 359

See also Kou Heiau; Loʻaloʻa Heiau; Pōpōiwi Heiau
Warther, Frances X., xvii
waterworn (*alā*) stones, 42, 48, 49
    as representations of deities, 42
waterworn gravel (*iliʻili*), 42
wealth, redistribution of, 31, 32
WF-AUW-100, 54, 90, 91, 109, 110, 150–152, **151**
WF-AUW-176A, 90, 91, 114, 152–154, **153, 155**
WF-AUW-338, 70, 90, 120, 130, 154–157, **156**
WF-AUW-343, 54, 120, 157–158
WF-AUW-359, 90, 91, 114, 120, 159–160
WF-AUW-391, 54, 90, 91, 120, 160–162, **161, 162**
WF-AUW-403, 162–165, **163**
    alignment upon Puʻu Pīmoe, 114
    associated with Kāne, 120
    axis of orientation of, 90, 91
    cist at, 50, 54
    pāhoehoe slabs at, 42, 50, 51, 121
    seating slab at, 51, 121
    sunrise viewed from, 120, 121–122, **121, 122, 165**
WF-AUW-493, 90, 114, 120, 122, 165–166
WF-AUW-574, 42, 43, 54, 55, 90, 120, 166–167

X-ray fluorescence, 73

ABOUT THE AUTHORS

Patrick Vinton Kirch is professor of anthropology at the University of Hawaiʻi at Mānoa, and professor emeritus at the University of California, Berkeley. Born and raised in Hawaiʻi, Kirch received his PhD from Yale University; he has also held positions at the Bishop Museum in Honolulu, and the Burke Museum at the University of Washington. During a research career that spans nearly five decades, Kirch has carried out archaeological fieldwork across the Pacific, from Papua New Guinea and the Solomon Islands, to Tonga and Samoa, the Cook Islands, French Polynesia, and Hawaiʻi. Among his many honors and awards, Kirch has been elected to the US National Academy of Sciences, the American Academy of Arts and Sciences, and the American Philosophical Society. He is the author of hundreds of scholarly articles and several dozen books and monographs, including *Feathered Gods and Fishhooks*, *On the Road of the Winds*, *Unearthing the Polynesian Past*, and the award-winning *A Shark Going Inland Is My Chief*.

Clive Ruggles is Emeritus Professor of archaeoastronomy in the School of Archaeology and Ancient History at the University of Leicester, United Kingdom. Trained as a mathematician and astrophysicist, Ruggles obtained his DPhil from Oxford University in 1978, but co-authored his first paper in archaeoastronomy, published in the journal *Nature*, while he was still an undergraduate. In 1979 he moved across to archaeology as a Research Fellow in Cardiff University, where he worked with Stonehenge's excavator Richard Atkinson. His early work focused on Scotland and Ireland, culminating in the publication of his award-winning book *Astronomy in Prehistoric Britain and Ireland* published by Yale in 1999. Over his forty-five-year career Ruggles has authored numerous books, papers and articles on subjects ranging from prehistoric Europe and pre-Columbian America to indigenous astronomies in Africa, as well as in other fields such as computer graphics and information systems. He was editor-in-chief of the three-volume *Handbook of Archaeoastronomy and Ethnoastronomy*, a definitive source on theory, method, and practice in these fields. In 2017 Ruggles was awarded the Royal Astronomical Society's new Agnes Clerke Medal for a "lifetime of distinguished work in the overlapping areas of archaeology, astronomy, and the history of science."